Dependable Computing and Fault-Tolerant Systems

Edited by
A. Avižienis, H. Kopetz, J. C. Laprie

Advisory Board
J. A. Abraham, V. K. Agarwal, T. Anderson, W. C. Carter,
A. Costes, F. Cristian, M. Dal Cin, K. E. Forward, G. C. Gilley,
J. Goldberg, A. Goyal, H. Ihara, R. K. Iyer, J. P. Kelly,
G. Le Lann, B. Littlewood, J. F. Meyer, B. Randell,
A. S. Robinson, R. D. Schlichting, L. Simoncini, B. Smith,
L. Strigini, Y. Tohma, U. Voges, Y. W. Yang

Volume 4

Springer-Verlag Wien New York

A. Avižienis, J. C. Laprie (eds.)

Dependable Computing for Critical Applications

Springer-Verlag Wien New York

Prof. Dr. Algirdas Avižienis, UCLA, Los Angeles, CA, USA
Dr. Jean-Claude Laprie, LAAS-CNRS, Toulouse, France

With 88 Figures

Printed on acid-free paper

ISBN-13: 978-3-7091-9125-5   e-ISBN-13: 978-3-7091-9123-1

DOI: 10.1007/ 978-3-7091-9123-1

FOREWORD

The International Working Conference on Dependable Computing for Critical Applications was the first conference organized by IFIP Working Group 10.4 "Dependable Computing and Fault Tolerance", in cooperation with the Technical Committee on Fault-Tolerant Computing of the IEEE Computer Society, and the Technical Committee 7 on Systems Reliability, Safety and Security of EWICS. The rationale for the Working Conference is best expressed by the aims of WG 10.4:

> " Increasingly, individuals and organizations are developing or procuring sophisticated computing systems on whose services they need to place great reliance. In differing circumstances, the focus will be on differing properties of such services – e.g. continuity, performance, real-time response, ability to avoid catastrophic failures, prevention of deliberate privacy intrusions. The notion of **dependability**, defined as *that property of a computing system which allows reliance to be justifiably placed on the service it delivers*, enables these various concerns to be subsumed within a single conceptual framework. Dependability thus includes as special cases such attributes as **reliability, availability, safety, security**. The Working Group is aimed at identifying and integrating approaches, methods and techniques for specifying, designing, building, assessing, validating, operating and maintaining computer systems which should exhibit some or all of these attributes. "

The concept of WG 10.4 was formulated during the IFIP Working Conference on Reliable Computing and Fault Tolerance on September 27-29, 1979 in London, England, held in conjunction with the Europ-IFIP 79 Conference. Profs A. Avižienis (UCLA, Los Angeles, USA) and A. Costes (LAAS-CNRS, Toulouse, France), who organized the London Conference and proposed the formation of the Working Group were appointed as Chairman and Vice Chairman, respectively, of the new WG 10.4 in 1980 and served until 1986, when Dr. J.C. Laprie (LAAS-CNRS, Toulouse, France) succeeded to serve as Chairman, and Profs J. Meyer (University of Michigan, Ann Arbor, USA), and Y. Tohma (Tokyo Institute of Technology, Japan) became Vice Chairmen of the

Working Group. The first meeting of the new WG 10.4 took place in Portland, Maine, USA, on June 22-23,1981. In attendance were 29 founding members of the Working Group. Since then, the membership has grown to 50 members from 15 countries. Sixteen WG 10.4 meetings have been held from 1981 through 1989 in various locations, including USA (6 meetings), France (3), and Australia, Austria, Canada, India, Italy, Japan, Tunisia (1 each). The main goal of WG 10.4 meetings is to conduct in-depth discussions of important technical topics. A principal theme since the first meeting has been the understanding and exposition of the fundamental concepts of fault-tolerant computing. Other major topics have been: distributed computing, real-time systems, certification of dependable systems, specification methods, design diversity, specification and validation of hard dependability requirements, methodologies for experiments, VLSI testing and fault tolerance, hardware-and-software testing and validation, fault tolerance in new architectures, telecommunication systems and networks. Beside the key themes, research reports by members and guests are presented at every meeting, and business meetings are held to plan future activities.

The program of the Working Conference is made up of 20 papers and three panel sessions. The papers were selected by the Program Committee from a very good set of submissions coming from academia, industry and government agencies. The submissions originated from 12 countries from America, Europe and Asia, and covered, as hoped, the broad range of topics concerned by dependable computing. Selection was greatly facilitated by an exceptional return rate of the requested reviews (83%), provided by external reviewers (3 or 4 per paper) as well as by the Program Committee members (2 per paper). As a Working Conference, the program was designed in order to promote the exchange of ideas by extensive discussions. All the paper sessions ended with a 30 minutes discussion on the topics dealt with in the session. In addition to the 9 paper sessions, three panel sessions were organized. The first panel was moderated by A. Avizienis; extending over two sessions, it concerned the *Successes and Limitations of Experimental Methods in Gaining Evidence on Dependability* of computing systems intended for critical applications; in one session it was people who have major responsibility for such systems as speakers, and researchers as discussants, in the other their roles were reversed. The second panel was moderated by J. Goldberg and B. Randell; it addressed the *Societal Issues involved in the utilization of Computers in Critical Applications*, such as the balancing of the risks incurred against the benefits to be gained, raised by the expanding deployment of computing systems in transport, energy production, military applications, etc. All our colleagues of the Program Committe deserve our thanks for selecting a first-rate program.

The Working Conference was being held on the beautiful grounds of the Santa Barbara campus of the University of California (UCSB), and we want to thank John P.J. Kelly from UCSB, the Local Arrangements Chair, for his enthusiastic support in the organization of the conference.

After the conference, the papers have undergone an additional reviewing process performed by the members of the Editorial Board of the series, aimed at advising the authors for the final presentation of their paper in the proceedings.

The continuous support of the Vice-General Chairman, Alain Costes from LAAS-CNRS, is acknowledged. Particular thanks go to Joëlle Penavayre from LAAS-CNRS for her diligent secretarial support during all the conference preparation; she performed, with the help of Jean Arlat, Christian Beounes and Yves Crouzet the reformatting of the papers for the proceedings.

General Chair	*Program Chair*
Algirdas Avižienis	Jean-Claude Laprie
UCLA	LAAS-CNRS
Los Angeles	Toulouse
USA	France

SPONSORS

IFIP Working Group 10.4 *Dependable Computing and Fault Tolerance*
In cooperation with:
  Technical Committee *Fault-Tolerant Computing* of the IEEE Computer Society
  Technical Committee 7 *Systems Reliability, Safety and Security* of EWICS

CONFERENCE ORGANIZATION

General Chair
A. Avižienis
UCLA
Los Angeles, California, USA

Program Chair
J.C. Laprie
LAAS-CNRS
Toulouse, France

Vice-General Chair
A. Costes
LAAS-CNRS
Toulouse, France

Local Organization Chair
J.P.J. Kelly
UCSB
Santa Barbara, California, USA

Program Committee

T. Anderson
The University of Newcastle upon Tyne
Newcastle upon Tyne, UK

A. Avižienis
UCLA
Los Angeles, California, USA

W.C. Carter
Bayley Island, Maine, USA

A. Costes
LAAS-CNRS
Toulouse, France

F. Cristian
IBM Almaden Research Center
San Jose, California, USA

G. Gilley
Aerospace Corporation
Los Angeles, California, USA

J. Goldberg
SRI International
Palo Alto, California, USA

H. Ihara
HITACHI
Yokohama, Japan

J.P.J. Kelly
UCSB
Santa Barbara, California, USA

H. Kopetz
Technische Universität Wien
Vienna, Austria

J.C. Laprie
LAAS-CNRS
Toulouse, France

B. Littlewood
The City University
London, UK

J.F. Meyer
The University of Michigan
Ann Arbor, Michigan, USA

B. Randell
The University of Newcastle upon Tyne
Newcastle upon Tyne, UK

J.M. Rata
Electricité de France
Clamart, France

L. Simoncini
IEI-CNR
Pisa, Italy

Y. Tohma
Tokyo Institute of Technology
Tokyo, Japan

U. Voges
KFK
Karlsruhe, Germany

Referees

V. Agarwal	H. Hecht	R.M. Needham
P. Amman	Heimann	P. Neumann
F. Anceau	A.J. Herbert	M. Pfluegl
J. Arlat	A.L. Hopkins	D. Powell
O. Babaoglu	W.E. Howden	G. Pucci
C. Beounes	O.C. Ibe	K. Richardson
T. Beth	J. Jacob	A.S. Robinson
P.G. Bishop	M. Joseph	J. Rushby
J.P. Blanquart	K. Kanoun	F. Saglietti
S. Bologna	R. Kemmerer	A. Sathaye
P. Bolzern	K. Kinoshita	G. Saucier
J. Brenner	Y. Koga	R.D. Schlichting
P. Caspi	A. Kouchakdjian	W. Schneeweiss
B. Courtois	J.H. Lala	J.P. Shen
Y. Crouzet	L. Lee	S.K. Shrivastava
A.T. Dahbura	P. Lee	D.P. Siewiorek
G. Dahll	M.M. Lehman	M. Sievers
M. DalCin	G. Le Lann	J. Sifakis
A. Damm	N.G. Leveson	J.L. Soler
Y. Deswarte	L. Lopriore	D. Taylor
J. Dobson	H. Lutnik	J. Torin
M. Dyer	E. Maehle	W. Toy
P. Ezichelvan	R. Marie	P. Traverse
J.C. Fabre	R. Maxion	K.S. Trivedi
K. Forward	E.J. McCluskey	R. Turn
M.C. Gaudel	J. McDermid	A. Villemeur
W. Gorke	D. Miller	J.P. Warne
G. Hagelin	M. Moulding	T. Williams
R. Hamlet	M. Mulazzani	R.A. Whitehurst

Post Conference Reviewers

J. Abraham	J. Kelly	D. Rennels
A. Avizienis	K. Kim	R. Schlichting
A. Costes	H. Kopetz	L. Simoncini
M. Dal Cin	B. Littlewood	T.B. Smith
J. Goldberg	J. Meyer	L. Strigini
A. Goyal	B. Randell	Y. Tohma
H. Ihara		

CONTENTS

Validation

Chair: M. Melliar-Smith (UCSB, Santa Barbara, California, USA)

DEPENDABLE COMPUTING

IN THE UK

Mel BARNES
Safety & Reliability Directorate, AEA Technology, Wigshaw Lane
Culcheth, Warrington, Cheshire, UK, WA3 4NE

Abstract

The increased usage and sophistication of computers applied to real time safety-related systems in the United Kingdom has spurred on the desire to provide a standard framework within which to design, build, and assess dependable computing systems. The recent introduction of the European Product Liability Laws has acted as a catalyst in this area.

This paper is a survey of the approaches being taken by various important organisations within the UK, in military, government, and industrial areas, and compares the strengths and weaknesses of the approaches. It also discusses frameworks for the design and assessment of dependable computer systems used in safety-critical applications, and embraces issues of reliability, safety, and security.

1. Introduction

The approach to producing dependable computing systems within the UK is inconsistent across the different user sectors. Various significant organisations have recognised this shortcoming and have taken initial steps to overcome the problem. The following text provides the author's interpretation of work being carried out by these organisations in areas of system reliability, availability, safety, and security.

Most of this work is focused on software, since this is the most onerous area, but hardware and systems aspects are also addressed.

2. The acard report

Pre-1986, it was recognised that the UK held a very small portion of a large and rapidly developing software market. In 1985, the Advisory Council for Applied Research and Development (ACARD) founded a working group to resolve a number of issues which needed to be addressed before the UK could increase its competitiveness. ACARD is a UK Government body, which provides advice to UK industry.

In 1986, ACARD published a report [1], which included the findings of that working group. An appendix of that report is entitled "Safety-critical Software"; it suggests and outlines a possible route that might be adopted over the next decade or so to:

- prevent (via good engineering practices)

- cope with (via management)

- prevent a recurrence of (via analysis and review)

a disaster occurring due to computer software failure. An outline of this imaginable approach is now presented, highlighting the salient points of [1]. The approach is based upon a formalised system of registration, certification, and licensing of:

- systems

- key personnel

- participating organisations.

2.1. Prevention of disasters

2.1.1. Registration of safety critical systems

All systems which are identified as safety critical (ie endangering lives or public safety) must be listed in a register of safety critical systems.

2.1.2 Licence to operate

A "Licence to Operate" (LTO) is required by an organisation before it can operate a safety-critical system. The LTO will be granted to an organisation when certain conditions have been fulfilled:

- Safety Certificate certificate must be issued, which will be valid for a limited period

- a Certified Software Engineer (CSE) trained in mathematical aspects of software engineering must be nominated as the person responsible for the system and its operation. The CSE must have undergone mathematical training to the appropriate level in software engineering

- the safety-critical system must be maintained in a manner which is appropriate to its safety role. The CSE will be responsible for the maintenance of the software.

2.1.3. Licence to construct

Safety-critical systems must be constructed by approved suppliers, who will require a "Licence to Construct" (LTC); this will be granted to suppliers who:

- can demonstrate a capability to construct systems to (not yet specified) approved standards

- use mathematically-based development methods

- use safety-certified software tools

- use safety-certified staff

2.1.4. Learning from experience

All systems which have been formally registered as safety-critical must have a data collection scheme and provide data for the purposes of:

- aiding research into system safety and reliability

- providing supporting evidence to assist Boards of Enquiry

2.2. Management of disasters

This aspect of the document focuses on the type of recovery action needed in the event of a failure which leads to a reduced level of safety.

2.2.1. Disaster management procedures

These should be defined and documented prior to system operation, and practised on a regular basis during operation.

2.2.2. Availability of emergency services

The emergency call-out of a Certified Software Engineer should be possible at any time at short notice. Where the emergency is of longer-term duration, procedures must exist for providing a specialist team of experts.

2.2.3. Preparation for learning from disasters

Provision should be made for equipment and procedures for the logging of data prior to and during a disaster, so that the subsequent Board of Enquiry will have the correct information to determine the cause and trend of the disaster. Such a system is analogous to the "black box" flight recorder on aircraft.

2.3. Analysis of disasters

After a disaster, it is necessary to determine its cause, so that precautions can be taken to reduce (or ideally eliminate) the probability of recurrence.

2.3.1. Official board of enquiry

A Board of Enquiry should be set up following a disaster, and should have the power to recommend and promulgate changes following the enquiry. These changes could be in many areas, for example:

- design changes

- development methods

- automated tools

- nominated staff

Any "close shave" (a serious incident which might have resulted in a disaster, but didn't), should be reported to the Licensing Authority, which then decides if the implications of the incident warrant a Board of Enquiry.

2.3.2. Incidents

It is desirable to formally investigate any fault in a safety-critical software system, no matter how trivial the error or its effect seems to be; thus the effectiveness of the software fault avoidance/removal procedures can be evaluated. It is implied that any faults other than trivial faults will result in disciplinary action to the staff responsible.

2.3.3. Data collection

The data collected during normal operation of the system must be submitted to the Licensing Authority. This will include data on performance and reliability. These data are additional to the operational data mentioned previously.

2.4. Safety certification

2.4.1. The certification process

The ability to provide certification of software system safety aspects must be developed, in the areas of construction and operation. This will include certification of:

- mathematical development methods

- the correct application development and repair methods

- automated tools used during development and maintenance

- software engineers who build/maintain the system

- the software product itself

2.4.2. Mathematical proofs

Methods which permit mathematical proofs to be performed should replace methods which are regarded as "good practice".

2.4.3. Certification of software engineers

In order to be certified, a software engineer will have to:

- satisfactorily complete a training course in formal mathematical techniques

- have an approved type and amount of experience

Certification will last for a limited period, after which the software engineer will need to requalify on existing and new techniques.

2.4.4. Certification of software products

In the past, methods for the certification of software have depended upon testing; the new approach will require proof obligations to be submitted in addition to testing.

2.4.5. Rigour of inspection

The extent and depth of testing will depend upon many factors, for example:

- the level of risk; risk is generally defined as the frequency at which a given level of harm will result from the realisation of a hazard . A hazard is generally defined as a physical situation with a potential for human injury, or damage to property or environment, or some combination of these

- the magnitude of the consequence

- the costs involved

2.4.6. Categorisation scheme

A categorisation scheme is visualised which will permit different levels of certification. The suggested categories and their associated protective measures required are presented below; it should be noted that there the categories are not complete. For example, Category A deals with the situation where more than 10 deaths could occur, and Category B where a single death could occur; it is not possible from the information presented in [1] to determine in which category, say, 5 deaths would lie. Similarly, in Category B, a person responsible for causing a death might be subjected to charges of criminal negligence; no such penalty appears in Category A. Hence the categories require firmer definition before usage.

Category A. *Disaster level*:

- where failure could cause more than 10 deaths

- where switch-off will not limit the damage.

- zero errors are required

Measures required:

- the entire software must be validated by a formal mathematical proof.

- independent checks of source code against object code for all automated tools used in the development

Category B. *Safety level*:

- where failure might cause a single death

- where switch off will prevent further danger

- no more than 1 fault expected per 100,000 lines of code

- shorter code will be expected to be fault-free

Measures required:

- the entire software should be developed by mathematical proof methods

- the proofs should be checked by a competent mathematician

- if a death occurs as a result of a human error, the Board of Enquiry must name the person(s) responsible for making the error and failing to detect it; such persons might be subjected to charges of criminal negligence

Category C. *High quality level*:

- commercial level software, where faults could result in financial loss for the user

- no more than 1 fault expected per 10,000 lines of code

- all software tools shall eventually be certified to this level

Measures required:

- all programmers must be competent in mathematical methods for software engineering

- sampling will be used to check their methods

- all corrected errors will require recertification using the methods in the "Safety Level" category

- certification will be withdrawn if the error rate is exceeded

Category D. *Normal quality level*:

- where faults could result in lower programmer productivity

Measures required:

- current "best software engineering practices"
- no more than 1 fault expected per 1,000 lines of code

3. IEE report on safety-critical software

The UK Institute of Electrical Engineers (IEE) responded on the ACARD report to the UK Department of Trade and Industry (DTI) , and made proposals to resolve some of the issues raised. In response to these proposals, the DTI placed a contract with the IEE to carry out a joint study with the British Computer Society (BCS) on Safety Critical Software Systems. A national open consultative meeting on Safety Related Systems (where the study was discussed), was held at the IEE, London, in September 1988. Following this meeting a draft report of the study [7] (incorporating comments from that meeting) was issued to the meeting attendees for comment. The following section provides some of the salient points discussed in the report.

In its proposals to the DTI, the IEE recommended areas for study, in:

- professional certification
- regulatory frameworks
- professional liability

One aim of the studies was to provide some recommendations in these areas, but the main objective of the work was to highlight gaps, overlaps, or any inconsistencies in guidelines, standards, codes of practice, and legislation in above areas.

The study revealed that there was a significant variation in the requirements and conditions between the sectors of industry studied; it is the desire of the DTI to harmonise standards across industry for safety related software systems.

Some important initial feelings from the study were:

- it is currently impossible to guarantee that software is free from faults

- formalism should not be over-relied upon

- legislation and regulations for safety-critical software should be no more difficult to effect than in other engineering branches

Certification is considered to be of prime importance, since it is necessary to provide evidence that the system is built and operated to the best accepted standards by adequately trained people. The areas which might well be certified in the future include:

- the as-built system, including any tools used

- the development and production processes

- the design organisations

- the development organisations

- individual professionals such as:

- the project manager

- the design team

- the safety assessor

(*Note*: It is the individual engineer who should be assessed, rather than the course of training and education, and this is a problem which needs to be addressed).

There was a large body of opposition to the use of certification; the UK industry mainly depends on various other approaches, such as:

- the "pedigree" of the design/manufacturing organisations,

- the competence of the design/manufacturing organisation's key personnel

- statutory law (eg European product liability laws)

- conditions of contract

There were 4 options proposed for a "way forward":

1. Harmonisation of existing industrial practices, but introduction of software specific or technology-oriented standardisation where considered appropriate

2. Where nothing exists, provide software-specific standardisation, and integrate it with existing practices

3. Where nothing exists, provide software-specific standardisation, but leave rest of industry "as-is"

4. Apply software-specific standardisation in all areas

5. Do nothing

The option proposed in the report is option 1, ie the harmonisation of current practices. Most industrial organisations have their own controls, and although there may be some shortcomings in the areas of software control, it is proposed that any new measures should be integrated with existing practices. Therefore the intent is to review existing practices, and to provide a package for their harmonisation.

4. HSE PES guidelines

The Health and Safety Executive (HSE) is a corporate body of 3 people appointed by the HSE Commission and has some 3600 employees, mainly inspectors, technical, scientific, and medical experts. It inspects and advises on all walks of industry, including mines, quarries, medical, chemical, engineering, etc, and enforces law. The HSE management board includes the Chief Inspectors of the various Inspectorates concerned with the enforcement of industrial safety and health. The Executive is the licensing authority for nuclear installations; the Chief Inspector of Nuclear Installations has this function delegated to him.

In the summer of 1987, the UK Health and Safety Executive (HSE) published its Guidelines for Programmable Electronic Systems (PES), in two parts, [4] and [5], for use in the construction and assessment of safety-related systems. Until that time, there were no nationally-recognised guidelines in existence in the UK industry, and although the guidelines are not perfect, they provide sound advice to enable the user to customise the guidelines to suit a particular application.

The PES guidelines comprise of 6 main steps:

1. Carry out a hazard analysis

2. Identify the PES systems whose failure could lead to the realisation of the hazard

3. Define the safety criteria

4. Design the system

5. Perform an assessment

6. Check adequacy

STEP 1. *Carry out a hazard analysis*: The first step is to carry out a hazard analysis of the plant, using techniques such as Event Tree Analysis. The output of such an analysis will be a number of different events, each leading to a consequence or hazard, together with a probability or frequency of occurrence.

STEP 2. *Identify the PES systems in which failure could lead to the realisation of the hazards*: The output from the hazards analysis forms an input into Step 2, where those PES systems in which failure could lead to loss of protection against the hazard (or indeed, which could trigger the hazard) are identified, using techniques such as Fault Tree Analysis. The top event of the fault tree is a hazard identified from Step 1, and the base event is the failure of a PES to operate, or spurious operation of a PES.

STEP 3. *Define the safety criteria*: In order to determine the target reliability of the PES systems, there has to be some safety criterion / criteria from which this is derived, (eg the frequency of release of radionuclides curves used in the Nuclear industry). From such a criterion, it is then possible to determine the target failure rate, or the probability of failure-on-demand of the PES.

STEP 4. *Design the system*. There are three systecharacteristics ("system elements") that govern the design and assessment strategy:

- The system configuration: the configuration should be such that no single channel failure (hardware or software) should cause a dangerous failure of the total system.

- The reliability of the hardware: this aspect addresses the safeguards applied to the system to reduce random hardware failures in a dangerous mode.

- The overall quality: this aspect focuses on the quality invested in the activities throughout the life-cycle of the equipment, ie from specification, through design, implementation, operation, to maintenance.

STEP 5. *Perform an assessment*: A quantitative assessment of the hardware design is required, together with an overall qualitative assessment, where considered appropriate.

A short-cut "component count" quantitative assessment method is suggested, which utilises "worst-case" guide figures. These guide figures are quoted for the various main building block components of a PES system. Three levels of guide figures are provided, ie "typical", "maximum", and "minimum". It is suggested that the "maximum" figures are used initially; if the calculated reliability does not meet the target figure, then justification is needed for the use of the lower figures, and a recalculation is made.

Guide figures are also provided to cater for the adverse affects of design dependencies (ie by the use of "beta-factors"), and for the advantages of using watch-dog timers and self-test features.

The qualitative assessment of the hardware and software is made by the use of checklists, one for each phase of the hardware and software life cycles. The checklists provide the assessor with a disciplined approach for checking that the best design and engineering practices have been carried out correctly and well documented, to assure a safe system.

STEP 6. *Check the adequacy*: This final step is simply to check whether the results of the quantified assessment have met the desired target. If not, then a more-detailed assessment is required. If, after the more-detailed assessment has been carried out, the safety target is still not met, a redesign will then be necessary.

5. HSE study on safety-related software

In 1987 the HSE commissioned a study into the procurement and development of software for safety-related systems; the results have just been published [6], and are currently under review by a number of UK experts.

The study included a survey of current development techniques used by a number of industries, university researchers, and government bodies, for safety-related software. The results of the study show that a non-unified approach exists, and that there is a need for an awareness programme and the establishment of a simple, pragmatic approach.

5.1. Objective of the study

The aim of this study is to recommend the software development techniques to be used in the future, and to propose a straightforward, pragmatic framework for their use.

5.2. The framework

The framework proposed is basically a database of recommended methods and standards, with a two-tiered index. The first tier relates to the activities to be carried out at each stage of the software life-cycle; the second tier relates to the recommended standards and activities. Overlaid onto these two tiers are Software Integrity Categories and Organisational Roles.

5.3. Software integrity categories

Four Software Integrity Categories are used; Category 1 signifies the highest level of trust, Category 4 the lowest. The categories are not defined explicitly, but two methods that could be used to determine safety integrity categories are provided, based upon:

- categorisation by risk, for example the number of deaths or injuries that could occur

- categorisation by hazard or criticality, for example the loss of various critical functions

5.4. Organisational roles

The study identifies 7 key Organisational Roles in the life-cycle of the safety-related software:

- the user

- the procurer

- the designer

- the developer

- the evaluator (assessor)

- the certifier

- the maintainer

5.5. Utilising the framework

The database will be accessed by each user and will inform the user of:

- the activities to be carried out

- the tools and methods to be used in carrying out the activities

for each phase, for each of the Organisational Users, and for each Software Integrity Category.

6. The UK ministry of defence standard
(DEF - STAN - 0055)

The impending Defence Standard DEF-STAN-0055 for Safety Critical Systems will undoubtedly have a major impact on the UK real-time software scene, and it therefore warrants more than a passing mention here.

DEF-STAN-0055 was due for promulgation by the UK Ministry of Defence (MoD) early in 1989, but the latest estimates suggest that it will not be available before early summer 1990. The standard will focus on software since this is the area in which, previously, there has been no unified approach. With the software issues in mind, a draft policy statement for the procurement and use of software used in safety critical applications was issued by the MoD in summer 1988 and circulated to selected organisations for comment.

It should be noted that a "Safety Critical System" is one in which a system failure due either to a design fault or a random hardware fault will result in a risk to human life. It is interesting to note that (at the time of writing) there are no categories within this definition; hence a system will be deemed safety critical whether there is risk to one human life or several hundreds.

The essence of the new Standard is one of formal (ie mathematical) methods for the development of the software, such that it will be possible to provide "formal proofs" of compliance between the software code and its requirements specification. Thus the MoD approach is one of "proof of 100% correctness".

A requirement for all defence projects is that a hazards analysis should be invoked to determine whether or not the system under review (or any of its parts) is safety critical. Thus all systems are initially to be considered as safety critical unless a hazards analysis has been carried out to demonstrate otherwise.

The policy for safety critical software will be a mandatory one, and will address the procedures to be applied throughout the entire software lifecycle, from specification and design, development and implementation, through to operation and maintenance.

The MoD has defined three authorities who between them bear the responsibility for the safety of the software:

- the SOFTWARE SAFETY AUTHORITY, who is the MoD project manager, and is responsible for ensuring that a Safety Plan is produced, and that a Safety Record is initiated and maintained.

- the DESIGN AUTHORITY (a nominated contractor to the MoD), responsible for managing software safety, and for verifying that the specified safety requirements have been met.

- the SAFETY ASSURANCE AUTHORITY, who also is a nominated contractor to the MoD, and is an independent software safety assessor. The Safety and Reliability Directorate (SRD) of the United Kingdom Atomic Energy Authority is such a Safety Assurance Authority. The main responsibility of the Safety Assurance Authority is to verify that the requirements of DEF-STAN-0055 have been met and continue to be met throughout the life of the system.

SPECIFICATION: The starting point for formal proofs is the method for specifying the requirements. The advantage of a mathematically expressed specification is that it is possible to check for completeness, ambiguity, and consistency. The main disadvantage with mathematically-based specifications is the lack of "visibility" of the software function, making it difficult to understand what the software does. The MoD has recognised this shortcoming, and the formal specification will be accompanied by a plain English version.

PROGRAMMING LANGUAGE: The MoD prefers high level languages to low level, but requires that the chosen language possesses a number of specified desirable characteristics. For example:

- it should be suitable for static analysis by automated tools

- it must be a mature language

- the language should have formally defined syntax and semantics

The languages favoured by the MoD are "safe" subsets of PASCAL (SPADE PASCAL) and ADA.

DEVELOPMENT: The development of the safety critical parts of the code will be carried out using rigorous codes of practice.

The MoD has recognised that there is a potential "weak link" in the development chain - ie that of the tools used. If the tools contain errors, then there is the possibility that errors may be introduced (or fail to be detected) by the tools themselves; thus the tools will also be subjected to the standard.

Thus all static analysis tools, testbeds, integrated project support environments, compilers, and the like, fall into this category; compilers which conform to the language definition will be used.

DESIGN: The design will be achieved by the use of structured design techniques, and formal mathematical techniques. Techniques to reduce the frequency or consequence of failure (eg on-line testing, fault detection, and defensive programming) are required in the design.

The size of modules will be limited to that commensurate with the application of formal methods and proofs. The current recommended maximum size is 8K of object code, with the absolute maximum being 16K.

QUALITY ASSURANCE: The MoD has specified that the requirements expressed in the NATO Defence Standards AQAP-1 [2], and AQAP-13 [3], shall be implemented for the Quality System. AQAP-13 has been well used and is firmly established on the UK MoD projects; it addresses all the important quality assurance aspects, including tools, techniques, methodologies, and configuration management. Configuration control is considered by DEF-STAN-0055 to be a very important part of Quality Assurance [8] and therefore part of the Quality

Assurance plan will be the strict application of an approved automated configuration control system.

VERIFICATION AND VALIDATION: These two activities will be formalised, and applied throughout the life cycle; static and dynamic analysis will be applied progressively throughout the life-cycle to reduce faults in the process.

CERTIFICATION: At the end of the project, a certificate will be issued which will state that the Safety Requirements have been achieved.

8. Author's comments

The widespread inconsistency that exists in the methods for developing and assessing dependable software within the United Kingdom has been recognised, and moves have been initiated to alleviate the problem.

The ACARD report proposes an ambitious (eg how can 1 fault per 10,000 lines of code be assured?) plan which is very formalised, based upon a system of registration, certification, and licensing. The detailed requirements of such a system will be extremely difficult to formulate; the costs to administer it would undoubtably be considerable, and would eventually be passed on to the end-user.

What is disturbing to note is the the apparent desire to charge with criminal negligence any person(s) who cause a death as a result of making an error. Surely more effort should be put into trying to ensure that "it never happens again" rather than initiating a "witch hunt".

It would be difficult in such an enquiry to determine who is to blame if software is installed with faults in it (as most software is). For example, an error initiated by the coder or designer would also have to be "missed" by many others, eg the peer reviewers, the safety analyst, and the person responsible for specifying test philosophy. Indeed senior management could be implicated if it could be shown that there were short-comings in the training or quality assurance requirements.

It is very comforting to note that part of the plan is a formalised method of "learning from past mistakes" ie the collection and analysis of data from past incidents and accidents.

At the review of the IEE Report on Safety Critical Software it was emphasised that the software engineer has been singled out for certification both in this study and the ACARD report; this has been very controversial, the main argument being that software is only one component of a system, and other components (eg hardware, human factors, etc) should be subjected to similar treatment.

The HSE PES Guidelines issued by the UK Health and Safety Executive (HSE) are limited with respect to software, since software is assessed qualitatively via checklists only, and no guidelines are provided on the acceptance criteria for the software. However, the Guidelines have gained favour as a pragmatic approach which can be built upon for specific applications. The HSE has recognised the shortcomings by initiating their separate study on Safety-related Software.

The HSE Study on Safety-related Software has received much debate in the UK. The reviewers of the study fall into two main camps. The first camp wish to quantify the return in the investment for adopting a particular set of techniques, and argue that the outcome of the study did not provide support for which set of techniques is better than another. The second camp wish to use "engineering judgement" rather than objective data, ie select current best practices, and determine which subsets of these are appropriate for a given level of criticality. Such subsets would be chosen by consensus and common agreement. This "argument of excellence" is gaining popularity both in the UK and in Europe. The study was considered complex, and is currently under review following invited criticism.

The MoD and ACARD approach is to encourage the development of safety-critical software which is "proven 100% correct", via the use of formal (mathematical) methods. Such proofs demonstrate compliance with the requirements specification; they cannot demonstrate that the requirement is correct in the first place. The techniques are not yet fully mature, and currently cannot show compliance with all requirements, for example accuracy and time response.

Formal methods are further limited in the size of module that can be formally proven; when this restriction is augmented by the requirement for specialist skills in formal methods and the inherent increased accompanying costs, formal methods are not likely to find favour in medium risk industrial applications.

Formal methods are not yet the panacea for all software ailments, and we should be well aware of this fact. However, research into formal methods should continue, because this is certainly a promising way to reduce the error content of software. However, it is conceit to think that it is possible to prove 100% compliance between software and requirements, except for the most trivial of examples. Formal proofs are long and complex themselves, and must surely be subjected to the same human error processes as the software itself. It would be folly to abolish "safety nets" (such as the traditional and well-established fault tolerant techniques) in favour of mathematical proofs.

The HSE approach is to develop packages of techniques for varying levels of risk. This is more likely to be accepted by industry than the formal approach, since industry has many different levels of risk, and the cost of implementing a solution will be mainly established on a commercial basis. The MoD has defined just one level of risk for safety-critical systems.

References

[1]   "Software - A Vital Key to UK Competitiveness" an ACARD report ISBN 011 630829, available from Her Majesty's Stationary Office, PO Box 276, London SW8 5DT.

[2]   AQAP-1: "NATO Requirements for an Industrial Quality Control System" (1984), available from Her Majesty's Stationary Office, PO Box 276, London SW8 5DT.

[3]   AQAP-13: "NATO Software Quality Control System Requirements" (August 1981), available from Her Majesty's Stationary Office, PO Box 276, London SW8 5DT.

[4]   "HSE Guidelines on Programmable Electronic Control Systems in Safety-related Applications, PART 1 ", ISBN 0 11 883906. Her Majesty's Stationary Office, PO Box 276, London SW8 5DT.

[5]   "HSE Guidelines on Programmable Electronic Control Systems in Safety-related Applications, PART 2 General Technical Guidelines", ISBN 0 11 883906. Her Majesty's Stationary Office, PO Box 276, London SW8 5DT.

[6]   "Safety-related Software Study", UK Health & Safety Executive.

[7]   Draft IEE report "Safety Critical Systems Employing Software", available from the Department for Trade and Industry, London, UK, reference: IT/24/27/39.

[8]   "Configuration Identification and Control of Software for Microprocessors" by A Lawrence, CEGB. Presented at the 4th Annual Symposium on Microprocessor-based Protection Systems (for the Institute of Measurement and Control), 10 December 1987.

SOFTWARE VALIDATION BY MEANS OF STATISTICAL TESTING: RETROSPECT AND FUTURE DIRECTION

Pascale THÉVENOD-FOSSE
Laboratoire d'Automatique et d'Analyse des Systèmes du C.N.R.S.
7, Avenue du Colonel Roche, 31077 Toulouse Cedex - France

Abstract

Statistical testing is a practical approach to software validation, involving both fault removal and fault forecasting. It consists in stimulating a program by test samples which are randomly selected based on a defined probability distribution of the input data. The first part of the paper provides a short view of the current state of investigation in statistical testing area. Then a comparison of the strengths and weaknesses of statistical testing with those of deterministic testing allows to put forward the complementary, rather than competing, features of these two methods of generating test data. Hence, a validation strategy organized in three steps is proposed, which mixes statistical and deterministic test data. The first two steps aim at revealing faults, and the third one provides an assessment of operational reliability. Future work to support the strategy is outlined.

1. Introduction

Nowadays, software validation remains an acute problem in the area of computing systems dependability. It involves two notions, namely fault removal and fault forecasting [25]. Substantial work has already been done in both fields, which are often investigated separately and through distinct approaches. Testing is the most used **fault removal** technique. Dynamic testing is usually performed at the end of the software development phase, with the aim of uncovering faults that exist in the program code to correct them before the release [32]. A program is tested by executing it on a computer with selected input data. Since exhaustive testing is not possible, one must choose a subset of the whole input domain. The test efficiency with respect to design fault exposure depends on the adequacy of the selected test data. Most often, **fault forecasting** calls for the utilization of software reliability models which assume that an adequate database of observed failures has been collected preferably during actual operation. It is recognized that the predictive quality of these models varies with the context (software type, available database), and no model has emerged as the most accurate one in all cases [1]. Moreover, such models are not well suited to reliability predictions during the software development phase, owing to the need for a database linked to failures during the system operational life [36].

Since the 1960s, a large amount of literature has evolved on software testing, which deals with the question of test data selection. Methods of generating test data proceed from two main principles: one is deterministic, the other probabilistic. In the first case, called **deterministic testing**, input data are derived through a careful analysis of the software specification or/and the program code. One has to devise a series of test cases that thoroughly exercises the software functionality (black box testing) or/and internal structure (white box testing). In the second case, called here[1] **statistical testing**, input data are randomly selected based on a defined probability distribution of the input domain, avoiding tedious investigations to produce test samples. Until nowadays, most of the studies have focused on deterministic approaches [30, 33].

This paper concerns software validation before the system operational life. *Section 2* provides a succinct survey on previous work related to statistical testing. The retrospect shows how the use of relevant random test data can be a practical and efficient means to reveal faults, and also to assess future reliability

---

[1] Other designations are found in some papers, such as *probabilistic testing* or *random testing*.

in operation by making inferences from the testing to operational environments. With regard to fault removal, deterministic testing and statistical testing go all over the program input domain in two basically different ways, so that they could be thought of as complementary rather than competing approaches. In *section 3*, a comparison of the strengths and weaknesses of these two methods of generating test data states their main complementary features. Hence, the validation strategy proposed in *section 4* is based on statistical testing combined with additional needful deterministic test data. It is organized in three steps aiming at uncovering faults and assessing the software reliability which will be perceived by users. Much work remains to be done in the proposed direction, and *section 5* outlines future studies to support the strategy.

2. Retrospect

Software testing procedures using randomly selected test samples have been studied since the early 70s. Some authors focused on the fault-revealing power of statistical testing, while others were interested in the fault-forecasting challenge. Main theoretical and experimental results they have published, most of them in the 80s, are summarized below. A detailed and more complete survey is reported in [37]. In previous work, different terms and notations are associated to similar notions, depending on the authors. This section is an attempt to give a nearly uniform presentation of the results.

2.1. Fault removal viewpoint

During a statistical test experiment, a program is run with a set of N randomly selected test data. The arising question is *how to quantify the efficiency of statistical testing as a verification tool?* That is to say, one has to evaluate the probability Q_N that a set of N random test cases detects at least one error in a faulty program or, inversely, the number N of test cases required to ensure a wanted target Q_N. The **detection probability** Q_N, also called **test quality** thereafter, is a measure of the test hardness. $1-Q_N$ is the probability that a faulty program does not fail during a test experiment, i.e. that an experiment states a faulty program is correct. Obviously, Q_N and N are closely linked through relations involving 1) *actual* faults and errors they can create, and 2) the *chosen* probability distribution of random input data. Section 2.1.1 focuses on test quality assessment.

As statistical testing involves a large number of test cases, we are specially faced with the well-known *oracle* question, i.e. *how to determine the correct response of the software to each input data?* In practice, a **back-to-back testing** scheme is often assumed. It requires the independent development of at least two variants of the program to be tested. Two (or more) variants can then be run with the same randomly generated test inputs, and the results are compared. A discrepancy indicates that one or several variants failed to produce the correct response. But the acute problem of **related faults**, which may lead two or more variants to produce the same wrong results for some inputs, can drastically decrease the test quality. Section 2.1.2 concerns back-to-back statistical testing with respect to related faults.

2.1.1. Test quality assessment

Girard and Rault from Thomson-CSF Research Laboratory (France), seem to pioneer the related theoretical research. In [21, 34], they investigate whether it would be possible to take advantage of the large body of hardware testing methods already devised and used in industrial environment, in order to design methods for software verification. They place a particular emphasis on two probabilistic procedures to generate test sequences, from which they think promising extensions could be found. In the first one, called *test case generation by simulation*, faults are purposely injected in a variant which is then tested against a "good" variant back-to-back, with random input data generated by a Monte-Carlo method. The process is stopped as soon as a given proportion of the seeded faults is revealed. Let R_i be the percentage of faults first revealed by the ith test case. An approximate value of the expected number N of input data required to reveal n faults is given by equation 1, assuming that R_i has a constant value R, $\forall$ i.

$$N \geq 1 - [\text{Log } n / \text{Log}(1-R)] \tag{1}$$

In the second procedure, called *statistical and comparative testing*, no fault is seeded but two variants of the program are tested back-to-back. Let p be the probability that a program fails on a random input data. If the probability that both variants produce the same wrong result is assumed to be nil, the number N of random data required to reach a detection probability Q_N is such that:

$$(1-p)^N \leq 1-Q_N \tag{2}$$

Hence,

$$N \geq Log(1-Q_N) / Log(1-p) \tag{3}$$

As in practice the exact calculation of p may not be feasible, the authors propose two approximate methods to assess a maximum value for 1-p, called the "worst case error condition". The first one, based on the program graph, is related to the frequency of use for each instruction. In the second one, p is determined through simulation by injecting faults in the program: the average detection probability R (defined in the first procedure) which is obtained gives the value of p.

Duran et al. [16-18] from the University of Texas at Dallas, present a theoretical study of statistical testing versus partition testing, illustrated with simulation results. The term partition testing refers to any test data generation method which partitions the input domain D into k subsets D_i (i = 1, ..., k) and forces at least one test case to come from each D_i. Hence, path testing is treated as a special case of partition testing. The detection probability Q_N of a statistical test experiment (input data are chosen randomly from the whole domain D) is compared with the detection probability, denoted thereafter q_N, of a partition test experiment in which n_i test cases are chosen randomly from each D_i. Equations 4 (similar to equation 2) and 5 give Q_N and q_N, respectively. p_i denotes the probability that a program fails on an input data chosen randomly from a subset D_i, and N = $\sum n_i$ with i = 1, ..., k.

$$Q_N = 1 - (1-p)^N \tag{4}$$

$$q_N = 1 - \prod (1-p_i)^{n_i} \quad \text{with } i = 1, ..., k \tag{5}$$

With respect to the same partitioning, and if an input data chosen randomly from the whole domain D has probability d_i of being selected from D_i, one has p = $\sum d_i p_i$. Hence, from equation 4:

$$Q_N = 1 - (1-\sum d_i p_i)^N \quad \text{with } i = 1, ..., k \tag{6}$$

Simulation results with different values of the (d_i, p_i) pairs suggest that statistical testing may often be more cost effective than partition testing strategies. For example [18], a 25 subset partition testing scheme simulated with one randomly chosen test data per subset (k = N = 25, n_i = 1) and a new set of d_i and p_i values picked for each trial, leads to $Q_{25} \geq q_{25}$ in 14 out of 50 trials, and the mean value of Q_{25}/q_{25} is .932. A similar experiment, but with $k = N = 50$ gives $Q_{50} \geq q_{50}$ in 7 out of 50 trials, and the mean value of Q_{50}/q_{50} is .949. Other theoretical results supported by simulations, are reported on the expected number of errors detected by statistical testing on the one hand, and by partition testing on the other hand. They reinforce the conjecture that statistical testing is often more cost effective than partition testing. One of the authors' conclusions from this work is that random testing of individual paths (partition testing assuring one test case for each path) could be less effective at revealing faults than statistical testing from the whole input domain.

Hamlet's papers investigate two models with the aim of assessing the value of Q_N [22-24]. *The first one* is the model defined by Girard and Rault (equation 2). Hamlet emphasizes that, as faults are put into the program, the space of faults is a textual one while the space of test data is the input domain, but the connection between inputs and textual coverage is not known. He suggests then to evaluate p by assuming that the number of faults is proportional to program length (on average, there are S faults per line), and that j distinct inputs, uniformly distributed over the input domain containing D points, reveal each fault. Hence in a program of L lines: $p = jSL/D$. This simple formula, based on textual sampling, is attractive. But parameters j, D and S remain to be known. *The second model* is derived from the sampling theory defined in a different context by Valiant [38]. Equation 7 is used to evaluate the number N of test data required to ensure that, when a program has not failed during a test experiment, the probability is better than $(1-p)$ that the failure probability is less than p. Hence, with our notation, $Q_N \geq 1-p$.

$$N > (2/p).(2+\ln(1/p)) \qquad\qquad (7)$$

For example, $p = .001 \Rightarrow N > 17815$. It means that after 17816 correct executions, the program failure probability is less than .001 with the confidence level .999. From this model, Hamlet compares statistical testing with partition testing, and numerical results strengthen Duran's conclusions. It must be noted that equation 7 involves an approximation that overestimates N for a given p. Indeed, it is more restrictive (because $Q_N = 1-p$) and pessimistic than equation 3 which gives N values more than twice lower [37]. For example, $p = .001$ and $Q_N = .999 \Rightarrow N \geq 6904$ from equation 3.

The following **empirical investigations** give a more convincing measure of statistical testing efficiency.

- In [12, 29], mutation analysis is used to provide an assessment of statistical testing. It consists in inserting simple changes in a program to find out if the test data reveal them. A mutant is a copy of the program under testing with a simple change. It is killed as soon as it produces a result which differs from the original program. The test data effectiveness is quantified by the proportion of killed mutants [13]. Experiments on a small sorting program show that statistical testing can kill most of the generated mutants, but its effectiveness depends on the interval from which the input values are drawn [12]. In Ntafos's study, seven programs (5496 mutants) are tested with from $N = 8$ to $N = 20$ random test data. Despite these few numbers of inputs, 79% of the mutants are killed by statistical testing, as compared to 84% for branch testing and 90% for required pairs testing. Statistical testing performs better than branch testing in four programs, and better than required pairs testing in one program. It is least efficient when special input values are needed to kill some mutants. That is the case in two triangle classification programs where equal values for two or three of triangle sides are important but difficult to generate randomly according to a uniform input probability distribution (see section 5.2).

- In [17, 18], experiments which encompass eight actual programs containing known "real-life" faults, show the high revealing power of statistical testing with respect to such design faults. A variety of test coverage measures, such as the proportion of segments or branches executed by sets of random test cases, is also proposed and used to evaluate how well a program is scanned. The experiments, carried on five programs with small sets of data ($20 \leq N \leq 120$) and various ranges for input variables, lead to good coverage percentages: on average 97% of segments, 93% of branches, 72% of required pairs [37].

- Finally, one has to mention the automated tool for test data generation presented in [31]. Random numbers are used to generate input data that exercise a given set of paths in a program graph. The implementation of a prototype test data generator for Fortran programs is described.

2.1.2. Back-to-back testing wrt related faults

Usual techniques to provide software fault tolerance at execution time rely on redundant software written to the same specification, i. e. on design diversity [39]. The main weakness of fault tolerant software arises from the presence of related faults in several program variants, leading to simultaneous failures so that the fault tolerance mechanism does not work for some inputs. **Back-to-back testing,** called *comparison testing* by some authors, is faced with the same problem. Hence it is an acute question since residual related faults will not be tolerated during the system operational life.

Brilliant et al. [2, 7], from the University of Virginia at Charlottesville, examine the potential fault-revealing power of back-to-back testing by comparing it to that of an *ideal testing system* defined to contain an *oracle* for determining the correctness of the outputs. They define the *failure subspace for a fault f* to be that subset of the input space in which f causes failure to occur. The capability of back-to-back testing to reveal f depends on the relationship between the failure subspace for f and the failure subspaces associated with faults in other variants. The portion of input space in which f causes failure and all compared variants contain faults which cause identical failures is called the *overlap subspace for f*. The *overlap ratio for f* is then the ratio of the probability of selecting a test input in the fault's overlap subspace to the one of selecting a test input in the fault s failure subspace. The authors refer to [6] in which a first analysis of the faults revealed in a previous N-version software experiment indicates that very few failure subspaces are identical, although intersections occur. Then, to search for the effects of such incomplete overlaps on back-to-back testing, several Markov models (one or several faults, two or more variants) are proposed and analyzed, assuming that successive test inputs are randomly selected according to a common distribution in a series of independent trials. They lead to analytical results the practical relevance of which depends on the overlap ratios that actually occur. These ratios are assessed for 34 "real-life" faults. The study shows that back-to-back testing is almost as effective as oracle-based testing in finding all faults except those having overlap ratios close to one. Thus it is a powerful method for a wide range of overlap ratios. Furthermore, the presumed performance improves rapidly as the number X of variants used for comparison increases, at least for typical small values of X and for the types of fault interrelationships one can expect to encounter in practice.

The theoretical work presented by **Littlewood and Miller** [26, 27], from City University (UK), also deals with the question of simultaneous failures but in a qualitative way. They examine the conceptual model of N-version software defined in [20], based on the assumption that a single methodology is applied to develop the versions and that execution of a program involves random selection of an input case from the input space. Hence, a generalization of the model is proposed, in which the versions are developed using diverse methodologies (different development environments, and/or languages, and/or testing methods, etc). The key idea is that each methodology induces its own measure on the population $P = \{\pi_1, \pi_2, ...\}$ of possible programs for a particular set of requirements. That is to say, the probability that a program version π_i is actually written depends on the applied methodology. If the methodologies are very diverse, one would expect a program π_i with a high selection probability under one methodology to have a low selection probability under others. Theoretical computation from the general model shows that the use of diverse methodologies to develop different versions decreases the probability of simultaneous failure. Empirical investigations are still needed to confirm the qualitative work, and quantify the advantages of methods proposed for allocating diversity between versions.

The **HALDEN project** consists of an European collective software diversity research in the nuclear field. It covers several projects, namely PODS (Project On Diverse Software) and STEM (Software Test and Evaluation Methods), involving United Kingdom, Norway and Finland. **PODS** aims at evaluating the merits of using N-version software. It focuses on the production of three diverse programs to the same requirement which is for a reactor over-power protection trip system [3, 39]. In the acceptance testing phase, each version is separately submitted to a common set of 672 test cases grouped into sequences of deterministic tests and random test data selected from a uniform distribution over the full range of each input parameter. Once the three programs have passed the acceptance tests, they are tested against each other back-to-back to track down residual faults: 2472 deterministic test cases are derived using equivalence partitioning, boundary value analysis and decision tables techniques; 662 816 statistical test data are generated which conform to four distinct distributions (uniform or gaussian over the input domain, rectangular or gaussian around domain boundaries). Seven different residual faults are revealed by discrepancies between the outputs of the three versions. Two faults are common to two programs and would have won a majority vote causing an overall system failure. To assess the probability p of failure per test case associated with each of the seven faults, a subsequent back-to-back testing is performed with uniformly distributed random input data. The **STEM** project, launched at the end of PODS, aims at examining some main validation techniques used during software development and operation. It makes use of the existing programs of the PODS trip software which contain known faults. Experimental investigations on back-to-back statistical testing efficiency at revealing faults, detailed in [4], encompass the four input probability distributions used in PODS. Deterministic and random test data are assessed against several criteria, such as percentage of statements executed, branches executed, undetected real faults, undetected seeded faults ... From the numerical results, it is clear that uniform random data over the input domain result in better coverage measures than any of the deterministic test data sets, and exhibit the most rapid growth in coverage. A more recent empirical study focuses on the failure characteristics of 46 faults revealed in the three programs developed in PODS [5]. The large amount of reported numerical results refers to two types of statistical tests called *uniform random tests* and *random walk tests*. In the first case, random input data are conformed to a uniform distribution over the input domain. In the second case, all the input parameters are altered by a random "step" on each execution, and the maximum step size for the random walk is typically 5% of the full range of the parameter. From the empirical results, it emerges that: 1) failures under uniform random input tests can conform to the assumption of constant failure probability p, but failures under random walk input tests do not because the failure probability on the ith test data is greater when the program has failed on the $(i-1)$th test case; 2) the failure rates of a population of faults in a program driven by uniform random data are nearly exponentially distributed; 3) the majority of non-identical fault pairs exhibits simultaneous failure characteristics close to the independence assumption.

Dunham's papers [14, 15] provide a detailed report on four data gathering experiments which are a part of a program pursued by NASA - Langley Research Center in order to develop a method for predicting operational reliability of flight control software. Two 3-version programming systems are studied and the software tools developed to support the experiments include two *N-version testers*, one for each system, which implement a back-to-back statistical testing scheme. Random test inputs are generated according to a prespecified program usage distribution. Each system is extensively tested for over 14 million test cases. Roughly speaking, the collected testing data seem to confirm conclusions 1 and 2 of the STEM project. In particular, software "error bursts", i.e. sequences of failures due to the current state of the system and the correlated nature of the inputs, are observed.

2.2. Fault forecasting viewpoint

Already experimented on several projects, **IBM's Cleanroom development method** [19, 28, 35] is a technical and organizational approach to produce software with certifiable reliability. It is organized around the incremental development of the product. The main purpose is to prevent the entry of faults during the software development. The focus of the method is imposing discipline on the development process by integrating formal methods for specification and design, nonexecution-based program development, and a certification procedure based on statistical testing. The *first priority*, namely *fault avoidance* (rather than fault removal), is achieved by using human verification in place of program debugging to prepare software for system test, so that there are few faults left by the time testing is done. The *second priority* is to provide valid *certification* of the software quality through representative-user testing at the system level. Hence the **certification procedure** explained in [10, 11] and briefly summarized below, aims at measuring the operational reliability which will be perceived by users. It is based on statistical testing which permits an accurate simulation of the software operating environments. Product inputs and their probability distributions are defined for all planned operating modes, and test samples are randomly selected based on these probabilities. The life cycle of incremental product allows a continuous quality assessment. The approach for measuring the reliability of the increments is through the use of *Mean Time To Failure* (MTTF) estimations. Considering a product increment under test, any observed failure is analyzed and engineering changes are proposed to correct the software. These changes likely increase the MTTF of the software and that increase should be captured. Let $MTTF_m$ be the MTTF after m changes. The model used for reliability prediction is

$$MTTF_m = MTTF_0 . R^m \qquad (8)$$

where $MTTF_0$ is the initial MTTF (before the first change) and R is the fractional improvement to the MTTF from each change. A corrected logarithm least squares technique is proposed to estimate $MTTF_0$ and R from the interfail times recorded during statistical test experiments. As changes affect the MTTF to greater or lesser extents depending on the rate at which the fixed fault caused failures, R is an average observed value. The practical approach to MTTF prediction is to record the execution time for each statistical test case run, sum the times between successive failures and input these interfail data into the theoretical model. MTTF predictions are then made on an increment basis from equation (8). A measure of the product reliability is computed as a weighted sum of the increment MTTFs predicted by the model. The weighting coefficients account for the increment contribution to the whole product functionality. It is noteworthy that if a target MTTF has been defined for a software, the computed reliability indicates the contributions from the released increments toward the target MTTF. The procedure is illustrated with an example in [10]. Numerical results show that the model projects the reliability growth trend observed in the interfail data, and provides accurate MTTF predictions. Moreover, the comparison of the results with those generated by three other well-known reliability prediction models is favorable to IBM's approach.

A different approach is presented by **Cho**, from COMPUTA Inc., in his last book [9] where he treats in great detail a software methodology first introduced in [8] and called *Quality Programming*. Emphasis is placed on deployment of statistical quality control during every stage of the software life cycle, from requirement specification up to software acceptance. The basic idea is to consider a program as a factory which processes raw materials (the data input to a program) into usable products (the data output by the program). Such an analogy between a factory and a program allows the use in software industry of statistical quality control techniques defined for hardware and from which numerous firms have benefited since World War II. The acceptance or rejection of a software is based on the results of statistical testing runs. If the software contains a "tolerable" number of faults (i.e. a target reliability is reached), it is accepted. *Two acceptance sampling plans* are described: *single* sampling plan and *sequential* sampling plan. In both cases, the plan is established with a criteria, called *the acceptance criteria*, involving a producer's risk α_1 (probability of rejecting an acceptable product) and a user's risk α_2 (probability of accepting a product below target reliability). The values of α_1 and α_2 (typically .05) must be agreed on by both parties before a plan is generated. The proper number of statistical test cases is then determined from (α_1, α_2) either before implementing the plan in case of single sampling, or during the experiments in case of sequential sampling. Several input probability distributions are analyzed, leading to five generation methods of random inputs: regular, weighted, boundary, invalid and special tests. A key concept is then the specification of the software input domain by means of the *Symbolic Input Attribute Decomposition (SIAD) tree* which is dual-purpose: it is a tool for describing the input domain, and a basis for construction of different kinds of statistical test inputs.

Two main differences between Cho's ideas and Cleanroom process arise. *First*, the Cleanroom certification model accounts for the reliability growth during development, while [9] does not involve the notion. *Second*, in Cho's book statistical testing aims at both fault removal (unit test and system test) and fault forecasting, while Cleanroom process uses it only with the purpose of fault forecasting.

2.3. Conclusion

Compared with deterministic testing, few theoretical studies have been performed on statistical testing. But experimental results have already led different authors to agree with the high fault-revealing power of statistical testing. Furthermore, it is noteworthy that several recent N-version software experiments use back-to-back statistical testing in addition to deterministic test cases during the certification process. Examples of such studies in safety-critical applications (nuclear, flight control, aerospace), recalled in [37], show that residual faults are often revealed by extra random inputs applied after thorough deterministic tests. Indeed, the complexity of actual software systems works in favor of statistical testing which is more and more helpful.

3. Statistical testing versus deterministic testing

Nowadays, current software validation strategies incorporate either deterministic test cases, or statistical test cases, or both. In practice, the choice of the used strategy is most often related to various factors, such as available testing tools, time limit, allocated budget, personnel's cultural background and usage, etc. We present here a comparison of the strengths and weaknesses of the two main test data generation methods which leads us to conclude that deterministic and statistical testing are two complementary, rather than competing, approaches. Figure 1 summarizes the main distinctions on which we briefly comment below. They justify our preference for the validation strategy designed in section 4. Numerical results given in section 5.2. illustrate some of the points involved in the comparison.

Statistical testing needs a high *number N* of test cases, but a random *input data generation* is easy, at least for a wide range of systems. On the contrary, N is reduced in deterministic approaches, but the input data selection requires tedious investigation. The evaluation of the *correct results* is more heavy for statistical testing (input data are numerous and not selected) than for deterministic testing. Concerning the *fault-revealing power*, one can reasonably think that the following relations are verified:

- errors due to faulty behaviors related to the structural and/or functional criteria used to select deterministic test data are most probably detected by such deterministic inputs; hence, statistical testing is expected to reveal a lower percentage V (V < U) of the faults creating these errors, called thereafter "regular" faults;

- for other faults, called thereafter "marginal" faults, no proper test inputs have been purposely put in the set of deterministic test inputs; the probability of revealing these faults is then an increasing function of the number N of executions; hence statistical testing involving a highest value N is expected to be more efficient (Z > Y), leading then to a more rapid reliability growth.

Finally, with regard to *reliability prediction* [10], the statistical method provides a scientific basis for making inferences from the testing to operational environments. It is not the case with the deterministic techniques as they tend towards a subjective selection of input data.

	DETERMINISTIC	STATISTICAL
# test cases N	low	high
Input data generation	often tedious	more easy
Output data calculation	from chosen inputs	very tedious
% of revealed faults: • regular faults • marginal faults	U Y	V < U Z > Y
Reliability prediction	*biased* by *selective* choice of an input data set	*unbiased* by using *operational* input profile

Figure 1. Complementary features of deterministic and statistical methods.

From figure 1, there emerge **three main strengths of statistical testing** in relation to deterministic testing:

- an automated generation of (unbiased) input data is often easy, so that large test samples are achievable;
- a better percentage of marginal faults should be revealed;
- its probabilistic nature is well suited to predict the future operational reliability with a defined confidence level.

In return, **two weaknesses** are unquestionable:

- some specific faults (linked to boundary input values, for example) can stay difficult to track down by means of random data, whatever be the chosen input probability distribution; such faults are easily revealed by a small number of appropriate deterministic input data;

- how to calculate the correct results for large random input samples is a fundamental question; in real complex software, such computation is not always feasible and, in any case, it can drastically decrease the cost effectiveness of statistical approach.

In the present state of the practice, a proper solution to turn away the second drawback seems to be the use of back-to-back testing scheme, which assumes that at least two program variants are available. Hence, although the global validation strategy proposed in the next section fits over any software, programs two or more variants of which are usable provide a practical application field. Let us note that a prototype is an example of program variant. But other solutions to the *oracle* problem must be investigated in future work.

4. A global validation strategy

To take advantage of the strengths of both input generation methods, one investigates a validation strategy which combines statistical and deterministic test cases, with the double aim of removing faults and of assessing the future operational reliability of the software. Three steps are necessary. The two first steps aim at fault removal, and the third one involves fault forecasting. For fault removal, various mix test strategies have already been suggested in previous work. In most of them, deterministic test cases are first applied and then enhanced by additional runs with random data. In others, testing begins with random data which detect a lot of errors without a great deal of effort, and goes on with deterministic data to ensure the coverage of extremal input values [4, 18]. In our mind, the second ordering is most cost effective provided that an appropriate input distribution is used to generate random data, as explained in § 5.3. The proposed strategy can be used for unit test and for system test.

4.1. Principle

The three following steps should constitute a careful and efficient software validation strategy.

Step 1: *Statistical testing for fault removal.*
Test cases are randomly selected based on an input probability distribution which lowers the number N of executions required to achieve a target test quality Q_N. A large number of faults are then expected to be revealed without a great deal of effort.

Step 2: *Deterministic testing for fault removal.*
Additional deterministic test cases are derived to ensure the coverage of some extremal/special input values the occurrence probability of which remains insufficient in step 1.

Step 3: *Statistical testing for fault removal and fault forecasting.*
Test cases are randomly selected based on an operational input profile which can vary depending on the user in case of a population of disparate users. Some residual faults are then uncovered and removed. A quantitative assessment of the user-perceived reliability is inferred from this final step.

4.2. Basic relation

Let us go back to equation 2 that we consider as the basic relation for both step 1 and step 3.

$$(1-p)^N \leq 1-Q_N \qquad\qquad (2)$$

From **fault removal** viewpoint, Q_N is the test quality, that is to say the probability that N random test inputs reveal any fault f such that the failure probability per test case associated with f is at least p. The number N of executions needed to reach a target Q_N for a presumed value of p is given by equation 3 (§ 2.1.1) deduced from equation 2. The problem is that the a priori knowledge of the value of p is assumed. For hardware, there are "classical" fault models for physical changes (stuck-at faults, couplings, pattern sensitive faults, etc), and several methods have been stated to evaluate the failure probabilities associated with these faults and deduce a relevant value of p. Unfortunately, in the present state of the art it does not seem feasible to define an accurate software design fault model. Section 5.3 outlines the approach we are presently studying to face up to this problem. As step 1 aims at detecting a high percentage of actual errors, it must track down the faults which cause the least frequent failures. Hence, in equation 3, one must take into account a **minimum value of p** in order to deduce the maximum number N needed to reach a test quality objective.

From **fault forecasting** viewpoint, the reliability R_N at discrete time N is the probability that no failure occurs during N executions. Hence, $R_N = 1-Q_N$ and equation 2 gives a lower bound for R_N. If the input probability distribution is representative of user profile, R_N provides an assessment of the operational reliability. Step 3 aims at exposing residual faults to achieve an experimental evaluation of the actual software failure probability p in operating environment, from which reliability prediction can be derived from equation 2. When running N random test cases, one first observes failures due to the residual faults which cause the most frequent failures. Hence, one must take into account a **maximum value of p** in order to reach a minimum reliability objective.

5. On-going research work

The main problem arising from step 1 is a suitable assessment of the parameter p. Due to the lack of a software design fault model, the failure probability per test case related to actual faults cannot be evaluated a priori. Hence, other directions must be investigated to approximate the value of p. Two first suggestions have been made by Girard and Rault, and a third one by Hamlet (§ 2.1.1). As deterministic test approaches intend to create data sets that cover various structural or/and functional criteria, in a similar way the coverage probability of such criteria could provide an a priori value for p. The test quality Q_N is then a measure of the test hardness wrt current software test criteria and p is a function of both the criterion to be covered and the input probability distribution. *Section 5.1* gives formal results in that direction, illustrated with a simple example in *section 5.2*. But the theoretical assessment of p from such criteria requires complex calculations for large programs. Moreover, the acute question must be raised whether the obtained value is representative of the actual software failure probability per test case. Thus, in *section 5.3*, we outline a different approach based on a functional model of the software.

5.1. Statistical coverage wrt test criteria

In order to select deterministic test input sets, several structural or/and functional test criteria have been defined and compared [30, 33].

For **example,** let us take the well known structural criterion of "branch testing", which requires that each program branch is executed. In deterministic testing, the criterion coverage is obtained by creating an input data set that executes each branch at least once. In statistical testing, we will say that the coverage is reached with probability Q_N if each branch has a probability equal to or greater than Q_N to be executed at least once during test runs with random inputs. Let a branch i, and p_i the probability that a random input executes i. p_i depends on the input probability distribution. Consider two branches i and j such that $p_i < p_j$. Equation 3, with $p = p_i$ (respectively $p = p_j$), gives the number N_i (resp. N_j) of executions required to ensure that i (resp. j) has been executed at least once with probability Q_N. Hence: $p_i < p_j \Rightarrow N_i > N_j$ for a given Q_N. That is to say, N_i executions allow to cover both i and j with probability at least Q_N. Consequently, if B is the whole set of branches in a program, the number N of executions required to reach the branch coverage with probability Q_N is given by equation 3 with $p = \min \{p_k, k \in B\}$.

First, we define the notion of **statistical coverage with respect to a test criterion**. Let C_i be a test criterion requiring that each element from a set SC_i is exercised (in the preceding example, C_i is "branch testing" and $SC_i = B$).

Definition. A criterion C_i is covered with probability Q_N (test quality wrt C_i) if each element $\in SC_i$ has a probability equal to or greater than Q_N to be exercised at least once during statistical test runs.

                                                                                                   ❏

Let p_k be the probability that a random input exercises $k \in SC_i$. The number NC_i of executions required to reach the coverage of criterion C_i with probability Q_N is given by equation 3 with $p = \min \{p_k, k \in SC_i\}$.

More generally, the notion of statistical coverage can be used for a **set C of several criteria**. Let C_i and C_j be two different criteria such that $pC_i = \min$ $\{p_k,\ k \in SC_i\}$ is lower than $pC_j = \min \{p_k,\ k \in SC_j\}$. Equation 3, with $p = pC_i$ (resp. $p = pC_j$), gives the number NC_i (resp. NC_j) of executions needed to reach the test quality Q_N wrt C_i (resp. C_j). Hence, $NC_i > NC_j$. In other words, after running NC_i random data, a test quality better than Q_N is reached wrt C_j. This leads to the following property which is due to the fact that, whatever be the criterion: 1) the test quality Q_N is an increasing function of the number of executions with random data, and 2) the choice of test data is not related to the criterion (only the minimum number of data for a target Q_N depends on the criterion).

Property. If a test quality Q_N is reached for a criterion C_i, then Q_N is also reached for any other criterion C_j such that $pC_j \geq pC_i$.

$\square$

In practice, to cover a set of criteria $C = \{C_k\}$, one has only to cover C_i such that $pC_i = \min \{pC_k,\ C_k \in C\}$. Then, the number NC of executions required to reach the coverage of any criterion $C_k \in C$ with probability at least Q_N is given by equation 3 with $p = \min \{pC_k,\ C_k \in C\}$. **With regard to actual faults**, NC executions lead to the same test quality Q_N wrt any fault set such that the failure probability per test case associated with each fault is at least p.

Finally, the following comments arise from numerical results we have obtained for several small programs. They are illustrated with a simple (academic) example in section 5.2.

- For a given criterion C_i, NC_i greatly varies with the input distribution. By choosing an appropriate distribution, one can then **lower the required NC_i** for a target Q_N.

- NC_i depends on the criterion C_i to cover. Inversely, for a given number N of executions, large variations are observed on the test qualities Q_N reached wrt different criteria. Hence, the problem of the **criterion adequacy wrt occurring faults** (leading to *actual* test quality) is obvious. However, it could be less acute for statistical testing than for deterministic testing because the method of generating test data does not depend on the criterion.

- Coverage probabilities associated to **structural criteria** are difficult, and even impossible, to calculate for large programs. Further investigations are needed to **define test criteria** proper to statistical approach (see § 5.3).

5.2. A simple example

Consider the Fortran program TRIANGL suggested in [31] which has been studied by several authors [12, 29, ...]. Given three integer values A, B, C with $A \geq B \geq C > 0$, the program classifies the triangle formed by them. It indicates which of the six following cases is satisfied by A, B, C (see figure 2).

- *Path 1:* they are not in nonincreasing order of their size.

- *Path 2:* they are the sides of a right angled triangle.

- *Path 3:* they are the sides of an isosceles triangle.

- *Path 4:* they are the sides of an equilateral triangle.

- *Path 5:* they are the sides of an obtuse angled triangle.

- *Path 6:* they are the sides of an acute angled triangle.

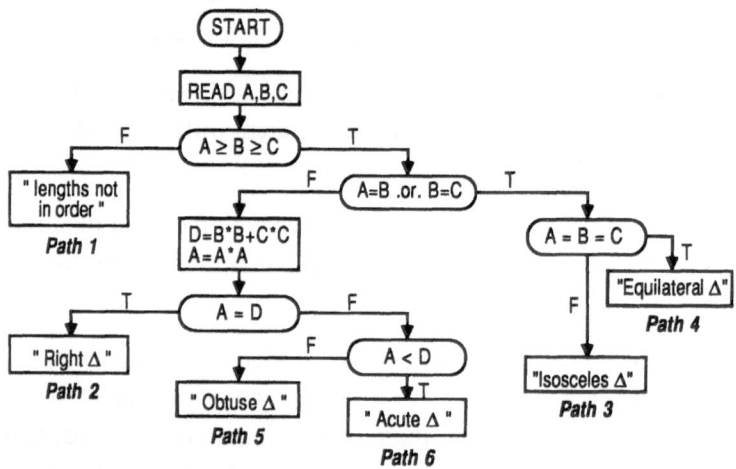

Figure 2. Flowchart of the program TRIANGL.

The program contains *a missing path fault*, noted f0 thereafter: it does not check that the lengths satisfy the triangle inequality A < B+C. Hence f0 produces an incorrect result for two data sets:

- A > B = C with A ≥ B+C (the incorrect result is "isosceles");
- A > B > C with A ≥ B+C (the incorrect result is "obtuse").

Three mutant programs [12] will also be studied:

- *M1:* the expression (A ≥ B ≥ C) in the first branch statement of TRIANGL becomes (A ≥ B) in M1. This fault noted f1 produces an incorrect result for inputs such that (A ≥ B .AND. B < C).

- *M2:* the same expression (A ≥ B ≥ C) in TRIANGL becomes (B ≥ C) in M2. This fault noted f2 produces an incorrect result for inputs such that (A < B .AND. B ≥ C).

- *M3:* the expression (A = B .OR. B = C) in the second branch statement of TRIANGL becomes (B = C) in M3. This fault f3 produces an incorrect result when (A = B .AND. B > C).

Let $p(i)$ be the probability of executing path i (i = 1, ..., 6) and $p(fj)$ be the probability of an incorrect result due to fj (j = 0, ..., 3) each time the program is run with random input data. CP will denote the criterion *path testing* with S_{CP} = {paths i}, and CF will denote the criterion *fault set* with S_{CF} = {faults fj}.

5.2.1. Uniform distribution over the input domain

Let us assume that the random data are chosen from a uniform distribution in the range [1-x] with x ≥ 1, i.e. $\forall$ k ∈ {1, ..., x}: Prob.(A=k) = Prob.(B=k) = Prob.(C=k) = 1/x. One finds the following relations, $\forall$ x ≥ 1:

$$p(1) = (x-1).(5x+2) / 6x^2; \quad p(3) = (x-1) / x^2; \quad p(4) = 1 / x^2;$$

$$p(2)+p(5)+p(6) = (x-1).(x-2) / 6x^2;$$

$$p(f0) = (x+2).(2x-1) / 24x^2 \quad \text{for even values of x;}$$

$$= (x^2-1).(2x+3) / 24x^3 \quad \text{for odd values of x;}$$

$$p(f1) = p(f2) = (x^2-1) / 3x^2; \quad p(f3) = (x-1) / 2x^2.$$

No simple relation gives the three probabilities p(2), p(5), p(6). To make possible values A, B, C which are the sides of a right triangle, one must take x ≥ 6, i.e. p(2) = 0 for x < 6. Figure 3 shows the values of p(i) and p(fj) with x ∈ [6, ..., 15], except for p(1) = max{p(i)} which goes from .741 with x = 6 up to .799 with x = 15, and p(f1) = p(f2) = max{p(fj)} which goes from .324 up to .332. Hence pCP = min {p(i)} = p(2) ∈ [.001, .005], and pCF = min {p(fj)} = p(f3) ∈ [.031, .069] ⇒ pCP < pCF.

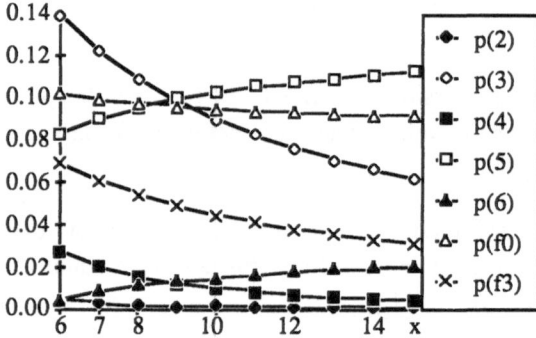

Figure 3. Probabilities p(i) and p(fj) with the inputs distributed uniformly in the range [1-x].

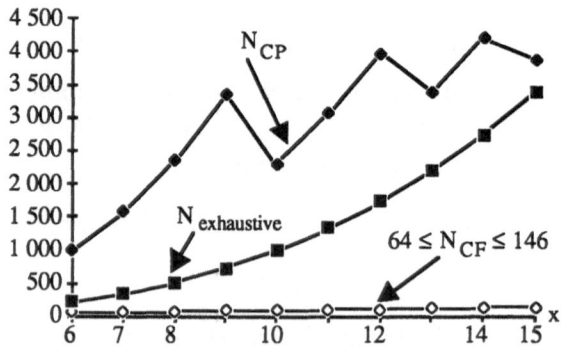

Figure 4. NCP and NCF required to reach QN = .99 with the inputs distributed uniformly in the range [1-x].

Then, for a target test quality Q_N equation 3 gives $N_{CP} > N_{CF}$, as shown in figure 4 for $Q_N = .99$. Inversely, N_{CP} executions lead to a detection probability higher than Q_N for any $fj \in S_{CF}$, as said in the property given in § 5.1. Figure 4 gives the number Nexhaustive $= x^3$ of different input cases (A, B, C). It is noteworthy that N_{CP} is greater than exhaustive testing, due to the fact that very few input data exercise path 2. From these numerical results, we conclude that the uniform distribution over the input domain is not well suited to exercise "rapidly" each part of the program. It is probably the reason why few TRIANGL mutants are killed by random data in Ntafos's experiments [29]. Hence, another input distribution is studied in section 5.2.2.

5.2.2. Uniform distribution over the path set

The input domain can be partitioned into six subsets D1, ..., D6 such that Di is the set of input cases (A, B, C) exercising path i. Let us assume that: 1) $\forall$ i, a randomly selected input case has a probability 1/6 of being chosen from Di, and 2) within a same subset Di the input cases are equally likely. That is to say, $\forall$ (A, B, C) $\in$ Di, the probability to select (A, B, C) is 1/(6.|Di|) where |Di| denotes the number of input cases $\in$ Di. |Di| depends on the range [1-x] from which the three values A, B, C are chosen, and $x \geq 6$ is required to have |Di| ≥ 0 $\forall$ i. Note that, as the test data generation method does not force at least one test case to come from each Di, it is not a "partition testing" as defined in [18]. For $x \geq 6$, one obtains: p(i) = 1/6 $\forall$ i, x; p(f1) = p(f2) = (x+1) / 3(5x+2); p(f3) = 1/12 $\forall$ x. No simple relation gives p(f0). Figure 5 shows the values of p(i) and p(fj) for x $\in$ [6, ..., 15]. Hence p_{CP} = p(i) = .167, and p_{CF} = min {p(fj)} = p(f1) = p(f2) which decreases from .073 (x = 6) to .069 (x = 15) $\Rightarrow$ $p_{CF} < p_{CP}$.

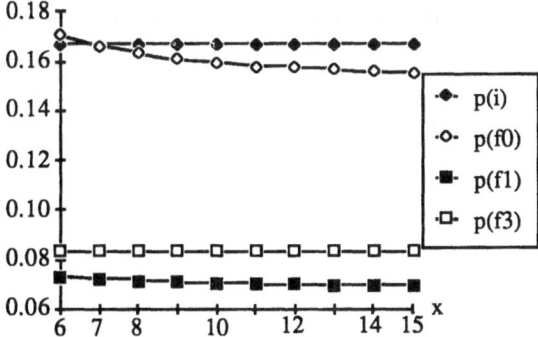

Figure 5. Probabilities p(i) and p(fj) with the inputs distributed uniformly over the path set.

Then $N_{CF} > N_{CP}$ for a target test quality. For $Q_N = .99$, equation 3 gives $N_{CP} = 26 \ \forall \ x$, while N_{CF} increases from 61 ($x = 6$) to 65 ($x = 15$). With the uniform distribution over the path set, the required numbers of executions are reduced. In particular, statistical testing wrt both criteria is much shorter than exhaustive testing. The test quality wrt C_F reached by running $N_{CP} = 26$ test cases is lower than .99. It decreases from .860 ($x = 6$) to .845 ($x = 15$), which is the detection probability related to f1 and f2. But the detection probability of the actual fault f0 is higher: it decreases from .992 ($> .99$) for $x = 6$ to .987 for $x = 15$.

5.3. A modeling tool under investigation

To carry on with theoretical work, a software modeling tool appropriate to analyze program behavior during random test runs is needed. Our present investigation involves two models. Starting from the system specification, a **functional model MF** is first defined. MF represents the software "elementary" functions and how they interact. The function granularity depends on the software complexity and on the test level (unit or system test). Then a **behavior model MB** is derived from MF. The model MB (e.g. a Markov chain) must be well suited to:

- study the influence of various input probability distributions on the function execution probabilities; a distribution such that each software function is properly scanned should be determined;

- infer reliability prediction for a given operational input profile.

Relevant input data (probability distribution and number N for a target Q_N) could then be settled from MB. The modeling tool will implicitly involve a test criterion related to the function execution probabilities. In conclusion, figure 6 sums up the approach applied to fault tolerant software. Theoretical and experimental work is still necessary to define a complete and usable modeling tool.

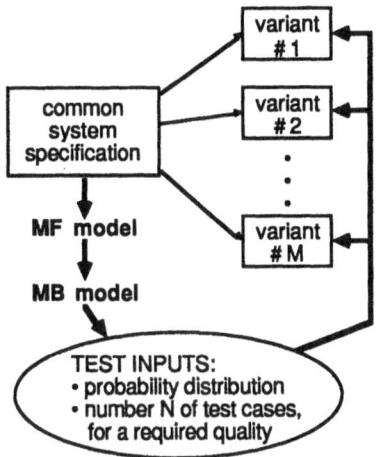

Figure 6. Statistical testing of fault tolerant software.

6. Conclusion

The complexity of real software systems and the lack of knowledge about actual design faults are reasons of the arising problems wrt software validation. The power of deterministic approach seems actually limited by these factors ("deterministic" and "lack of knowledge" are conflicting notions). The paper attempts to give a global, although surely incomplete, view of statistical approach which can be helpful to face up to such difficulties. We have tried to point out some basic ideas, and a promising validation strategy is deduced. Obviously, statistical testing is not the ideal method, in the sense that it cannot guarantee the removal of all residual faults. But is it possible to devise an ideal method? To our knowledge, no deterministic test can give such a confidence, except exhaustive testing which is impossible in practice. Therefore, the proposed validation strategy combines both deterministic and statistical data. Further theoretical and experimental studies are needed to support the strategy. Our present research work is going in that direction.

Acknowledgment

I thank Alain COSTES and Jean-Claude LAPRIE at LAAS for their constructive comments and suggestions during the preparation of this paper.

References

[1]   A. A. Abdel-Ghaly, P. Y. Chan, B. Littlewood, "Evaluation of competing software reliability predictions", *IEEE Transactions on Software Engineering*, Vol. SE-12, No. 9, September 1986, pp. 950-967.

[2]   P. E. Ammann, S. S. Brilliant, J. C. Knight, "Using multiple versions for verification", *proc. 4th NSIA Annual National Joint Conference on Software Quality and Productivity*, Washington, USA, March 1988, pp. 220-223.

[3]   P. G. Bishop & al., "PODS - A project on diverse software", *IEEE Transactions on Software Engineering*, Vol. SE-12, No. 9, September 1986, pp. 929-940.

[4]   P. G. Bishop & al., "STEM - A project on software test and evaluation methods", *proc. Conference SARS'87*, Altrincham, UK, November 1987, pp. 100-117.

[5]   P. G. Bishop, F. D. Pullen, "PODS revisited - A study of software failure behaviour", *proc. 18th Symposium on Fault-Tolerant Computing*, Tokyo, Japan, June 1988, pp. 2-8.

[6]   S. S. Brilliant, "Analysis of faults in a multi-version software experiment", Master's Thesis, University of Virginia, USA, May 1985.

[7]   S. S. Brilliant, "Testing software using multiple versions", Doctoral Dissertation, University of Virginia, USA, January 1988.

[8]   C-K. Cho, *An introduction to software quality control*, Wiley, New york, 1980.

[9]   C-K. Cho, *Quality programming: developing and testing software with statistical quality control*, John Wiley & Sons, 1987.

[10] P. A. Currit, M. Dyer, H. D. Mills, "Certifying the reliability of software", *IEEE Transactions on Software Engineering*, Vol. SE-12, No. 1, January 1986, pp. 3-11.

[11] P. A. Currit & al., "Correction to Certifying the reliability of software", *IEEE Transactions on Software Engineering*, Vol. SE-15, No. 3, March 1989, pp. 362.

[12] R. A. DeMillo, R. J. Lipton, F. G. Sayward, "Hints on test data selection: help for the practicing programmer", *IEEE Computer Magazine*, Vol. 11, No. 4, April 1978, pp. 34-41.

[13] R. A. DeMillo & al., "An extended overview of the Mothra software testing environment", *proc. 2nd IEEE Workshop on Software Testing*, Banff, Canada, July 1988, pp. 142-151.

[14] J. R. Dunham, "Experiments in software reliability: life-critical applications", *IEEE Transactions on Software Engineering*, Vol. SE-12, No. 1, January 1986, pp. 110-123.

[15] J. R. Dunham, "Software errors in experimental systems having ultra-reliability requirements", *proc. 16th Int. Symposium on Fault-Tolerant Computing*, Vienna, Austria, July 1986, pp. 158-164.

[16] J. W. Duran, J. J. Wiorkowski, "Quantifying software validity by sampling", *IEEE Transactions on Reliability*, Vol. R-29, No. 2, June 1980, pp. 141-144.

[17] J. W. Duran, S. C. Ntafos, "A report on random testing", *proc. 5th Conference on Software Engineering*, San Diego, USA, March 1981, pp. 179-183.

[18] J. W. Duran, S. C. Ntafos, "An evaluation of random testing", *IEEE Transactions on Software Engineering*, Vol. SE-10, No. 4, July 1984, pp. 438-444.

[19] M. Dyer, H. D. Mills, "The Cleanroom approach to reliable software development", *proc. Validation Methods Research for Fault-Tolerant Avionics and Control Systems Sub-Working-Group Meeting: Production of Reliable Flight-Crucial Software*, Research Triangle Institute, NC, USA, November 1981.

[20] D. E. Eckhardt, L. D. Lee, "A theoretical basis for the analysis of multi-version software subject to coincident errors", *IEEE Transactions on Software Engineering*, Vol. SE-11, No. 12, December 1985, pp. 1511-1517.

[21] E. Girard, J-C. Rault, "A programming technique for software reliability", *proc. 1st IEEE Symposium on Computer Software Reliability*, New York, USA, 1973, pp. 44-50.

[22] R. G. Hamlet, "Testing for probable correctness", *proc. 1st IEEE Workshop on Software Testing*, Banff, Canada, July 1986, pp. 92-97.

[23] R. G. Hamlet, "Probable correctness theory", *Information Processing Letters*, Vol. 25, No. 1, April 1987, pp. 17-25.

[24] R. G. Hamlet, "Testing for trustworthiness", *proc. Symposium on Directions & Implications of Advanced Computing (DIAC-87)*, Washington, USA, July 1987, pp. 87-93.

[25] J-C. Laprie, "Dependable computing and fault tolerance: concepts and terminology", *proc. 15th Int. Symposium on Fault-Tolerant Computing*, Ann Arbor, USA, June 1985, pp. 2-11.

[26] B. Littlewood, D. R. Miller, "A conceptual model of multi-version software", *proc. 17th Int. Symposium on Fault-Tolerant Computing*, Pittsburgh, USA, July 1987, pp. 150-155.

[27] B. Littlewood, D. R. Miller, "A conceptual model of the effect of diverse methodologies on coincident failures in multi-version software", *proc. 3rd Int. GI/ITG/GMA Conference on Fault-Tolerant Computing Systems*, Bremerhaven, RFA, September 1987, pp. 263-272.

[28] H. D. Mills, M. Dyer, R. C. Linger, "Cleanroom software engineering", *IEEE Software magazine*, September 1987, pp. 19-25.

[29] S. C. Ntafos, "On testing with required elements", *proc. COMPSAC'81*, November 1981, pp. 132-139.

[30] S. C. Ntafos, "A comparison of some structural testing strategies", *IEEE Transactions on Software Engineering*, Vol. SE-14, No. 6, June 1988, pp. 868-874.

[31] C. V. Ramamoorthy, S-B. F. Ho, W. T. Chen, "On the automated generation of program test data", *IEEE Transactions on Software Engineering*, Vol. SE-2, No. 4, December 1976, pp. 293-300.

[32] C. V. Ramamoorthy, A. Prakash, W-T. Tsai, Y. Usuda, "Software engineering: problems and perspectives", *Computer*, Vol. 17, No. 10, October 1984, pp. 191-209.

[33] S. Rapps, E. J. Weyuker, "Selecting software test data using data flow information", *IEEE Transactions on Software Engineering*, Vol. SE-11, No. 4, April 1985, pp. 367-375.

[34] J-C. Rault, "Extension of hardware fault detection models to the verification of software", Chapter 19 in *Program test methods*, Edited by W. C. Hetzel, Prentice-Hall, Inc., Englewood Cliffs, New Jersey, USA, 1973, pp. 255-262.

[35] R. W. Selby, V. R. Basili, F. T. Baker, "Cleanroom software development: an empirical evaluation", *IEEE Transactions on Software Engineering*, Vol. SE-13, No. 9, September 1987, pp. 1027-1037.

[36] M. L. Shooman, "Software reliability: a historical perspective", *IEEE Transactions on Reliability*, Vol. R-33, No. 1, April 1984, pp. 48-55.

[37] P. Thévenod-Fosse, "Statistical testing of software: a survey", LAAS Research Report No. 88.355, December 1988.

[38] L. G. Valiant, "A theory of the learnable", *Communications of the ACM*, Vol. 27, No. 11, November 1984, pp. 1134-1142.

[39] U. Voges (ed.), *Software diversity in computerized control systems*, Series on Dependable Computing and Fault-Tolerant Systems, Vol. 2, Springer-Verlag, Wien, Austria, 1988.

Design Diversity Assessment

Chair: J. Arlat (LAAS-CNRS, Toulouse, France)

ERROR MASKING: A SOURCE OF FAILURE DEPENDENCY IN MULTI-VERSION PROGRAMS

P.G. BISHOP, F.D. PULLEN
National Power Technology and Environmental Centre,
Kelvin Avenue, Leatherhead, KT22 7SE - England.

Abstract

This paper presents some empirical measurements of failure dependencies between the known faults detected in an earlier software diversity experiment (PODS). The results showed that some apparently unrelated pairs of faults had high (and very similar) levels of dependency. This has been explained in terms of a *error masking* process. It is shown that this process is likely to occur in many software applications, including the missile launcher application used in the Knight and Leveson experiment. Error masking behaviour can be predicted from the specification (prior to implementation), and simple modifications to the program design can minimize the error masking effect and hence the observed dependency.

Keywords: Multi-version programming, failure dependency, error masking, mapping functions.

1. Introduction

N-version programming [1,2] has been advocated as a means of enhancing the reliability of software. The basis of the method is to develop *N* diverse programs to the same specification. During the subsequent operation of the software, majority voting is used to mask failures occurring in a minority of software versions.

The method relies on the assumption that the faults in diverse programs are dissimilar. Ideally, diverse programs should conform to the independence assumption. Under this assumption, if two independent faults exist in separate programs with failure probabilities P_a and P_b the probability of co-incident failures P_{ab} will be:

$$P_{ab} = P_a . P_b$$

However, if dependencies exist between faults, then the probabilities of co-incident failure could range from completely positively correlated to completely negatively correlated. This is shown schematically in Figure 1.

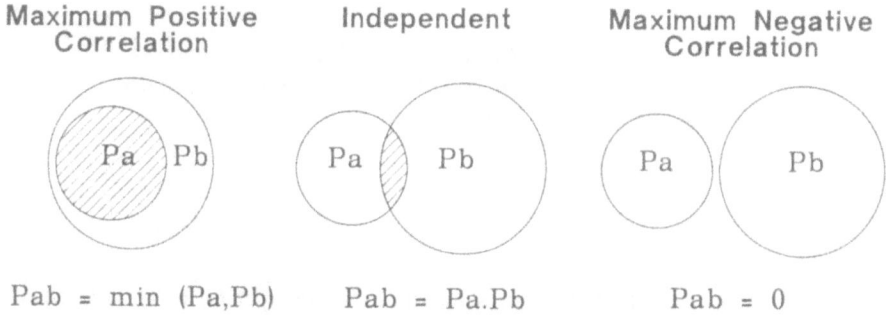

Figure 1. Examples of failure dependency

Clearly the independence assumption will not apply to common faults caused by problems in the initial specification. This is a general problem that affects all software development approaches and can be regarded as a separate issue. Recently however, doubts have been cast on the independence of *design-level* faults. Failure dependencies have been observed experimentally [3] and predicted theoretically [4,5].

Even if software dependencies do exist, significant improvements in software reliability may be achieved if the degree of dependency is not too large. The initial objective of the authors was to determine empirically the actual levels of dependency occurring between software faults. Subsequent analysis of the observed dependency effects led to a general theory of *error masking*. This paper expands on the theory presented in an earlier paper [6], and considers the implications of the theory for other applications.

2. Measurement of failure dependency

Dependency measurements were made on three diverse programs (and their known faults) produced in an earlier project [7]. A test harness was developed which could apply different types of random data to a "golden" program in parallel with a faulty version of the same program. The faulty program was specially instrumented so that individual faults could be switched on or off by the test harness. The harness compared the outputs of the "golden" program with those of the faulty program on every test cycle. If the outputs disagreed, the cycle number and the faulty and "golden" output values were stored in a failure history file. These failure history files could then be analysed singly and in combination to determine:
- the failure rate for each individual fault,
- the co-incident failures between faults in different programs.

In order to compare the number of coincident failures with that expected under the independence assumption, the observed coincident failure rate P_{ab} was plotted against the failure rate expected under the independence assumption $(P_a.P_b)$. Figure 2 presents the results of all the fault pairs analysed.

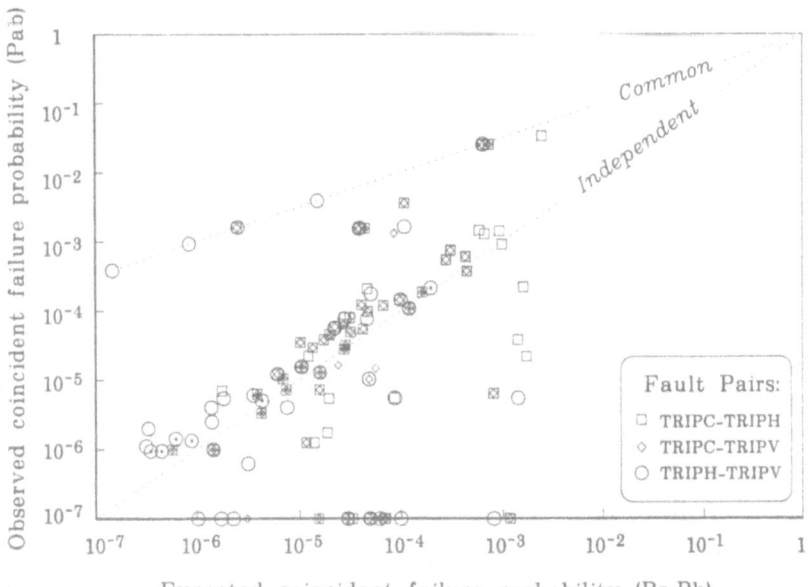

Figure 2. Coincidence of fault pairs with Pa, Pb < 0.1

Figure 3 presents the same information in a different way by plotting the dependency factor:

$$\frac{P_{ab}}{(P_a.P_b)}$$

for all fault pairs. A factor of unity implies that the failures are independent. A ratio greater than unity implies positively correlated failures, while a ratio less than unity implies negative correlation. It can be seen that the distribution of dependency factors ranges from strong positive correlation to strong negative correlation.

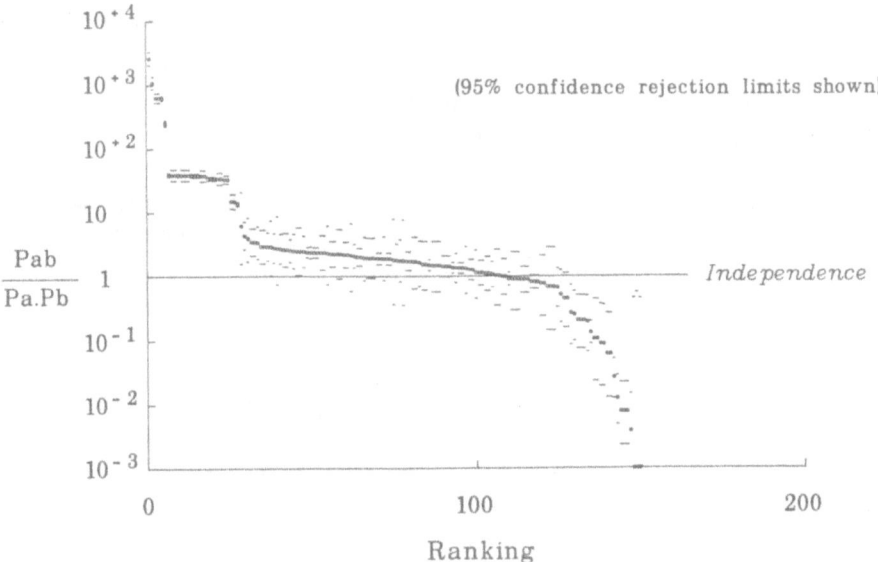

Figure 3. Dependency factor for fault Pairs with Pa, Pb < 0.1

3. Sources of dependency

In order to determine the cause of the dependencies between the faults, the fault pairs were analysed in some detail. Most of the high dependency factors were due to known common mode faults. However some strange clusters of high dependency fault pairs were observed which could not be accounted for by common mode faults. These fault pairs had dependency factors which were clustered around three particular values (14.0, 34.0 and 38.8). These clusters are the high dependency 'plateau's in Figure 3. The three faults in the 14.0 cluster were clearly related. The failures were associated with a common computational function and there were significant overlaps in the sets of input values that initiated failure.

The two remaining high dependency clusters were associated with failures on the same single-bit output. However when the faults were examined there was no obvious reason for the dependency since there was little or no commonality between the input data or computation function of the faulty code sections.

A satisfactory explanation was eventually found, which showed that the dependency effects were caused by "error masking". In this paper, "error" is defined to be an incorrect behaviour of some internal component of the software. The single-bit output value was determined by an "OR" of several logical conditions as shown in Figure 4. An incorrectly calculated condition value could only be detected at the binary output when all the other logical conditions were zero. Thus only a fraction of the internal errors caused failures at the output of the OR function.

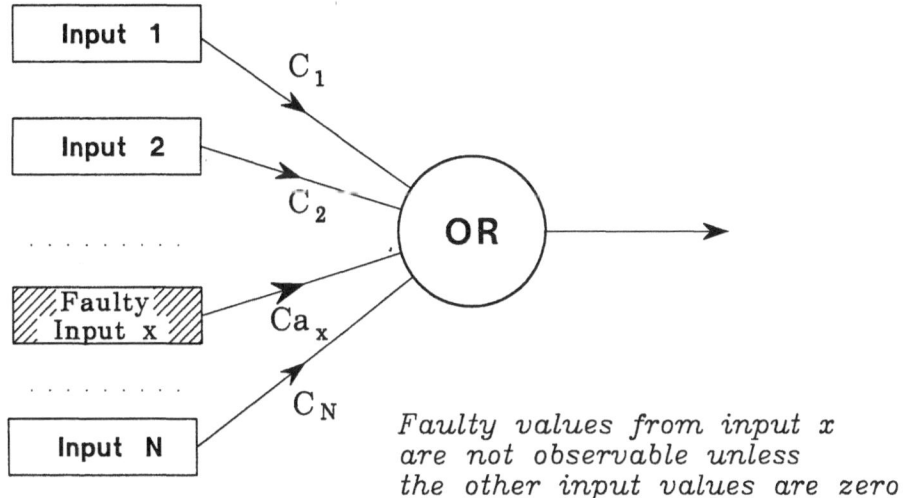

Figure 4. Error masking in OR logic

It was this error masking effect that gave rise to the observed failure dependency. This principle is illustrated schematically in Figure 5 for a particular case where the individual and coincident error rates are reduced by the same fraction.

More generally, for any error masking function, if $fdet_a$, $fdet_b$ and $fdet_{ab}$ are the fractions of the individual and simultaneous errors resulting in failures at the output, the dependency factor computed from the observed failure rates at the output (P'_a, P'_b, P'_{ab}) will be:

$$\frac{P'_{ab}}{P'_a.P'_b} = \frac{fdet_{ab}}{(fdet_a.fdet_b)} \cdot \frac{P_{ab}}{P_a.P_b}$$

Thus if the internal error probabilities P_a and P_b are independent, the *observed* dependency factor will be:

$$\frac{fdet_{ab}}{(fdet_a.fdet_b)}$$

This is a "functional dependency factor" caused by the error masking properties of the common function interposed between the internal errors and the output (rather than by any dependency between the internal errors).

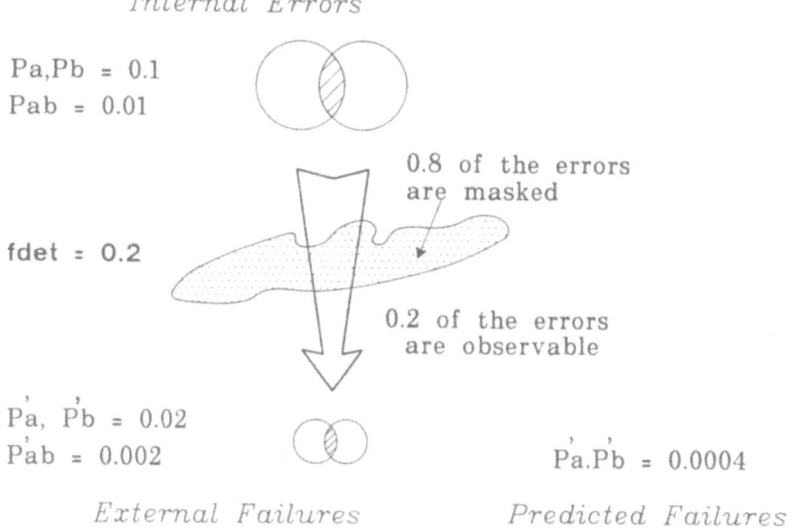

Internal Errors

Pa,Pb = 0.1
Pab = 0.01

0.8 of the errors are masked

fdet = 0.2

0.2 of the errors are observable

Pa, Pb = 0.02
Pab = 0.002

Pa.Pb = 0.0004

External Failures   *Predicted Failures*

Figure 5. Illustration of dependency due to error masking

4. Functional dependency in OR logic

For the OR logic function, the functional dependency factors were determined for a range of cases where a single input condition is faulty in each version. In this analysis:

N is the number of inputs to the OR logic,

C_x is the correct value of condition x,

C_y is the correct value of condition y,

Ca_x is the value of faulty condition x in version a,

Cb_y is the value of faulty condition y in version b,

$p0_x$ is the probability of generating zero for "golden" input condition x,

$P0$ is the probability of generating zero at the "golden" output

$$(P0 = \prod_{i=1}^{N} p0_i)$$

For a faulty input condition x, the fraction of erroneous values detected in version a will be:

$$fdet_a = \prod_{i=1, i \neq x}^{N} p0_i$$

From the definition of P0, this can be represented as:

$$fdet_a = \frac{P0}{p0_x}$$

Similarly:

$$fdet_b = \frac{P0}{p0_y}$$

For the case where x and y represent the same input condition in the two versions, the same fraction of simultaneous errors will survive the masking process:

$$fdet_{ab} = \frac{P0}{p0_x}$$

Hence by substitution, the functional dependency factor is:

$$\frac{P0_x}{p0}$$

The situation is more complex when x and y represent different conditions in versions a and b. Detectable simultaneous failures are only observed when detectable failures occur in versions a and b, i.e.:

$$(C_x \neq Ca_x) \wedge (C_y = 0) \wedge (C_z = 0) \dots$$

$$\wedge (C_x = 0) \wedge (C_y \neq Cb_y) \wedge (C_z = 0) \dots$$

This condition is only satisfied when all golden inputs are zero ($C_x = 0$, $C_y = 0$ etc). However simultaneous errors can occur *internally* for all combinations of input values. Thus the fraction simultaneous errors detected is:

$$fdet_{ab} = P0$$

By substitution, the dependency factor when $x \neq y$ is :

$$\frac{p0_x.p0_y}{p0}$$

A similar analysis may also be applied to faults which generate only one erroneous value (e.g. $C_{ax}=1$ when $C_x = 0$). This restriction makes no impact on the $(x = y)$ case, but in the $(x \neq y)$ case, simultaneous failures are impossible for three out of the four possible combinations of error mode, thus *fdet*$_{ab}$ =0. In the remaining case, simultaneous failures only occur when the intended input values are: $C_x = 0$, $C_y = 0$, and conditions a and b generate '1'. This event is detectable providing the remaining inputs are zero, hence:

$$fdet_{ab} = \frac{P0}{p0_x.p0_y}$$

Thus by substitution, the functional dependency factor is:

$$\frac{1}{P0}$$

The results of the dependency analyses are summarized below:

FAILURE VALUE		FUNCTIONAL DEPENDENCY FACTOR	
Ca_x	Cb_y	$x \neq y$	$x = y$
-	-	$\dfrac{p0_x - p0_y}{P0}$	$\dfrac{p0_x}{P0}$
1	1	$\dfrac{1}{P0}$	$\dfrac{p0_x}{P0}$
1	0	0	
0	1	0	
0	0	0	$\dfrac{p0_x}{P0}$

Table 1. OR logic functional dependency factors

It is clear that positive dependency effects will occur since $P0 < p0_x < 1$. As the number of inputs N increases, P0 decreases; this can result in high observed dependency factors.

It can also be seen that the OR logic is capable of producing *negatively* correlated failures (dependency factors of zero, in fact).

5. Analysis of data in terms of error masking

It might be expected that typical faults would generate erroneous '1' *and* erroneous '0' values. Surprisingly, all the faults examined had single erroneous values. Typically this resulted from a failure to set or clear a variable under certain conditions (e.g. due to the omission of an IF clause).

It was found that all the fault pairs in the 38.8 cluster corresponded to the $(1, 1, x \neq y)$ case of Table 1. Under uniform random test conditions, $P0 = 0.02577$. The predicted dependency factor of 38.9 agreed with the measured values within twice the standard deviation.

To check that this effect was due to error masking, a subset of the 38.8 cluster of fault pairs was re-tested using a different input distribution with a reduced level of error masking ($P0=0.591$). The measured dependency factor was slightly below that predicted by the error masking theory. This indicates the internal error rates are almost independent. A comparison of the measured and predicted dependencies is shown in Table 2.

FAULT PAIR	P0	DEPENDENCY FACTOR	
		Predicted	*Measured*
V06,C16	0.0257	38.9	38.22 ± 0.4
	0.591	1.69	1.65 ± 0.02
V13,C16	0.0257	38.9	38.2 ± 0.4
	0.591	1.69	1.41 ± 0.02

Table 2. Variation of predicted and measured dependencies with P0

An analysis of the 34.0 cluster of fault pairs showed they were combinations of masked single-bit faults and a bit transposition fault. The functional dependency factor for this case approximated to $(P0 + P0_2)^{-1}$ where $P0_2$ was the probability of zero on the second output bit. Based on measurements of $P0$ and $P0_2$, the functional dependency factor was calculated to be 33.85. This agreed with measured dependency factor within the standard deviation.

Further analysis of the negatively correlated fault pairs showed that many of them were $(0, 0, x \neq y)$ or $(0, 1, x \neq y)$ error cases where conditions for detectable failure in the two versions are disjoint.

An examination of dependency factors in the 1.0 to 10.0 range showed that some of the dependency effects could be explained by the "range-clipping" of an erroneous output to a correct final value. This range-clipping effect was found to give a functional dependency factor of 2.18. Since this effect applied to the 14.0 cluster, the dependency factor for this cluster arising from correlation of the internal errors was actually less than 7.

In summary therefore, the high dependency factors between fault pairs were determined either by common problems in the initial specification, or by error masking. Negatively correlated fault pairs could also be explained in terms of a masking effect which prevents some combinations of uni-directional error conditions from being observed simultaneously through the OR logic.

In general, the observed dependency factors tended to align with the main identified levels of functional dependency (38.8, 2.18, 0.0) or with unity (independence). The intermediate values of dependency may also be explicable in terms of error masking if the fault pairs affect several outputs subject to different levels of error masking. The observed dependency factor would then represent an average of the dependencies for each output, weighted according to failure probabilities on each output.

The intermediate fault pairs have not been analysed to check this supposition, but the effect was observed in the 34.0 cluster.

6. General applicability of error masking

Error masking is likely to occur to some degree in any application program since any N to 1 mapping function is capable of masking errors (Figure 6). An error can be regarded as a *displacement* to a new value in the input domain. If the new input value maps to the correct output value (i.e. is in the same *equivalence class*), no failure is observed and the error is masked.

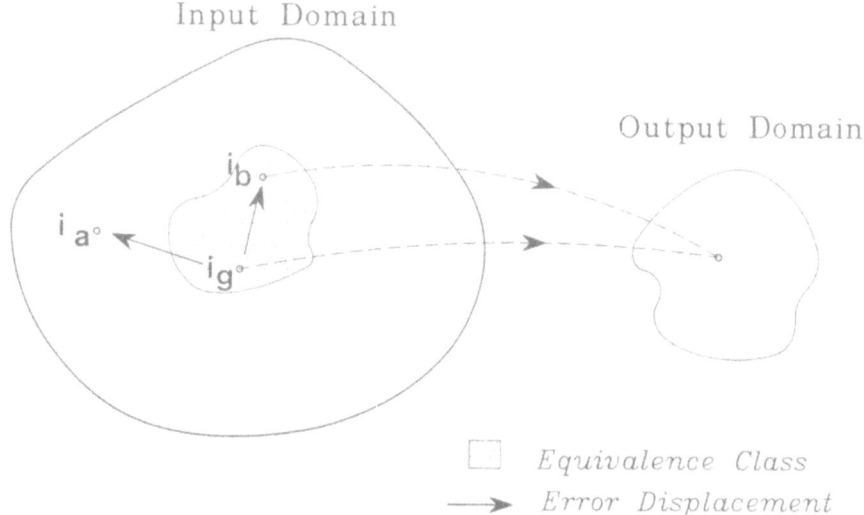

Figure 6. Error masking: N to 1 mapping function

In general, the dependency factor observed associated with an error masking function will be:

$$\frac{fdet_{ab}}{fdet_a.fdet_b}$$

In mapping function terms, $fdet_a$, $fdet_b$, and $fdet_{ab}$ represent the probabilities of displacement outside the equivalence class, for both individual and simultaneous errors. For example, with a mapping function f, and an input domain I:

$$fdet_a = \frac{P(f(i_a) \neq f(i_g))}{P(i_a \neq i_g)}$$

where i_a are input values in version a, and i_g are the golden input values.

The detection probabilities will depend on:

- the distribution of intended input values i_g over the input domain I

- the distribution of the displacements from the golden value ($i_a - i_g$, $i_b - i_g$) occurring in the two versions.

- the size and orientation of the equivalence classes.

In practice it is likely that the error masking will be greatest where:

- the equivalence classes are large.

- equivalent input values can be selected by a displacement involving a single input variable (i.e. by a movement along a single dimension in the input space).

Typical functions which satisfy these requirements are:

- AND, OR,

- MAX, MIN,

- range clipping,

- quantization,

- selector functions.

A function such as '+' has relatively small equivalence classes, and they are oriented diagonally. To mask an error, the displacement must also be diagonal. This requires complementary error displacements for the two dimensions in the input domain, so there is likely to be a low probability of masking.

In many cases, $fdet_{ab}$ will be of the same order of magnitude as $fdet_a$ and $fdet_b$. However, for error displacements that occur along *specific axes* in the input domain, $fdet_{ab}$ can approach zero.

This can occur, for example, in the OR logic function (Table 1), where $fdet_{ab} = 0$ for some error displacement combinations. In general there will be *no* detectable coincident failures if the subsets from which the error displacements are detectable in the two versions are disjoint. This condition is illustrated in Figure 7 for failures to the $(0,0, x \neq y)$ case of the OR logic function.

Another example of disjoint failure occurs in the *selector function*. A selector function is analogous to a multi-way switch which selects a value from one of several possible inputs. Simultaneous failures cannot occur if different inputs are faulty in the two versions, since they cannot be selected at the same time.

It can be seen that $fdet_{ab} \ll fdet_a fdet_b$ for some functions and some error modes. This can result in lower levels of dependency (or even complete negative correlation). However for predictive purposes, the error displacement modes which give the worst-case dependency should be assumed.

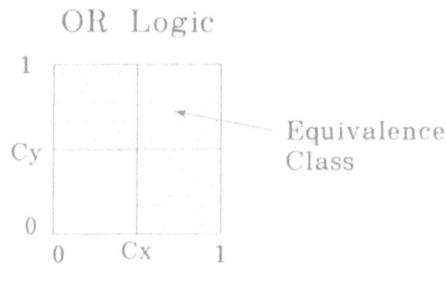

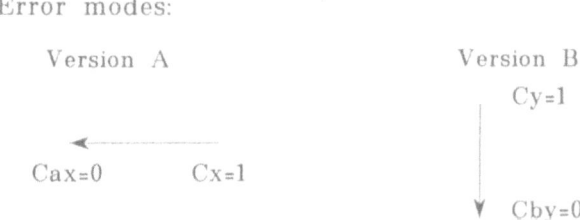

Figure 7. Disjoint detectable failure subsets

7. Error masking in the missile launcher experiment

It has been asserted that error masking effects are a general phenomenon. It has also been stated that dependency effects due to error masking should be predictable from the specification with no *a priori* knowledge of the implemented programs. To support these assertions, the missile launcher specification used in the Knight and Leveson [3] dependency experiment was analyzed. The Knight and Leveson experiment developed 27 diverse programs to this common specification. A statistical test of failure independence between these programs was rejected to high confidence.

A diagrammatic representation of the specification is shown in Figure 8. A launch interceptor condition (LIC) calculation is applied to a sequence of different input data values (representing radar tracking coordinates). Each calculation yields a Boolean value, and the results are effectively OR'd together to produce a single Boolean value.

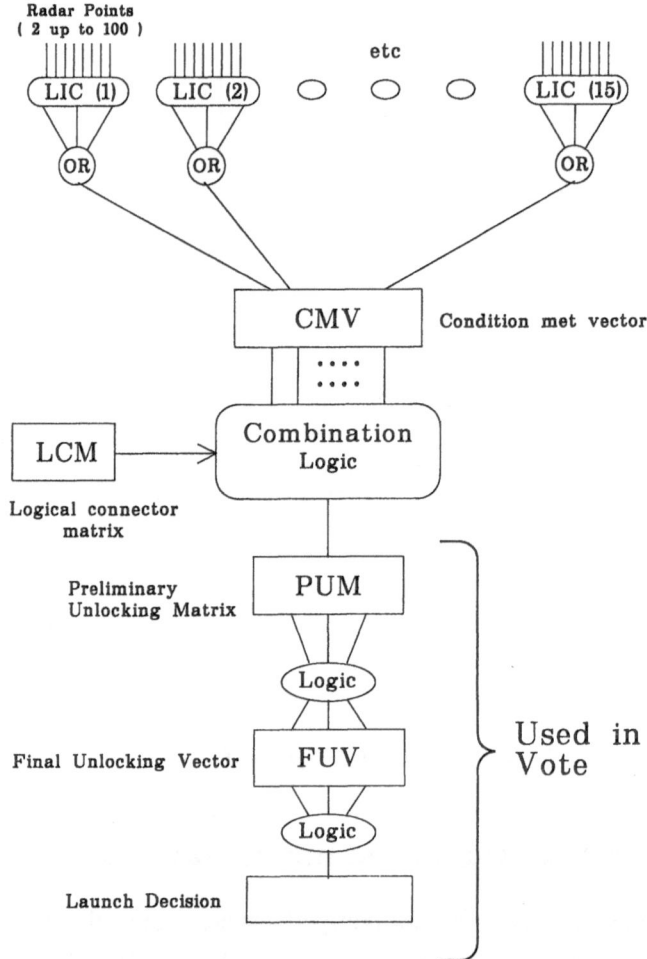

Figure 8. Missile launcher application

Fifteen different types of LIC computation are applied to the same sequence of radar track coordinates. The results of these computations determine the value of the 15 bits in the Conditions Met Vector (CMV). The CMV is then processed according to the logic specified in the logical connection matrix (LCM). The LCM logic processing combines bit values in the CMV using AND, OR and IGNORE logic to generate a 15 by 15 preliminary unlocking matrix (PUM).

The PUM and FUV and the final trip decision value are all included in the final vote comparison between the programs. However it is clear that there is considerable scope for error masking in the OR of individual LIC values, and in the processing of CMV bits to compute the PUM.

The error masking process for the LIC calculations is somewhat different to the earlier analysis of OR logic masking. The *same* LIC computation function is used on all inputs but with *different* radar track coordinates data. However any correct '1' value is still capable of masking errors on other inputs.

The observed failure rate for N OR'd LIC calculations is:

$$Pfail(N) = perr(N).fdet(N)$$

where *Perr(N)* is the error rate and *fdet(N)* is the fraction of errors surviving OR logic masking. Note that subscripts a, b, ab will be used to distinguish the equations associated with individual and simultaneous error masking in versions a and b.

For a probability of zero per input, $P0$, error probabilities per input, d_a, d_b, and N inputs:

$$Perr_a(N) = (1 - (1 - d_a)^N)$$

For small d_a, it can be shown that:

$$fdet_a(N) \approx p0^{(N-1)}$$

Note that while *Perr* increases with N, the fraction detected $fdet$ falls exponentially. This implies that the failure rate will rise to a maximum then *decrease* as the number of faulty inputs to the OR logic increases (Figure 9).

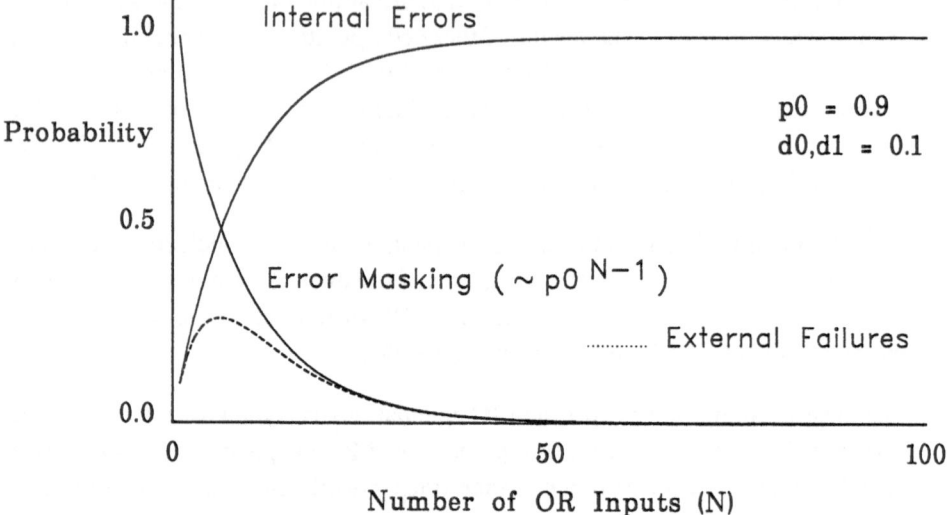

Figure 9. OR logic masking of a common function

For simultaneous errors, it can be shown that:

$$Perr_{ab}(N) = Perr_a(N).Perr_b(N)$$

$$fdet_{ab}(N) \approx \frac{(1 - p0)\, p0^{N-1}}{N} + p0^N$$

Hence the dependency factor $D(N)$ for N randomly failing inputs will be:

$$D(N) \approx N^{-1}(1 - p0)p0^{1-N} + p0^{2-N}$$

For large values of N, the dependency factor could easily exceed 10^6. However in the missile launcher experiment, the number of OR'd LIC calculations varied randomly between 1 and 100 (depending on the number of input coordinates). This significantly reduces the mean level of dependency observed. The mean fraction of errors detected through the OR logic with randomly selected values of N is:

$$\overline{fdet} = \frac{\sum\limits_{N=1}^{100} Perr(N).fdet(N)}{\sum\limits_{N=1}^{100} Perr(N)}$$

This equation can be used to derive $\overline{fdet_a}$, $\overline{fdet_b}$ and $\overline{fdet_{ab}}$ and hence determine the mean functional dependency factor. The factors are insensitive to the error rates since, for small *Perr*:

$$Perr_a(N) \approx N.d_a,$$

$$Perr_b(N) \approx N.d_b,$$

$$Perr_{ab}(N) \approx N^2.d_a.d_b$$

To a first approximation, the error rate terms cancel when calculating the detection probabilities. Some example mean dependency factors are shown in Table 4.

$p0$	MEAN DEPENDENCY FACTOR	
	$N = 1..10$	$N = 1..100$
0.99	1.32	2.23
0.9	1.91	17.05
0.5	4.95	49.93
0.1	8.26	82.65

Table 4. Example LIC mean dependency factors $(d_a, d_b = 0.001)$

It is therefore credible that the functional dependency factors for each bit in the CMV could have values between unity and 100. Furthermore, the logic used to compute the PUM (employing ANDs, ORs, and IGNORE's) would introduce additional levels of error masking.

Some indication of the actual levels of error masking occurring in the missile launcher application can be found in a comment made in the original paper. It was noted the two major faults only had detected failure probabilities of about 10^{-3}. This may indicate that functional dependency factors of up to 100 are credible in the missile launcher application.

If error masking is a significant source of dependency, then it is possible to make a number of predictions about the failure behaviour of the missile launcher programs under different test conditions. In particular:

- The use of more "realistic" test data where $p0 \approx 1$ will minimize masking and hence reduce the observed dependency effect.

- Reducing the number of radar track co-ordinates for each test will also minimize masking and hence decrease the observed dependency effect.

- Including all the launch interceptor condition values (LICs) and the conditions met vector (CMV) in the external vote would:

  - Increase the observed failure rate.

  - Decrease the observed dependency.

8. Discussion

8.1. Relationship to earlier research

Theoretical studies by Eckhardt and Lee [4] and by Littlewood and Miller [5] has shown that dependency effects will be observed if different input values are more prone to failure than others. This is a general theory which applies whatever the cause of "failure-proneness". The relevance of the present work, is that it identifies a *specific and predictable* mechanism which accounts for at least some of that variability.

With regard to the empirical test of independence performed by Knight and Leveson [3], it is considered that the missile launcher example was a 'worst case' example in terms of error masking. Other applications [8,9] have far lower levels of potential functional dependency and hence N-version programming are likely to yield very different results. Indeed changes to the voting and test strategies in the missile launcher experiment are predicted to produce lower levels of dependency.

Of course this does not mean that error masking is the only mechanism causing failure variability. The specification is a known source of dependency. Knight and Leveson [3] have reported common faults due to "thinking traps" in specific problem areas. In addition, different implementation methods may also have an impact on "failure-proneness". It would be instructive to re-analyze the Missile Launcher Experiment programs (and other software diversity experiments) to determine the contribution of error masking as a source of variability compared with other sources.

8.2. Impact on N-version programming

What impact do these observations have on N-version programming in general?

The pessimist might consider that this further minimizes the potential benefits of diversity since some dependencies will always occur.

The optimist may say that although the dependencies occur, they only exist because the individual *and* coincident failure rates have been reduced by error masking. In other words the problem only exists because the software is *inherently fault tolerant*.

The pragmatist would make use of the knowledge to maximize the benefits of N-version programming. The main advantage of this failure dependency mechanism is that it can be predicted *prior* to implementation, and this information can be utilized to optimize program development. The original program specification can be analyzed to determine the sources of error masking. For any given program design, it then should be possible to:

- Design test strategies to minimize masking, and hence maximize the probability of detecting faults during development.

- Determine potential dependency effects using operational input data distributions. This can place an upper bound on the level of dependency to be expected due to error masking effects.

Equally it should be possible to modify the program design such that:

- the internal values that feed error masking functions are made visible to the external vote logic, or

- the internal values are presented to a separate "discrepancy checker", but not included in the voting logic.

The first approach enhances the probability of detecting faults in the development phase, but may reduce availability during operation. The second approach could give an "early warning" of internal problems without affecting the system availability during operation.

More speculatively, it may be possible to claim some kind of "scaling factor" between internal error rates and external failure rates. This scale factor could be determined by an analysis of the failure masking functions or experimentally (e.g. using known development faults or "seeded" faults). Assuming that the failure masking factor is known, then it might be possible to scale up the observed interval between discrepancies to obtain an estimate for the mean time between voting failures.

One final observation. This paper has been solely concerned with dependency effects in software. However the observations relating to error masking should be equally applicable to transient failure mechanisms in hardware systems. It is therefore likely that independent internal errors in error-masking hardware designs will result in dependent external failure behaviour, unless similar precautions are adopted.

9. Conclusions

1. In this experiment:

- The majority of non-identical fault pairs exhibited co-incident failure characteristics approximately in accord with the independence assumption (once error masking effects had been taken into account).

- Negative correlation effects were observed (which could also be accounted for in part by error masking).

2. The error masking theory used to explain the observed dependency effects is considered to be of general application. Dependencies caused by error masking effects can be predicted in advance on the basis of the program specification. This information can be used to:

- Optimize test strategies.

- Place an upper bound on the degree of dependency which may be expected from this effect.

- Identify intermediate computation values to be presented as part of the program output to reduce the dependency effects.

3. The error masking theory has been applied to the Knight and Leveson missile launcher specification and it has been shown that high levels of masking are possible. If error masking is a significant source of dependency it is predicted

that: a reduction of the number of coordinate points per test *or* the inclusion of intermediate computation values in the voting logic will:

- Increase the failure rate.

- Decrease the observed level of dependency.

Acknowledgments

This work was carried out at the National Power Technology and Enviromental Centre and is published with the permission of National Power plc The authors also wish to thank the members of the STEM Project for providing support and technical assistance.

References

[1] A. Avizienis, "Fault Tolerance and Fault Intolerance: Complementary Approaches to Reliable Computing", *Proc. Int. Conf. on Reliable Software*, Los Angeles, California, 1975.

[2] L. Chen, A. Avizienis, "N-Version Programming: A Fault Tolerance Approach to Reliability of Software Operation", *Proc. 8th IEEE Int. Symposium on Fault Tolerant Computing*, Toulouse, France, 1978.

[3] J.C. Knight, N.G. Leveson, "An Experimental Evaluation of the Assumption of Independence in Multiversion Programming", *IEEE Trans on Software Engineering*, Vol SE-12, Jan 1986.

[4] D.E. Eckhardt, L.D. Lee, "A Theoretical Basis for the Analysis of Multi-version Software Subject to Coincident Failures", *IEEE Trans. Software Engineering*, Vol SE-11, No. 12, 1985.

[5] B. Littlewood, D. Miller, "A conceptual Model of the Effect of Diverse Software Methodologies on Coincident Failures in Multi-Version Software", *Proc. Third Int. Conference on Fault-Tolerant Computing Systems*, Bremerhaven, 1987, Informatik-Fachberichte 147, Springer Verlag, Berlin, 1987.

[6] P.G. Bishop, F.D. Pullen, "PODS Revisited - A Study of Software Failure Characteristics", *Proc. Eighteenth Int. Symposium on Fault Tolerant Computing (FTCS-18)*, Tokyo, June 27-30, 1988, IEEE Computer Society Press.

[7] P.G. Bishop et al, "PODS - A Project on Diverse Software", *IEEE Trans. Software Engineering*, Vol SE-12, No. 9, 1986.

[8] A. Avizienis, M.R. Lyu, W. Schütz, "In Search of effective Diversity: A Six Language Study of Fault Tolerant Flight Control Software", *proc. Eighteenth IEEE Int. Symposium on Fault Tolerant Computing (FTCS-18)*, Tokyo, June 27-30, 1988.

[9] J.P.J. Kelly, D.E. Eckhardt, M.E. Vouk, D.F. McAllister, A. Caglayan, "A Large Scale Second Generation Experiment in Multi-version Software: Description and Early Results", *Proc. Eighteenth IEEE Int. Symposium on Fault Tolerant Computing (FTCS-18)*, Tokyo, June 27-30, 1988.

RECOVERY BLOCK RELIABILITY ANALYSIS WITH FAILURE CLUSTERING

Attila CSENKI*)
Centre for Software Reliability, The City University,
Northampton Square, London EC1V 0HB - Great Britain

ABSTRACT

New discrete- and continuous-time models of the recovery block are presented using Markov chain techniques. The time to failure distributions are fully described by their probability generating functions and Laplace transforms respectively. A discrete-time model is developed accounting for the commonly observed clustering of failure points in the input space. The Markov framework is a useful tool to solve this model even though the system's behaviour is non-Markovian: additional states are artificially introduced to model the degraded state of individual modules.

Keywords: clusters of failure points; fault-tolerant software; Markov modelling; recovery block; software reliability; time to failure; transform methods.

---

*) Present address: Department of Computer Science and Applied Mathematics, Aston University, Aston Triangle, Birmingham B4 7ET, Great Britain.

1. Introduction

The recovery block scheme, described by Randell [17] and subsequently discussed by Anderson and Lee [2], is a fault-tolerant software construct using diversity which consists of a primary module, one or more stand-by spare modules, and an acceptance test. The primary module and the stand-by spares are not replicas of the same piece of code but are based on different algorithms for the same problem and are preferably implemented by (teams of) programmers who, for the sake of a high degree of independence of module failure behaviour, do not communicate with each other. For a recent theoretical discussion of problems associated with achieving a completely independent failure behaviour of modules see, however, Littlewood and Miller [16].

The primary module is executed first when entering a recovery block and the result is then subjected to an acceptance test. If the acceptance test is passed control is then returned to the next higher level but the first alternate module is executed otherwise. This process is repeated with the remaining alternate modules in turn until one of them is found to produce a result which passes the acceptance test or until all of them are found to produce erroneous outputs in which case the recovery block is considered to have failed. An appropriate syntax for a recovery block with m-1 alternate modules is ([2], [17]) as follows:

```
ensure  <acceptance test>
by      <primary module>
else by <alternate module 1>
else by <alternate module 2>
        .
        .
        .
else by <alternate module m-1>
else error
```

Randell [17] and Anderson and Lee [2] dicuss various points concerning the implementation of a recovery block such as the use of local variables, the reinstatement of nonlocal variables to their initial value prior to entering a new alternate module, etc. In this paper, we shall concentrate on probabilistic models of the recovery block structure, and questions of its implementation will not be dealt with.

Markov models in computer science are a widely used tool for the probabilistic analysis of program behaviour and queueing systems [13], [21]. In the next Section, simple Markov chain models of a recovery block are formulated leading to the joint probability distribution of the number of successfully processed input points by each of the m modules. Furthermore, moments of the failure time distribution are obtained using Laplace transforms. The behaviour of the mean time to failre (MTTF) and the variance of the failure time distribution are discussed for small values of the (conditional) failure rates of the primary (and all the other) alternates. The Laplace transform and moments of the response time (i.e. the CPU-time for any one input point) will also be derived.

It should be noted that papers addressing the performance evaluation of a recovery block do exist but they are restricted to calculating mean times to failure, see [3], [11] and [14], or the probability of failure for any given (randomly chosen) input point, see [6], [18] and [19]. In the present paper a complete description of all relevant distributions is given in terms of their Laplace transform and probability generating function.

In Section 3, a model is proposed accounting for randomly encountered clusters of failure points in the input space, a well-known problem [1] which does not seem to have been modelled previously.

For reasons of mathematical tractability, the acceptance test is assumed infallible throughout this paper. Even though this restriction reduces the practical applicability of the models, it is thought that the present approach is a good first approximation to recovery block reliability along which later refinements could achieve more realism. Work is currently being carried out to incorporate more realistic features into the models.

2. Simple Markov models of the recovery block

The technique of modelling the failure behaviour of modular software by a Markov process has been known for more than a decade [15]. Siegrist [20] discussed, most recently, a discrete-time Markov chain model of a modular software structure whose execution is terminated by arriving at one of its two absorbing states standing for mission success and mission failure respectively. General software tools for the analysis of such systems admitting a large number of states have also been reported. (See, for example, [10] for a review of some of them). The purpose of this Section is to show that one need not resort to powerful, all-purpose tools to model a recovery block but the special structure of the transition probability and transition-rate matrices may be exploited to essentially hand-calculate a number of useful quantities.

The notation and model assumptions are as follows. The recovery block consists of m modules numbered in descending order: m refers to the primary module and m-1, ..., 1 to the respective alternate modules. The execution times of the acceptance test and of module i are exponentially distributed with expectations $1/\mu$ and $1/\lambda_i$ respectively. The acceptance test is infallible and it signifies a correct output on exit from module i with probability $1-\rho_i > 0$. Upon producing a correct output control is returned to the primary module which takes the next point in the sequence of input values and processes it. The recovery block is considered to have failed if an input point causes all modules to produce erroneous results.

2.1. A simple discrete-time Markov chain model

The average number of input points which the recovery block will be able to process successfully is obtained by application of a simple discrete-parameter Markov chain model. The state-transition diagram is shown in Fig. 1; its states are the m (primary and alternate) modules and the (absorbing) failed state denoted by m, ..., 1 and 0 respectively.

The transition probability matrix is given by

$$
\begin{bmatrix}
1 & 0 & 0 & \ldots & 0 & 0 & 0 \\
\rho_1 & 0 & 0 & \ldots & 0 & 0 & 1-\rho_1 \\
0 & \rho_2 & 0 & \ldots & 0 & 0 & 1-\rho_2 \\
. & . & . & \ldots & . & . & . \\
0 & 0 & 0 & \ldots & \rho_{m-1} & 0 & 1-\rho_{m-1} \\
0 & 0 & 0 & \ldots & 0 & \rho_m & 1-\rho_m
\end{bmatrix}
$$

and thus has the form

$$
\begin{bmatrix}
\begin{array}{c|cccc}
1 & 0 & 0 & \ldots\ 0 & 0 \\
\hline
* & & & & \\
* & & & S & \\
* & & & & \\
* & & & &
\end{array}
\end{bmatrix}
$$

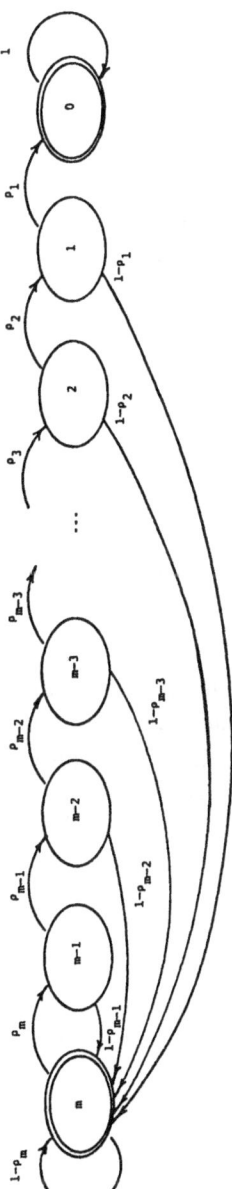

Figure 1. State-transition diagram for the simple discrete-time Markov chain model of a recovery block

with a substochastic matrix **S** defined in an obvious manner. The fundamental matrix **U** ([12], [21]) which is defined as

$$U = (I - S)^{-1}, (I = \text{identity matrix}),$$

is easily verified to be

$$
U = \begin{bmatrix}
\dfrac{1}{\rho_1} & \dfrac{1-\rho_1}{\rho_1\rho_2} & \dfrac{1-\rho_1}{\rho_1\rho_2\rho_3} & \dfrac{1-\rho_1}{\rho_1\rho_2\rho_3\rho_4} & \cdots & \dfrac{1-\rho_1}{\rho_1\cdots\rho_m} \\[2ex]
\dfrac{1}{\rho_1} & \dfrac{1}{\rho_1\rho_2} & \dfrac{1-\rho_1\rho_2}{\rho_1\rho_2\rho_3} & \dfrac{1-\rho_1\rho_2}{\rho_1\rho_2\rho_3\rho_4} & \cdots & \dfrac{1-\rho_1\rho_2}{\rho_1\cdots\rho_m} \\[2ex]
\dfrac{1}{\rho_1} & \dfrac{1}{\rho_1\rho_2} & \dfrac{1}{\rho_1\rho_2\rho_3} & \dfrac{1-\rho_1\rho_2\rho_3}{\rho_1\rho_2\rho_3\rho_4} & \cdots & \dfrac{1-\rho_1\rho_2\rho_3}{\rho_1\cdots\rho_m} \\[2ex]
\vdots & \vdots & \vdots & \vdots & \ddots & \vdots \\[2ex]
\dfrac{1}{\rho_1} & \dfrac{1}{\rho_1\rho_2} & \dfrac{1}{\rho_1\rho_2\rho_3} & \dfrac{1}{\rho_1\rho_2\rho_3\rho_4} & \cdots & \dfrac{1}{\rho_1\cdots\rho_m}
\end{bmatrix}.
$$

$u_{i,j}$ can be shown to be the expected number of visits to state j until the absorbing state 0 is reached when starting from state i ([12], [21]). Furthermore, the entries of the i-th row of the matrix $U^{(2)}$, which is defined by

$$U^{(2)} = U(2 U_{dg} - I),$$

are the respective second moments of the number of these visits when starting from i ([12]). (U_{dg} stands for the diagonal matrix which has the same diagonal as U). Let L_0 denote the number of successful passes through the recovery block. The average number of successfully processed input points is thus

$$E(L_0) = u_{m,m} - 1 = 1/(\rho_1\rho_2\cdots\rho_m) - 1.$$

Combining this with

$$E(L_0 + 1)^2 = u^{(2)}_{m,m} = u_{m,m} [2/(\rho_1\cdots\rho_m) - 1] = (2 - \rho_1\cdots\rho_m)/(\rho_1\cdots\rho_m)^2,$$

we obtain respectively the second moment and the variance of L_0 as

$$E(L_0^2) = (2 - \rho_1\cdots\rho_m)(1 - \rho_1\cdots\rho_m)/(\rho_1\cdots\rho_m)^2,$$

$$Var(L_0) = (1 - \rho_1\cdots\rho_m)/(\rho_1\cdots\rho_m)^2.$$

In fact, L_0 can be thought of as the number of repeated independent trials on which the first failure occurs when the probability of failure for each individual input point is $\rho_1...\rho_m$. L_0 therefore has a modified geometric distribution as follows

$$P(L_0 = l\ ) = \rho_1...\rho_m(1 - \rho_1...\rho_m)^l, \quad l = 0,1,... \tag{1}$$

Let K_i denote the number of input points for which module i produces the correct result. The average number of times module i is entered and produces a correct result is then

$$E(K_i) = u_{m,i} - u_{m,i-1} = (1 - \rho_i) / (\rho_1\rho_2...\rho_i\ ). \tag{2}$$

($u_{m,0}$ is to be put unity.) This shows, for instance, that when assuming a roughly equal chance of $1 - \rho$ of a successful completion by each of the m modules the utilization of module i is decreasing in a geometric fashion as i decreases:

$$u_{m,i} - u_{m,i-1} = (1 - \rho) / \rho^i\ .$$

The joint distribution of $K_1, ...,K_m$ is obtained as follows. For every fixed input point which is not in the failure region we have

P(correct result is produced by module i | no failure) =

$$= \rho_m\rho_{m-1}...\rho_{i+1} (1 - \rho_i) / (1 - \rho_1...\rho_m) , \quad i = 1, ..., m\ . \tag{3}$$

Given that the total number of successfully processed input points was l, the joint conditional probability of the i-th module producing k_i of the l correct results, $k_1 +...+ k_m = l$, is a multinomial probability with individual event probabilities (3):

$$P(\ K_1 = k_1, ..., K_m = k_m \mid L_0 = l) = \frac{l!}{k_1!...k_m!} \prod_{i=1}^{m} \left(\frac{(1 - r_i)r_1...r_m}{(1 - r_1...r_m)r_1...r_i} \right)^{k_i} . \tag{4}$$

Since $\{\ K_1 = k_1, ..., K_m = k_m \} \subset \{\ L_0 = l\ \}$, (1) and (4) give

$$P(\ K_1 = k_1, ..., K_m = k_m\ ) = P(\ K_1 = k_1, ..., K_m = k_m \mid L_0 = l\ )\ P(\ L_0 = l\ ) =$$

$$= \frac{(k_1+...+k_m)!}{k_1!...k_m!}\ (\rho_1...\rho_m)^{k_1+...+k_m} \prod_{i=1}^{m} \left(\frac{1 - \rho_i}{\rho_1...\rho_i} \right)^{k_i} . \tag{5}$$

(5) shows that the joint distribution of K_1, ..., K_m is a negative multinomial distribution; see, e.g., [7]. The i-th marginal distribution of (5) is seen to be a modified geometric distribution

$$P(K_i = k) = q_i (1 - q_i)^k, \quad k = 0, 1, 2, \ldots,$$

with

$$q_i = \left(\prod_{j=1}^{i} \rho_j \right) \left(1 - \rho_i \left(1 - \prod_{j=1}^{i-1} \rho_j \right) \right)^{-1}.$$

The mean of this distribution is indeed the previously calculated value (2).

2.2. A simple continuous-parameter Markov chain model

The state diagram of the corresponding homogeneous continuous-time Markov chain $\{ X_t : t \geq 0 \}$ is shown in Fig. 2. It has $2m+1$ states: the odd state E_{2i-1} stands for the acceptance test following the execution of the i-th module while the even state E_{2i} denotes execution of module i itself, $i = 1, \ldots, m$. E_0 is an absorbing state and it stands for recovery block failure. The $2m+1$ components of the probability vector $P(t)$, $t \geq 0$, are defined by

$$P_j(t) = P(X_t = E_j), \quad j = 0, \ldots, 2m.$$

The equations for the probability flow rate into individual states are obtained by inspection from Fig.2, (see, e.g. [13]), and they are as follows

$$\left. \begin{aligned} &P_0'(t) \quad = \rho_1 \, \mu \, P_1(t); \\ &P_{2i-1}'(t) = \lambda_i \, P_{2i}(t) - \mu \, P_{2i-1}(t) && \text{for } i = 1;\ldots;m; \\ &P_{2i}'(t) \quad = \rho_{i+1} \, \mu \, P_{2i+1}(t) - \lambda_i \, P_{2i}(t) && \text{for } i = 1;\ldots;m-1; \\ &P_{2m}'(t) \quad = \mu \sum_{i=1}^{m} (1 - \rho_i) \, P_{2i-1}(t) - \lambda_m \, P_{2m}(t). \end{aligned} \right\} \tag{6}$$

The transpose of the transition rate matrix will be denoted by $\mathbf{Q}$ and it is therefore given by

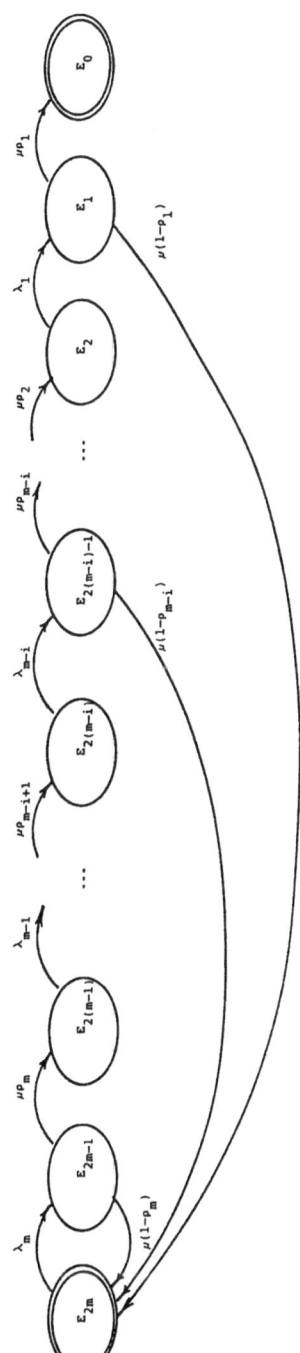

Figure 2. State-transition-rate diagram for the simple continuous-time Markov chain model of a recovery block

$$
Q = \begin{bmatrix}
0 & \mu p_1 & 0 & 0 & 0 & 0 & 0\ldots & 0 & 0 & 0 \\
0 & -\mu & \lambda_1 & 0 & 0 & 0 & 0\ldots & 0 & 0 & 0 \\
0 & 0 & -\lambda_1 & \mu p_2 & 0 & 0 & 0\ldots & 0 & 0 & 0 \\
0 & 0 & 0 & -\mu & \lambda_2 & 0 & 0\ldots & 0 & 0 & 0 \\
0 & 0 & 0 & 0 & -\lambda_2 & \mu p_3 & 0\ldots & 0 & 0 & 0 \\
\cdot & \cdot & \cdot & \cdot & \cdot & \cdot & \cdot\ldots & \cdot & \cdot & \cdot \\
0 & 0 & 0 & 0 & 0 & 0 & 0\ldots-\lambda_{m-1} & \mu p_m & 0 \\
0 & 0 & 0 & 0 & 0 & 0 & 0\ldots & 0 & -\mu & \lambda_m \\
0 & \mu(1-p_1) & 0 & \mu(1-p_2) & 0 & \mu(1-p_3) & 0\ldots & 0 & \mu(1-p_m) & -\lambda_m
\end{bmatrix}.
$$

The system of differential equations (6) can now be written as

$$P'(t) = Q\, P(t). \tag{7}$$

Execution of the recovery block is assumed to start at time $t = 0$ by entering the primary module, i.e., the initial conditions for (6) are

$$P_{2m}(0) = 1, \quad P_j(0) = 0, \quad j = 0, \ldots, 2m - 1. \tag{8}$$

Denoting by $P^*(s)$ the Laplace transform of $P(t)$ we get from (7) and (8)

$$(Q - s\, I)\, P^*(s) = (0, \ldots, 0, -1)^T. \tag{9}$$

The matrix $(Q - s\, I)$ is of the form

$$
M = \begin{bmatrix}
a_0 & b_0 & 0 & 0 & \ldots & 0 & 0 \\
0 & a_1 & b_1 & 0 & \ldots & 0 & 0 \\
0 & 0 & a_2 & b_2 & \ldots & 0 & 0 \\
\cdot & \cdot & \cdot & \cdot & \ldots & \cdot & \cdot \\
0 & 0 & 0 & 0 & \ldots & a_n & b_n \\
c_0 & c_1 & c_2 & c_3 & \ldots & c_n & c_{n+1}
\end{bmatrix},
$$

with n = 2m-1, and such that its entries are defined by

$$
a_j = \begin{cases} -s & \text{if } j = 0; \\ -(s + \lambda_j/2) & \text{if } j \in \{1; ...; n\} \text{ is even;} \\ -(s + \mu) & \text{if } j \in \{0; ...; n\} \text{ is odd;} \end{cases} \tag{10a}
$$

$$
b_j = \begin{cases} \mu \rho_{1+j/2} & \text{if } j \in \{0; ...; n\} \text{ is even;} \\ \lambda_{(j+1)/2} & \text{if } j \in \{0; ...; n\} \text{ is odd;} \end{cases} \tag{10b}
$$

$$
c_j = \begin{cases} 0 & \text{if } j \in \{0; ...; n\} \text{ is even;} \\ -(s + \lambda_m) & \text{if } j = n+1; \\ \mu(1 - \rho_{(j+1)/2}) & \text{if } j \in \{0; ...; n\} \text{ is odd.} \end{cases} \tag{10c}
$$

Expanding the determinant of **M** by its last row shows that

$$
\det(\mathbf{M}) = \sum_{j=0}^{n+1} (-1)^{n+1+j} c_j \left(\prod_{i=0}^{j-1} a_i \right) \left(\prod_{i=j}^{n} b_i \right).
$$

It is easily verified that the (unique) solution of the system

$$
\mathbf{M} (x_0, x_1, ..., x_n, x_{n+1})^T = (0, ..., 0, d)^T
$$

is given by

$$
x_k = \frac{(-1)^k d \left(\prod_{i=0}^{k-1} a_i \right) \left(\prod_{i=k}^{n} b_i \right)}{\sum_{j=0}^{n+1} (-1)^j c_j \left(\prod_{i=0}^{j-1} a_i \right) \left(\prod_{i=j}^{n} b_i \right)}, \qquad k = 0, ..., n+1, \tag{11}
$$

if $\det(\mathbf{M}) \neq 0$. Substituting k = 0 and d = -1 in (11), and using the values defined by (10a) - (10c), we get from (9)

$$
P_0^*(s) = \frac{\alpha}{\sum_{i=0}^{m} \beta_i s(s + \mu)^i \prod_{j=1}^{i} (s + \lambda_j)},
$$

with the following coefficients α and β_i

$$\alpha = -\mu^m \prod_{i=1}^{m} \lambda_i \rho_i \,,$$

$$\beta_i = \begin{cases} \mu^{m-i} (1 - \rho_{i+1}) \left(\prod_{j=i+2}^{m} \rho_j \right) \left(\prod_{j=i+2}^{m} \lambda_j \right) & \text{for } i \in \{0; \ldots; m-1\}; \\ -1 & \text{for } i = m. \end{cases}$$

The (local CPU-) time to failure, T, has density $P_0'(t)$ and its Laplace transform $T^*(s)$ is therefore given by

$$T^*(s) = s\, P_0^*(s) = \frac{\alpha}{\displaystyle\sum_{i=0}^{m} \beta_i (s + \mu)^i \prod_{j=1}^{i} (s + \lambda_j)} \,. \tag{12}$$

The denominator in (12) is a polynomial of degreee 2m in s

$$\gamma_0 + \gamma_1 s + \gamma_2 s^2 + \ldots + \gamma_{2m} s^{2m},$$

say, such that $\gamma_0 = \alpha$. It is more conveniently expressed in the form

$$\alpha + \sum_{i=1}^{m} \left(s^{2i-1} \sum_{j=i}^{m} \beta_j \, \delta_{j;2i-1} + s^{2i} \sum_{j=i}^{m} \beta_j \, \delta_{j;2i} \right),$$

with coefficients

$$\delta_{j;r} = \sum_{\substack{0 \le i; \, i' \le j \\ i+i'=r}} \binom{j}{i} \mu^{j-i} \, \varepsilon_{j;j-i'} \,,$$

where

$$\varepsilon_{j;h} = \begin{cases} \displaystyle\sum_{1 \le k_1 < \ldots < k_h \le j} \lambda_{k_1} \ldots \lambda_{k_h} & \text{for } h \in \{1; \ldots; j\}; \\ 1 & \text{for } h = 0. \end{cases}$$

(See [8] for details.) The moments of T are now given by

$$E(T^k) = k! \, \alpha \, (-1)^k \, \tau_k, \quad k = 0, 1, 2, \ldots \tag{13}$$

where τ_k is calculated by recursion from

$$\tau_0 = 1/\alpha$$

$$\tau_k = - \sum_{i=\max\{0;k-2m\}}^{k-1} \gamma_{k-i} \, \tau_i / \alpha \qquad \text{for } k \geq 1. \tag{14}$$

Using (13); the first moment of T works out to be

$$E(T) = \frac{\frac{m}{\mu} + \sum_{i=1}^{m} \frac{1}{\lambda_i}}{\rho_1 \cdots \rho_m} - \sum_{j=1}^{m-1} \frac{1 - \rho_{j+1}}{\rho_1 \cdots \rho_{j+1}} \left(\frac{j}{\mu} + \sum_{i=1}^{j} \frac{1}{\lambda_i} \right). \tag{15}$$

The MTTF (15) can also be arrived at by the following informal argument using the discrete-time Markov chain result (2): for any given input point which is not in the failure region of the recovery block the expected CPU-time is; when the correct output is produced by module i, obviously

$$\frac{m-i+1}{\mu} + \sum_{j=i}^{m} \frac{1}{\lambda_j} , \tag{16}$$

and hence by (2) the expected CPU-time expended on all such input points totals

$$\frac{1 - \rho_i}{\rho_1 \cdots \rho_i} \left(\frac{m-i+1}{\mu} + \sum_{j=i}^{m} \frac{1}{\lambda_j} \right). \tag{17}$$

Summing over all modules i and also adding the expected time spent on the last input point; which initiates failure; we arrive at

$$E(T) = \sum_{i=1}^{m} \frac{1 - \rho_i}{\rho_1 \cdots \rho_i} \left(\frac{m-i+1}{\mu} + \sum_{j=i}^{m} \frac{1}{\lambda_j} \right) + \sum_{i=1}^{m} \left(\frac{1}{\mu} + \frac{1}{\lambda_i} \right), \tag{18}$$

an expression that can be shown to be identical to (15). In fact; both (15) and (18) are easily seen to be equivalent to the following (finite Taylor series) representation of $\rho_1 \ldots \rho_m \, E(T)$:

$$\rho_1 \cdots \rho_m\, E(T) = \left(\frac{1}{\mu} + \frac{1}{\lambda_m}\right) + \sum_{i=1}^{m-1}\left(\frac{1}{\mu} + \frac{1}{\lambda_i}\right)\left(\prod_{j=i+1}^{m}\rho_j\right). \tag{19}$$

Notice that (18) holds irrespective of the assumed exponentiality of holding times: the transition from (16) to (17) is justified by a standard result on the expectation of random sums, [9], which is valid also in the non-exponentially case. (19) therefore also holds for semi-Markov processes with waiting time distributions not dependent upon the state to be entered.

The second moment and the variance of T are obtained from (13) and (14) as

$$E(T^2) = 2\,[E(T)]^2 - 2\,\gamma_2\,/\,\alpha;$$

$$\left.\begin{array}{l} \\ \\ \end{array}\right\} \tag{20}$$

$$Var(T) = [E(T)]^2 - 2\,\gamma_2\,/\,\alpha;$$

with

$$\rho_1 \cdots \rho_m\, \gamma_2\,/\,\alpha = \sum_{j=1}^{m}\left\{\frac{1}{\mu\lambda_j} + \left(\frac{1}{\mu} + \frac{1}{\lambda_j}\right)\sum_{i=1}^{j-1}\left(\frac{1}{\mu} + \frac{1}{\lambda_i}\right)\right\}\left(\prod_{i=j+1}^{m}\rho_i\right). \tag{21}$$

(19), (20) and (21) show that for small values of $\rho_1 \cdots \rho_m$ the second moment and the variance of T are closely approximated by $2[E(T)]^2$ and $[E(T)]^2$ respectively.

(12) will now be used to gain some further insight into the structure of the failure time distribution. The Laplace transform of T can be expanded into a geometric series to give

$$T^*(s) = \sum_{k=0}^{\infty} \rho_1 \cdots \rho_m (1-\rho_1 \cdots \rho_m)^k \left(\frac{\mu}{s+\mu}\right)^m \left(\prod_{i=1}^{m}\frac{\lambda_i}{s+\lambda_i}\right) \times$$

$$\times\left\{\sum_{i=1}^{m}\frac{(1-\rho_i)\rho_{i+1}\cdots\rho_m}{1-\rho_1\cdots\rho_m}\left(\frac{\mu}{s+\mu}\right)^{m-i+1}\left(\prod_{j=i}^{m}\frac{\lambda_j}{s+\lambda_j}\right)\right\}^k. \tag{22}$$

(22) shows that the distribution of T is obtained from the initial exponential distributions, $EXP(\mu)$, $EXP(\lambda_1)$, ..., $EXP(\lambda_m)$; through a combination of convolution and mixture operations:

$$\sum_{k=0}^{\infty} p_1 \cdots p_m (1-p_1 \cdots p_m)^k \, EXP(\mu)^{(m)} \otimes \left(\overset{m}{\underset{i=1}{\circledast}} \, EXP(\lambda_i) \right) \otimes$$

$$\otimes \left\{ \sum_{i=1}^{m} \frac{(1-p_i)p_{i+1} \cdots p_m}{1-p_1 \cdots p_m} EXP(\mu)^{(m-i+1)} \otimes \left(\overset{m}{\underset{j=i}{\circledast}} \, EXP(\lambda_j) \right) \right\}^{(k)} .$$

Another, and perhaps even more interesting variable is the response time R of the recovery block, i.e., the CPU-time needed to respond to an input point either by a correct output or an error message. The Laplace transform of R could now be again derived from an appropriate Markov chain model of a single pass through the recovery block; it is obvious, however, that the Coxian "service facility", [13], shown in Fig. 3, is an appropriate model of the distribution of R. This allows us to immediately determine the Laplace transform or R by inspection as follows

$$R^*(s) = p_2 \cdots p_m \left(\frac{\mu}{s+\mu} \right)^m \left(\prod_{i=1}^{m} \frac{\lambda_i}{s+\lambda_i} \right) +$$

$$+ \sum_{i=1}^{m-1} (1-p_{i+1})p_{i+2} \cdots p_m \left(\frac{\mu}{s+\mu} \right)^{m-i} \left(\prod_{j=i+1}^{m} \frac{\lambda_j}{s+\lambda_j} \right). \tag{23}$$

The distribution of R is therefore a finite mixture of hypoexponential distributions. Moments of R are thus obtained as convex combinations of the respective moments of those hypoexponential distributions:

$$E(R) = p_2 \cdots p_m \left\{ \sum_{i=1}^{m} \left(\frac{1}{\mu} + \frac{1}{\lambda_i} \right) + \sum_{i=1}^{m-1} \frac{1-p_{i+1}}{p_2 \cdots p_{i+1}} \sum_{j=i+1}^{m} \left(\frac{1}{\mu} + \frac{1}{\lambda_j} \right) \right\} =$$

$$(= p_1 \cdots p_m \, E(T)) , \tag{24}$$

(see (18)), and

$$E(R^2) = p_2 \cdots p_m \left\{ \frac{m}{\mu^2} + \sum_{i=1}^{m} \frac{1}{\lambda_i^2} + \left(\frac{m}{\mu} + \sum_{i=1}^{m} \frac{1}{\lambda_i} \right)^2 + \right.$$

$$\left. + \sum_{i=1}^{m-1} \frac{1-p_{i+1}}{p_2 \cdots p_{i+1}} \left(\frac{m-i}{\mu^2} + \sum_{j=i+1}^{m} \frac{1}{\lambda_j^2} + \left(\frac{m-i}{\mu} + \sum_{j=i+1}^{m} \frac{1}{\lambda_j} \right)^2 \right) \right\}.$$

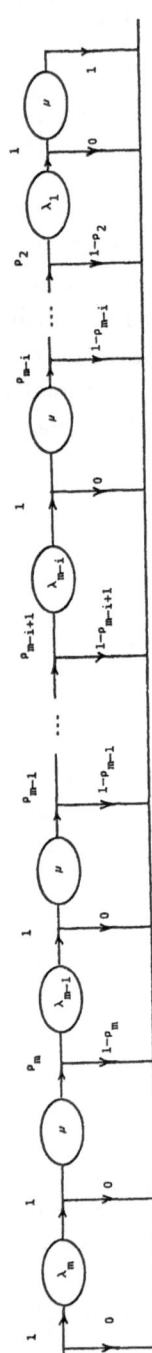

Figure 3. Coxian "service facility" for the response time R in the simple continuous-time Markov chain model of a recovery block

(24) may be used as an objective function when choosing the order of modules in a recovery block to achieve a minimum expected response time.

Finally; the following interrelationship between the Laplace transforms of T and R is seen to hold from (22) and (23)

$$T^*(s) = \frac{\rho_1 \cdots \rho_m \, \psi(s)}{1 - R^*(s) + \rho_1 \cdots \rho_m \, \psi(s)} , \qquad (25)$$

where $\psi(s)$ is defined by

$$\psi(s) = \left(\frac{\mu}{s+\mu}\right)^m \left(\prod_{i=1}^{m} \frac{\lambda_i}{s+\lambda_i}\right).$$

(25) provides an alternative to (13) and (14) for obtaining moments of T once those of R are known. By differentiating (25) we get

$$E(T) = E(R) / (\rho_1 \cdots \rho_m) ,$$

$$E(T^2) = \frac{\rho_1 \cdots \rho_m \, E(R^2) + 2 \, [E(R)]^2 + 2 \, \rho_1 \cdots \rho_m \, E(R) \, \psi'(0)}{(\rho_1 \cdots \rho_m)^2} ,$$

where, of course,

$$\psi'(0) = -\frac{m}{\mu} - \sum_{i=1}^{m} \frac{1}{\lambda_i} .$$

(The above representation of E(T) is already known from (24).)

3. Modelling the effects of failure point clustering in the input space

Software experiments indicate that failure points in the input space are seldom isolated but tend to occur in clusters [1]. It will be assumed that the sequence of input points evolves in a continuous fashion, i.e., that each input value is reasonably close to the preceding one. Consequently, the probability of an erroneous output by a given (primary or alternate) module will depend upon the entire failure history of that module for previous input values. It will be shown in this section that such non-Markovian behaviour still can be modelled by a Markov process if the state space is suitably enlarged.

3.1. Development of a theoretical model

We shall consider here recovery blocks containing only one single alternate module. Two kinds of failure events of the primary module will be distinguished: upon entering the failure region of the primary module in the input space the module suffers a failure which can be termed *random* or (perhaps more aptly) *spontaneous*. There follows a finite sequence of *certain* failures the length of which will depend upon the size of that particular cluster of failure points and upon the sequence of input points which the recovery block is required to process; see Fig. 4. The number of certain failures subsequent to any spontaneous failure is a random variable x. It will be assumed that the number of certain failures does not exceed a fixed value σ, say, i.e.,

$$P(\xi = i) = p_i; \quad i = 0; 1; ...; \sigma; \quad \sum_{i=0}^{\sigma} p_i = 1.$$

This temporary restriction allows the theory of finite Markov chains to be applied, but it will be removed later. The alternate module has, of course, its own characteristic clusters of failure points in the input space. This, however, need not be included in the model explicitely since exercising such an input by the alternate module terminates the execution of the recovery block, i.e., the alternate module can only fail in the spontaneous mode.

The state-transition diagram of the corresponding discrete-time Markov chain is shown in Fig. 5. 2_0 and 1_0 stand for the primary and secondary module respectively. A dummy state "d" and 2σ additional states (subscripted by $1,..., \sigma$) have been included to model the deteriorated state of the primary module: 2_i is followed by 1_i with probability 1 which then behaves as another instance of the alternate module. The number of successfully processed input points, L, is the total number of transitions $2_0 \rightarrow 2_0$, $1_0 \rightarrow d$, and $1_i \rightarrow 2_{i-1}$, $i = 1, ..., \sigma$. The chain has one absorbing state, 0, and it starts in 2_0. Fig. 6.a shows the state-transition diagram of a Markov chain which is obtained from Fig. 5 by merging pairs of states where transitions occur with certainty ($2_i \rightarrow 1_i$, $i = 1, ..., \sigma$), and by eliminating the dummy state. Reducing the Markov chain further gives Fig. 6.b: L now becomes the total number of transitions between any of the transient states of this Markov chain; they are $2_0, ..., 2_\sigma$. Consequently, $N = L + 1$ is the number of state transitions until absorption.

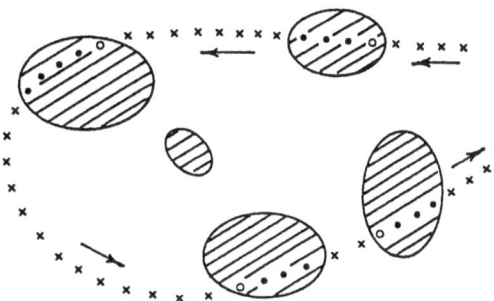

Figure 4. Evolving sequence of inputs with clusters of failure points
of the primary module.
("o") = random failure; "•" = certain failure; "x" = no failure)

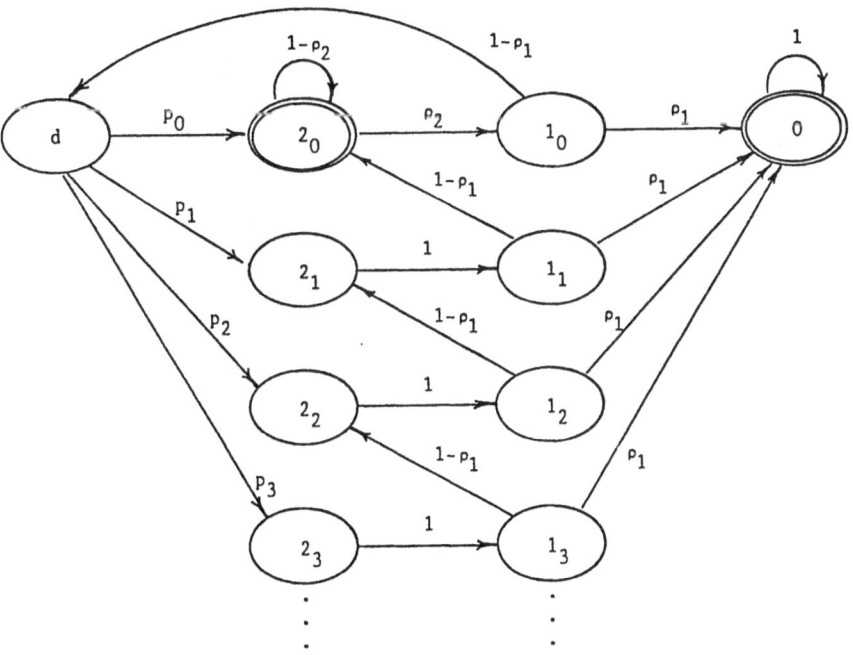

Figure 5. State-transition diagram for the Markov chain model of the
recovery block when clusters of failure points are admitted.

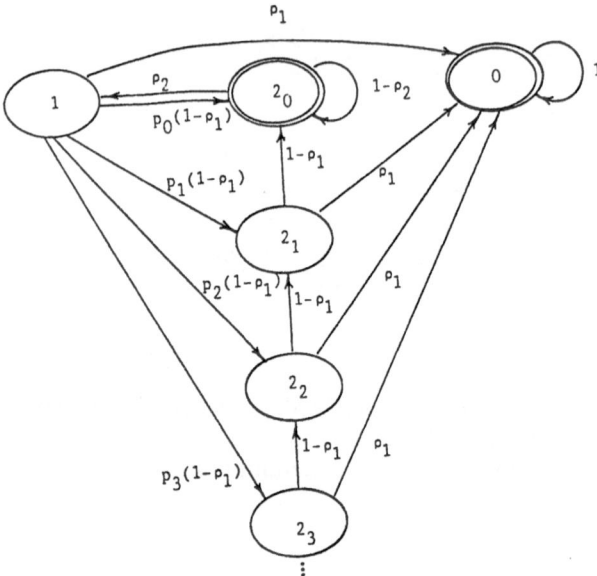

Figure 6.a. The reduced state-transition diagram; stage one

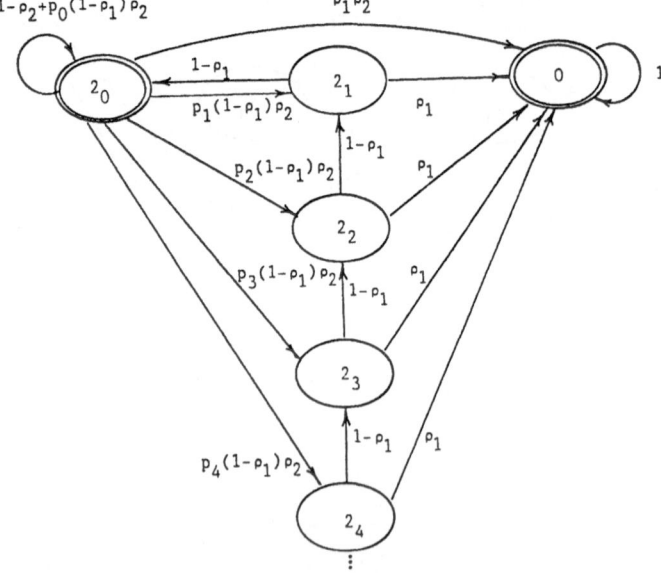

Figure 6.b. The reduced state-transition diagram; stage two

The distribution of N will now be described in terms of its moments and its probability generating function N^*. The $(\sigma+2)$ x $(\sigma+2)$ transition probability matrix Q has the form

$$0 \quad 2_0 \ 2_1 \ldots 2_{\sigma-1} \ 2_\sigma$$

$$Q = \begin{bmatrix} 1 & 1 & 0 & 0 & \ldots & 0 & 0 \\ \text{---} & & & & & & \\ & 1 & & & & & \\ r & 1 & & S & & & \\ & 1 & & & & & \\ & 1 & & & & & \end{bmatrix} \begin{matrix} 0 \\ 2_0 \\ \cdot \\ \cdot \\ \cdot \\ 2_s \end{matrix}$$

with

$$S = \begin{bmatrix} (1-\rho_2)+\rho_0(1-\rho_1)\rho_2 & \rho_1(1-\rho_1)\rho_2 & \rho_2(1-\rho_1)\rho_2 & \ldots\rho_{\sigma-1}(1-\rho_1)\rho_2 & \rho_\sigma(1-\rho_1)\rho_2 \\ 1-\rho_1 & 0 & 0 & \ldots & 0 \\ 0 & 1-\rho_1 & 0 & \ldots & 0 \\ 0 & 0 & 1-\rho_1 & \cdots & 0 \\ \cdot & \cdot & \cdot & \cdots & \cdot \\ 0 & 0 & 0 & \cdots & 1-\rho_1 & 0 \end{bmatrix},$$

and

$$r^T = (\rho_1\rho_2, \rho_1, \ldots, \rho_1)^T = \rho_1 \ 1^T - \rho_1 \ (1-\rho_2) \ e_0^T,$$

where e_0 denotes the first element of the standard basis of the $((\sigma+1)$-dimensional) Euclidean space; and 1 stands for the column vector with all entries unity. For an initial probability vector p_0 whose first entry is zero, i.e.,

$$p_0^T = (0; a_0; a_1; \ldots; a_\sigma)^T,$$

it is, see, e.g., [5],

$$P(N = n) = a^T \ S^{n-1} \ r, \ n = 1,2, \ldots \ .$$

The probability generating function of N can therefore be shown to be as follows ([4], [5]):

$$N^*(z) = (a^T \ (I - z \ S)^{-1} \ r) \ z. \tag{26}$$

Differentiating (26), we get (see [8])

$$E(N) = \mathbf{a}^T (\mathbf{I} - \mathbf{S})^{-1} \mathbf{1}, \tag{27}$$

$$E(N^2) = \mathbf{a}^T (\mathbf{I} + \mathbf{S}) (\mathbf{I} - \mathbf{S})^{-2} \mathbf{1}. \tag{28}$$

In the present application, $\mathbf{a} = e_0$ (starting point is at state 2_0). Furthermore, the matrix $\mathbf{B} = (\mathbf{I} - z \mathbf{S})$ has the form

$$
\mathbf{B} =
\begin{bmatrix}
b_0 & b_1 & b_2 & \dots & b_{\sigma-2} & b_{\sigma-1} & b_\sigma \\
-c & 1 & 0 & \dots & 0 & 0 & 0 \\
0 & -c & 1 & \dots & 0 & 0 & 0 \\
. & . & . & \dots & . & . & . \\
0 & 0 & 0 & \dots & -c & 1 & 0 \\
0 & 0 & 0 & \dots & 0 & -c & 1
\end{bmatrix}
\tag{29}
$$

with

$$
\left.
\begin{aligned}
b_0 &= 1 - z\,[(1-\rho_2) + \rho_0(1-\rho_1)\rho_2]; \\
b_i &= -z\,\rho_i(1-\rho_1)\rho_2 \qquad \text{for } i = 1;\ \dots;\ \sigma\,; \\
c &= z\,(1-\rho_1).
\end{aligned}
\right\}
\tag{30}
$$

It is easily verified, [8], that the inverse of a matrix $\mathbf{B}$ whose structure is given by (29) is

$$\mathbf{B}^{-1} = \frac{1}{b_0 + b_1 c + b_2 c^2 + \dots + b_\sigma c^\sigma} (1, c, \dots, c^\sigma)^T \mathbf{1}^T \mathbf{D} - \mathbf{C}, \tag{31}$$

where $\mathbf{D}$ is a diagonal matrix defined by

$$\mathbf{D} = \mathrm{diag}\left(1, \frac{b_0}{c}, \frac{b_0 + b_1 c}{c^2}, \dots, \frac{b_0 + b_1 c + \dots + b_{\sigma-1} c^{\sigma-1}}{c^\sigma}\right),$$

and C is given by

$$C = \begin{bmatrix} 0 & c^{-1} & c^{-2} & \dots & c^{-(\sigma-2)} & c^{-(\sigma-1)} & c^{-\sigma} \\ 0 & 0 & c^{-1} & \dots & c^{-(\sigma-3)} & c^{-(\sigma-2)} & c^{-(\sigma-1)} \\ 0 & 0 & 0 & \dots & c^{-(\sigma-4)} & c^{-(\sigma-3)} & c^{-(\sigma-2)} \\ . & . & . & \dots & . & . & . \\ 0 & 0 & 0 & \dots & 0 & 0 & c^{-1} \\ 0 & 0 & 0 & \dots & 0 & 0 & 0 \end{bmatrix}$$

Combining (26) and (31), we get

$$N^*(z) = a^T B^{-1} r z =$$

$$= \left\{ \left(\sum_{i=0}^{\sigma} b_i c^i \right)^{-1} e_0^T (1; c; \dots; c^\sigma)^T 1^T D r - e_0^T C r \right\} z$$

$$= \left\{ \left(\sum_{i=0}^{\sigma} b_i c^i \right)^{-1} 1^T D \left(\rho_1 1 - \rho_1(1-\rho_2)e_0 \right) - e_0^T C \left(\rho_1 1 - \rho_1(1-\rho_2) e_0 \right) \right\} z$$

$$= \left\{ \left(\sum_{i=0}^{\sigma} b_i c^i \right)^{-1} \left(\rho_1 \operatorname{tr}(D) - \rho_1(1-\rho_2) \right) - \rho_1 \sum_{i=1}^{\sigma} c^{-i} \right\} z. \tag{32}$$

Using geometric series and applying (30), it can be shown that ([8])

$$\operatorname{tr}(D) - \left(\sum_{i=0}^{\sigma} b_i c^i \right) \left(\sum_{i=1}^{\sigma} c^{-i} \right) = 1 + (1 - c)^{-1} \sum_{i=0}^{\sigma} b_i (c^i - 1) =$$

$$= 1 + (1 - \rho_1)\rho_2 \frac{1 - \xi^*((1-\rho_1)z)}{1 - (1-\rho_1)z} z, \tag{33}$$

where ξ^* stands for the probability generating function of ξ. Substituting the above into (32) gives

$$N^*(z) = \rho_1\rho_2 \frac{1 - \xi^*((1-\rho_1)z)\,(1-\rho_1)z}{[1 - (1-\rho_2)\,z - \xi^*((1-\rho_1)z)(1-\rho_1)\rho_2\,z]\,[1 - (1-\rho_1)z]}\,z. \tag{34}$$

(See [8] for details.) Using (27) and (33), the first moment of N is seen to be

$$E(N) = e_0^T\, B^{-1}\, \mathbf{1}$$

$$= e_0^T \left(\frac{1}{b_0 + b_1c + b_2c^2 + \ldots + b_\sigma c^\sigma}(1, c,\ldots, c^\sigma)^T\, \mathbf{1}^T\, D - C \right)\mathbf{1}$$

$$= \frac{\mathrm{tr}(D)}{b_0 + b_1c + b_2c^2 + \ldots + b_\sigma c^\sigma} - (c^{-1} + \ldots + c^{-\sigma})$$

$$= \frac{\rho_1 + (1-\rho_1)\rho_2\,[1 - \xi^*(1-\rho_1)]}{\rho_1\rho_2\,[1 - (1-\rho_1)\,\xi^*(1-\rho_1)]}$$

$$= \frac{1}{\rho_1\rho_2} - \frac{(1-\rho_1)(1-\rho_2)}{\rho_1\rho_2}\,\frac{1 - \xi^*(1-\rho_1)}{1 - (1-\rho_1)\,\xi^*(1-\rho_1)}. \tag{35}$$

(In the course of the derivation of (35), all quantities depending on z are to be evaluated at $z = 1$.) Using (28), it can be shown in a similar fashion, [8], that the second moment of N is given by

$$E(N^2) = \frac{(1-\rho_2)\,[\rho_1 + \rho_1\rho_2\,(1-\rho_1)^2\,\xi^{*\prime}(1-\rho_1) + \rho_2 - \rho_2\,(1-\rho_1)\,\xi^*(1-\rho_1)]}{\rho_1\rho_2^2\,[1 - (1-\rho_1)\,\xi^*(1-\rho_1)]^2} \times$$

$$\times [2 - \rho_2 + (1-\rho_1)\,\rho_2\,\xi^*(1-\rho_1)] + \frac{(1-\rho_1)^2\,(1-\rho_2)\,\xi^{*\prime}(1-\rho_1)}{1 - (1-\rho_1)\,\xi^*(1-\rho_1)} + \frac{2 - \rho_1\rho_2}{\rho_1^2}. \tag{36}$$

Letting $\sigma \to \infty$, a suitable approximation argument shows that (34) - (36) hold even if ξ is not bounded; see [8].

The results on the moments of L_0 in Section 2.1 are easily rederived from (35) and (36): for isolated failure points in the input space put $\xi \equiv 0$ and hence $\xi^* \equiv 1$ and $\xi^{*\prime} \equiv 0$. The other extreme, namely permanent degradation of the primary module upon encountering an input point in its failure region is modelled by moving the probability mass of ξ to infinity: in the limit, $\xi^*(1-\rho_1) = \xi^{*\prime}(1-\rho_1) = 0$. Subscripting the corresponding quantities with ∞, we have from (35)

$$E(L_\infty) = E(N_\infty) - 1 = 1/\rho_1 + 1/\rho_2 - 2 \;\; (= E(L_0) - (1-\rho_1)(1-\rho_2)/(\rho_1\rho_2) \;) . \tag{37}$$

The actual value of $E(L)$ lies in between the two extremes; from (35) and (37) it is seen ([8]) that $E(L)$ ($= E(N) - 1$) is a convex combination of $E(L_\infty)$ and $E(L_0)$:

$$E(L) = \frac{1 - \xi^*(1-\rho_1)}{1 - (1-\rho_1)\,\xi^*(1-\rho_1)} E(L_\infty) + \frac{\rho_1\,\xi^*(1-\rho_1)}{1 - (1-\rho_1)\,\xi^*(1-\rho_1)} E(L_0). \tag{38}$$

3.2. An example

3.2.1. Random cluster sizes

The case of "entirely random" cluster sizes is modelled by assuming that ξ has a modified geometric distribution, i.e., for some $q \in (0, 1]$

$$P(\,\xi = i\,) = q\,(1-q)^i, \;\; i = 0, 1, \dots \;. \tag{39}$$

Due to the memoryless property of the geometric distribution, (39) models the failure behaviour of the recovery block correctly if it is appropriate to assume that subsequent to each failure of the primary module it returns to its normal mode of operation with probability q irrespective of its past failure history. The current failure behaviour of the primary module is thus influenced only by what has happened to the input point immediately preceding the present one. Using (34), the probability generating function of L now becomes, [8],

$$L^*(z) = \frac{\rho_1\rho_2}{1 - [1 + (1-q)(1-\rho_1)(1-\rho_2) - \rho_1\rho_2]\,z + (1-q)(1-\rho_1)(1-\rho_2)\,z^2} .$$

L^* may be shown to be the product of the probability generating functions of two modified geometric distributions with parameters $q_1, q_2 \in (0, 1]$ such that

$$q_{1,2} = 1 - \frac{2\,(1-q)(1-\rho_1)(1-\rho_2)}{1 + (1-q)(1-\rho_1)(1-\rho_2) - \rho_1\rho_2 \pm \sqrt{[1+\rho_1\rho_2-(1-q)(1-\rho_1)(1-\rho_2)]^2 - 4\rho_1\rho_2}}.$$

(See [8] for details.) Furthermore, using (37) and (38), it is shown in [8] that

$$E(L) = \frac{1}{\rho_1} + \frac{1}{\rho_2} - 2 + q\frac{(1-\rho_1)(1-\rho_2)}{\rho_1\rho_2}. \tag{40}$$

It is also shown in [8] that the variance of L can be represented as follows:

$$Var(L) = \left(\frac{1 - (1-q)(1-\rho_1)(1-\rho_2)}{\rho_1\rho_2}\right)^2 - \frac{1 + (1-q)(1-\rho_1)(1-\rho_2)}{\rho_1\rho_2}. \tag{41}$$

(40) and (41) may be used to *estimate* the model parameters by the method of moments: ρ_1, which is an unconditional failure probability, may be available from unit testing while estimates of $E(L)$ and $Var(L)$ are obtained from integration testing. Solving (40) and (41) for ρ_1 and q gives the required estimates.

3.2.2. An illustrative example

Consider two modules with rather mundane reliability levels: both the primary and the alternate will be assumed to have the same spontaneous failure probabilities

$$\rho_2 = \rho_1 = 10^{-3}.$$

(Note, however, that as far as ρ_1 is concerned, it is really a *conditional* failure probability: the input space of the alternate is the failure region of the primary module.) The parameter q is in terms of the *expexted cluster size* $E(\xi)$

$$q = [1 + E(\xi)]^{-1}.$$

The moments of the number of correctly processed inpot points, L, are shown below for various values of $E(\xi)$:

$E(\xi)$	$E(L)$	$Var(L)$	$COV(L)$
0	1.000×10^6	1.000×10^{12}	1.0000
1	5.010×10^5	2.510×10^{11}	1.0000
2	3.347×10^5	1.120×10^{11}	1.0000
3	2.515×10^5	6.325×10^{10}	1.0000
10^1	9.272×10^4	8.596×10^9	1.0000
10^2	1.188×10^4	1.392×10^8	0.9929
10^3	2.995×10^3	6.979×10^6	0.8821
10^4	2.097×10^3	2.407×10^6	0.7398
∞	1.998×10^3	1.998×10^6	0.7075

where COV(L) denotes the coefficient of variation of L. The expected number of sucessfully processed input points is seen to decrease from the *product* of the mean number of correct outputs by each individual module to that of the *sum* of these means as the average cluster size increases; the actually achieved figure will be somewhere in between these two extremes. The coefficient of variation of L is seen to approach a limit as the mean cluster size increases: this is indeed what we expect knowing that for $E(\xi) \to +\infty$ the situation approaches that of two irrecoverable units in series.

4. Conclusions and future work

Probabilistic models of the recovery block have been presented which enables the distribution of the time to failure to be represented in terms of its Laplace transform and its probability generating function. Application of the theory of absorbing Markov chains resulted in a reliability model for the recovery block accounting for effects of failure point clustering in the input space.

Work currently in progress includes the following: imperfect acceptance tests; cluster models for *any number* of alternate modules; conditional performance measures; continouos-time cluster models; models with several failure modes to increase model realism; questions of parameter estimation; more examples of ξ; joint distribution of module utilization.

Acknowledgment

This work was supported by the Alvey Directorate and Science and Engineering Research Council., UK.

References

[1] P. E. Amman and J. C. Knight, "Data Diversity: An Approach to Software Fault Tolerance", *Proc. 1987 IEEE Int. Symp. Fault-tolerant Computing*, pp. 122 - 126, July 1987. Also published in *IEEE Trans. Comp.*, vol. 37, pp. 418-425, April 1988.

[2] T. Anderson and P. A. Lee, *Fault Tolerance: Principles and Practice*, Englewood Cliffs, NJ: Prentice/Hall International, 1981.

[3] J. Arlat, K. Kanoun, and J.-C. Laprie, "Dependability Evaluation of Software Fault-Tolerance", *Proc. 1988 IEEE Int. Symp. Fault-Tolerant Computing*, pp. 142 -147, June 1988.

[4] M. S. Bartlett, *An Introduction to Stochastic Processes with Special Reference to Methods and Applications*, Third Edition, Cambridge: Cambridge University Press, 1978.

[5] B. R. Bhat, "Some Properties of Regular Markov Chains", *The Annals of Mathematical Statistics*, vol. 32, pp. 59 - 71, 1961.

[6] S. D. Cha, "A Recovery Block Model and Its Analysis", *Proc IFAC SAFECOMP 1986*, Sarlat, France, pp. 21 - 26, 1986.

[7] C. L. Chiang, *Introduction to Stochastic Processes in Biostatistics*, New York: Wiley, 1968.

[8] A. Csenki, "Reliability Models of Fault-Tolerant Software", Technical Report, Centre for Software Reliability, The City University, London, 1989.

[9] W. Feller, *An Introduction to Probability Theory and Its Applications*, Volume II, New York: Wiley, 1966.

[10] R. M. Geist and K. S. Trivedi, "Ultrahigh Reliability Prediction for Fault-Tolerant Computer Systems", *IEEE Trans. Comp.*, vol. C-32, pp. 1118-1127, December 1983.

[11] A. Grnarov, J. Arlat, and A. Avizienis, "On the Performance of Software Fault-Tolerant Strategies", *Proc. 1980 IEEE Int. Symp. Fault-tolerant Computing*, pp. 251 - 253, October 1980.

[12] J. G. Kemeny and J. L. Snell, *Finite Markov Chains*, New York, Berlin, Heidelberg: Springer-Verlag, 1976.

[13] L. Kleinrock, *Queueing Systems*, Volume I: Theory, New York: Wiley, 1975.

[14] J.-C. Laprie, "Dependability Evaluation of Software Systems", *IEEE Trans. Software Eng.*, vol. SE-10, pp. 701 - 714, November 1984.

[15] B. Littlewood, "A Reliability Model for Systems with Markov Structure", *Appl. Statist.*, vol. 24, pp. 172 -177, 1975.

[16] B. Littlewood and D. R. Miller, "A Conceptual Model of Multiversion Software", *Proc. 1987 IEEE Int. Symp. Fault-Tolerant Computing,* pp. 150 - 155, July 1987.

[17] B. Randell, "System Structure for Software Fault Tolerance", *IEEE Trans. Software Eng.,* vol. SE-1, pp. 220 - 231, June 1975.

[18] R. K. Scott, J. W. Gault, D. McAllister, and J. Wiggs, "Experimental Validation of Six Software Reliability Models", *Proc. 1984 IEEE Int. Symp. Fault-Tolerant Computing,* pp. 102 - 107, June 1984.

[20] K. Siegrist, "Reliability of Systems with Markov Transfer of Control", *IEEE Trans. Software Eng.,* vol. SE-14, pp. 1049 - 1053, July 1988.

[21] K. S. Trivedi, *Probability and Statistics with Reliability, Queuing, and Computer Science Applications,* Englewood Cliffs, NJ: Prentice-Hall, 1982.

Application-Oriented Systems

Chair: H. Kopetz (Technische Universität, Wien, Austria)

SAFETY AND FAULT-TOLERANCE

IN COMPUTER-CONTROLLED

RAILWAY SIGNALLING SYSTEMS

Katsuji AKITA, Hideo NAKAMURA
Safety and Telecommunications Lab. - Railway Technical Research Institute
8-38, 2-Chome, Hikaricho - Kokubunji-City - Tokyo - Japan

Abstract

This paper presents safety and fault-tolerance technology of railway signalling controls. First,the type of faults to be handled and the required level of safety and reliability are given. Next,two types of microcomputer architectures with hardware redundancy are shown. Safety methods for a video display unit and data transmission are also given. Fault-tolerant system structures with triple-modular-redundancy, duplex and single,which have been applied to signalling systems in JR are also described. Further, software approach to fail-safety and fault-tolerance is discussed. Field data on 4 kinds of microcomputer-controlled signalling systems are shown. Since 1985 about 650 sets of fail-safe microcomputers have been operating in JR. An endangering failure has never occurred and the safe-side failure rate up to now including an initial error period has been in the order of 10^{-5} to 10^{-6}/h.

1. Introduction

Recently, the use of microcomputers has been increasing in very vital areas. A typical example is their use in railway signalling systems.

Since 1985 several kinds of microcomputer-controlled signalling systems have been introduced into Japan Railways, JR (Table 1). They are a solid-state interlocking system, named SMILE which controls signals and point-machines at a station [1], an electronic blocking system which prevents a collision of trains on a single track line [7], an Automatic Train speed Control system (ATC) on Shinkansen lines,an Automatic Train Stop system (ATS) on narrow-gauge lines,a level crossing alarm controller [6] and so on. These systems require not only high availability, but must be fail-safe to prevent collisions and train derailments. To satisfy these requirements, both hardware and software must have fail-safety and fault-tolerance. Furthermore, a system must be able to withstand the extreme environmental conditions encountered in the railway field, including electromagnetic noise and electrical surges. In signalling control, safe-side state of a system can be uniquely defined. If any errors are detected, the control output must infallibly give a stop signal,close the crossing gate and so on. To improve the dependability of a whole system, management of safety and fault-tolerance is indispensable throughout a system's life cycle. Guidelines for introducing an electronic signalling system have been established and are followed in JR.

2. Types of faults and countermeasures against them

(Table 2)

Several kinds of faults are liable to happen in a computer-controlled system. They originate from the following circumstances:

(1) Wrong design, production and construction due to the human factor.

(2) Wrong operation and maintenance also due to the human factor and illegal input from other systems.

(3) Deterioration of hardware elements due to physical conditions.

(4) Environmental interference like electro-magnetic noise and high or low temperature also due to physical factor.

Year	System	Level 1)	Redundancy	Comparison 2)
1981	Prototype of solid-state interlocking 3)	A	TMR	H
1983	Axle counter for level crossing control	B	Dual 4)	H
1985	Centralized traffic control system CTC	C	Duplex	
1985	Solid-state interlocking system SMILE	A	TMR	H
1985	Automatic train control ATC for Shinkansen	A	[Dual*[TMR]	H
1986	Electronic blocking system for single track	A	Dual 4)	S
1987	Solid-state interlocking system μ-SMILE	A	[Dual*[Hot-standby]	H
1987	Automatic train stop system ATS-P	A	Dual 4)	H
1987	Transponder	A	Dual 4)	H
1987	Track circuit without insulation joint	A	Dual 4)	H
1987	Constant warning control for level crossing 3)	A	Dual 4)	H
1988	ATC for Seikan tunnel	A	[Dual*[Hot-standby]	H
1988	Safety data transmission in Seikan tunnel	A	Dual 4)	S
1989	Solid-state signal/point-machine controller	A	Dual 4)	H

1) A:Safety system   B:Backup system for safety control   C:System not required safety
2) H:Hardware comparison at bus level   S:Software comparison on Task level
3) Field test stage   4) Dual without backup

Table 1. Microcomputer-controlled signalling systems in Japan

Fault			Countermeasures	
			Fault-avoidance	Fault-tolerance
Human factor	Design		Design review	
	Production	Hardware	System test	Built-in acceptance test by software
			Safety analysis	Latent error detection by software
			System test	Watch-dog timer
				Double-link-file structure
		Software	Structured software	Same as measures against hardware error
			System test	Interlock technique
	Construction			
	Operation/Maintenance		Training	Integrity check against input error
Physical factor	Hardware deterioration		Accelerated test	Error detection by software
				Hardware redundancy
	Overload		Overload test	Watch-dog timer
			by simulator	
	Noise/Surge		Noise suppressor	Hardware redundance
			Noise injection test	Time redundance

*) Malicious interference and disaster are excluded.

Table 2. Type of faults and countermeasures in Japan railways

Additional factors which are not addressed here are natural disasters like earthquakes and fires,also mischievous and malicious damage.

Both fault-tolerance by design/software diversity [3 - 9] and fault-avoidance are effective methods to guard against the faults of type (1). However, the former has not been adopted yet in JR, except in the application of a double link structure in part of the files. Our technology for removing such faults basically relies on fault-avoidance, that is, design review and exhaustive testing during system development and production stages. The reasons why we depend on fault-avoidance are as follows:

    a. Cost of computer-controlled signalling systems is required to be lower than that of conventional ones. We think diversity software will be expensive.

    b. The control algorithm of a microcomputer-controlled signalling system is based on that of a conventional one, which has been established on the view point of fail-safety.

    c. Design faults and software faults are expected to be eliminated through long term exhaustive testing in the laboratory and in the field. Usually synthetic system testing can be done by operating a computer system in parallel with the conventional system in actual use and by comparing the outputs between them.

Fault-tolerance technology is applied against the faults of type (2). Against wrong handling by an operator, an interlock technique is adopted so that an illegal input can be rejected depending on the internal state of the system. For example, a train route-setting command from the operator is never accepted when other trains are running on the track ahead, or after an order for other trains to enter the track has been issued. However, other illegal data or data at unexpected timing may enter the system. Acceptance test at check points assigned by software is carried out,and the software reverses internal states and control outputs to safe-side,when a serious error is detected (described in chap.6).

Hardware redundancy is used against the faults of types (3) and (4) in JR. Two or three computers,operating independently,compare the processing data with each other on either a hardware basis or a software basis (described in chap. 4). The following considerations are indispensable in this architecture:

    a. To avoid a double error by increasing the frequency of data comparison between computers and by detecting a single error instantly.

b. To avoid a double error in common mode due to environmental interference by ensuring independence of power supply, clock, installation of printed board, arrangement of wiring, and so on between computers.

3. Required level of safety and reliability

Most signalling systems are required to operate round the clock, so preventive maintenance is practiced.

Dangerous failure rate of computer-controlled signalling systems is required to be less than 10^{-10}/h, which is not more than that of conventional relay-type ones. On the other hand mean system down time of safe-side is required to be less than 30 minutes per 10 years (unavailability: less than 5.7×10^{-6}).

4. Fail-safe technology

4.1. Fail-safety in computer control

Hitherto, the safety of railway signalling systems has been ensured by means of electromagnetic relays with fail-safe characteristics as their logical elements.

On the other hand, a system consisting of electronic components without asymmetrical failure mode is required to ensure its safety by adopting redundant structure and error detection mechanisms. We have developed two types of architectures based on hardware redundancy in JR (Fig.1). One is comparison at bus level and the other is that of software level. Either of the two is selected according to the mission of a given system.

4.1.1. Comparison at bus level

Two computers installed with identical software are synchronized with a master clock,and data transferred on the inner buses are compared bit by bit for every machine cycle on a hardware basis. Error and frequency fluctuation of the clock is diagnosed based on a window-timer method, where a software timer is compared with a hardware timer accumulating the pulses of an additional clock (Fig.2). If any difference is detected between them, computers are switched to halt the state by software. This system architecture has the following advantages:

(1) Errors of processors,memories and input circuits can be reliably revealed on the buses.

(2) The probability of a double error occurring in the same mode and at the same instant can be made negligibly small by frequent data comparison.

(3) Software need not be basically conscious of redundancy and can be simplified. Overhead time of software for error detection and system reconfiguration is extremely short.

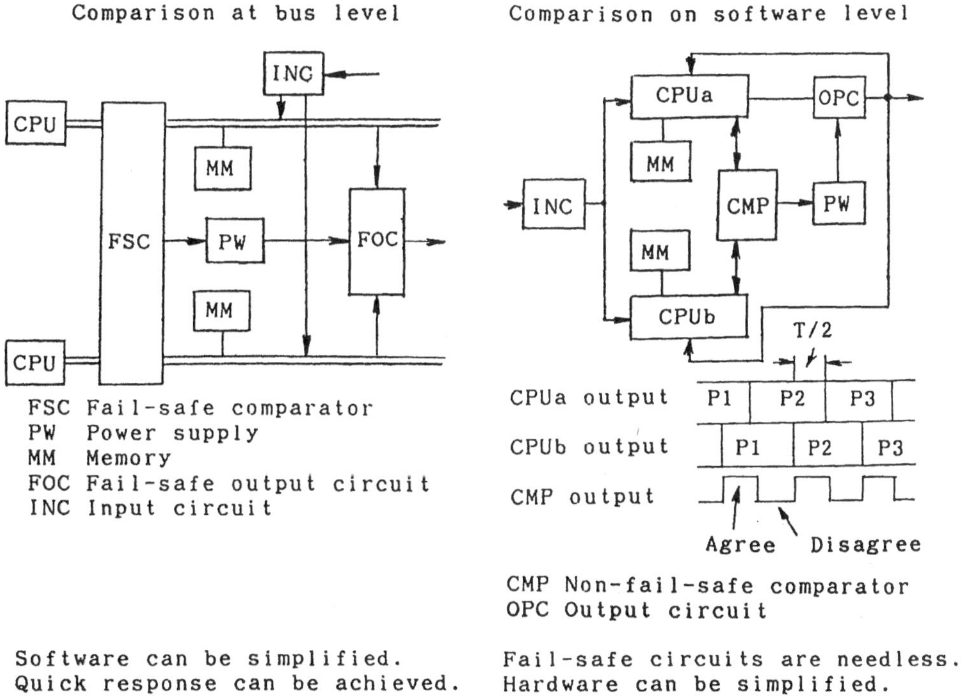

Comparison at bus level   Comparison on software level

FSC Fail-safe comparator
PW   Power supply
MM   Memory
FOC Fail-safe output circuit
INC Input circuit

CMP Non-fail-safe comparator
OPC Output circuit

Software can be simplified.   Fail-safe circuits are needless.
Quick response can be achieved.   Hardware can be simplified.

Figure 1. Fail-safe microcomputer with dual structure

Further, the detection of latent errors in important circuits is periodically done by software (described in Sec. 6.3). In order to avoid a common-mode error due to environmental interference, which is feared in this type of architecture, d.c. power supply, installation of printed boards and arrangement of wiring on mother-boards are separated and independence is ensured between the two computers. As the result of experimental tests by inserting electromagnetic noise from a noise generator into a system, errors were positively detected and control outputs were switched to safe-side. The comparator must be furnished with fail-safe and high-speed features. We have developed a fail-safe comparator (FSC) consisting of PISO (parallel-in-serial-out) registers, 2-bit bi-directional shift register and so on (Fig.3) [1]. FSC latches the data on buses, when computers issue "read or write" control signals. FSC has an inner clock which gives a pulse (left shift signal) to the shift register in every comparison cycle. Following this, a pair of latched data taken out in serial form are sent to the shift register through an exclusive OR(EOR) element. This operation is accomplished in synch with the inner clock of the FSC.

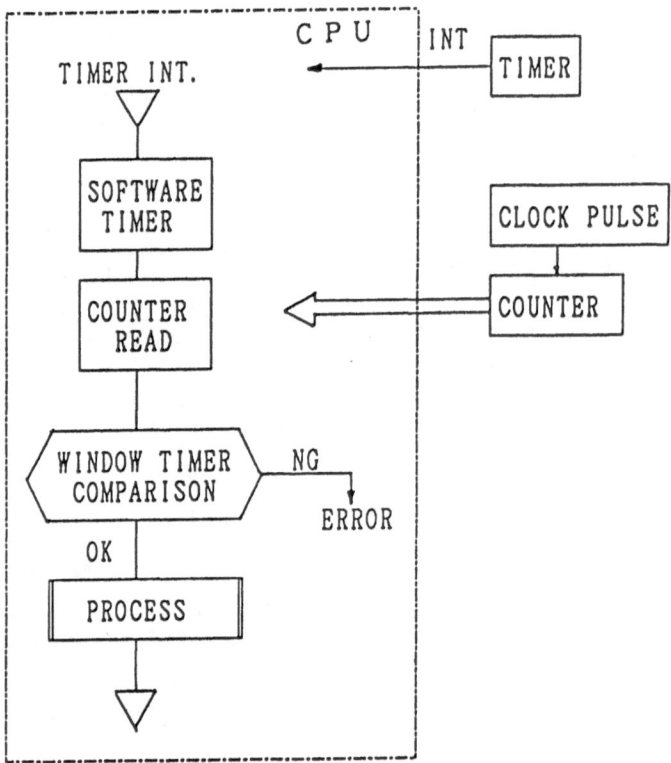

Figure 2. Concept of window limer method

In normal operation FSC produces 1 and 0 outputs periodically, which are rectified to drive a d.c. relay. If any part of FSC fails or a disagreement between computers occurs, an alternative output is never issued and so power to the output circuit is switched off.

Errors in an output circuit cannot be detected by FSC, therefore the output circuit must be also fail-safe. We have developed a fail-safe output circuit (FOC) consisting of fail-safe AND, photo-coupler and fail-safe driver. If outputs from the computers happen to disagree or FOC itself fails, the system output is switched to safe-side state.

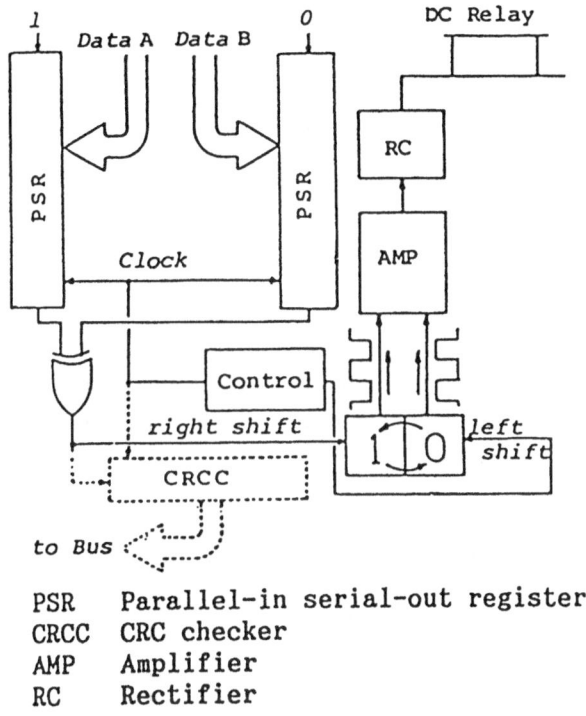

PSR	Parallel-in serial-out register
CRCC	CRC checker
AMP	Amplifier
RC	Rectifier

Figure 3. Concept of fail-safe comparator (FSC)

4.1.2. Comparison on software level

Identical software is employed in two computers driven by their own clocks and the processing in one of the two is performed with a time lag of T/2 (T=1ms). Accordingly the comparator outputs an agreement signal and a disagreement signal alternately, when each computer sends its processing results and diagnostic data (Table 3) to the comparator at a time interval of T. This dynamic signal drives a d.c. relay in normal operation. If errors occur in one computer, the comparator never issues a dynamic signal and the power supply to the output circuit is switched off.

Each computer, which is well aware in advance whether the comparator output agrees or not, can instantly detect a comparator error and the computer itself as a whole through feedback of the contents of the comparator. Further, an interrupt signal is exchanged between the two computers in every period of T/2, and each computer checks the time difference between its own timer and the other's.

Diagnostic program	Test Pattern for check of comparator Results of ROM check(summing up value) Results of RAM check(read/write check) Results of instruction check Important areas of RAM asigned
Application program	Task numbers sequentially executed Results of processing at check points

Table 3. Diagnostic data between two computers with software level comparison

Control outputs are issued from a computer in alternate signal form on a software basis and rectified in the output circuit. When a computer detects any errors,it stops issuing the output signal and switches the system to safe-side state. This architecture has the following advantages:

(1) Off-the-shelf low-cost components are available for comparator and other circuits, therefore complicated fail-safe circuits are needless.

(2) System structure can be simplified.

(3) A double error in common mode due to electromagnetic noise is avoidable by means of time difference processing.

The disadvantage of this architecture is that as much as 50% more processing time is required for software diagnosis. When the volume of application software required is large, it becomes difficult to select the check points to ensure system safety without some oversight.

4.2. Safety in data transmission

In a decentralized signalling system where fail-safe computers are located around a railway, data transmitted between computers must have safety features. Our technology for protecting data transmission from critical errors is as follows:

(1) Error detection and diagnosis are done by fail-safe computers on an end-to-end basis.

(2) Check-code ensuring hamming-distance of not less than 4 is attached, and information-code is provided by one of the two transmitting fail-safe computers and check-code is made by the other.

(3) Time-redundant transmission is done periodically and when a receiving fail-safe computer detects illegal data or an interruption of the transmission, it switches its inner state and control output to safe-side state.

5. Fault-tolerance technology

5.1. Redundant Structure

Fault-tolerance technology excluding fail-safety is shown here. A railway signalling system is required to operate round the clock,but requirements for availability and cost differ depending on the mission of a given system.

5.1.1. Triple-Modular-Redundancy (TMR)

TMR is one method for ensuring non-stop operation and transient error correction. It is applied to ATC in Shinkansen and SMILE (Fig.4) at large stations. These systems controlling heavy traffic trunk lines can not tolerate a system shutdown, whether it is instant or permanent. The heart of SMILE,

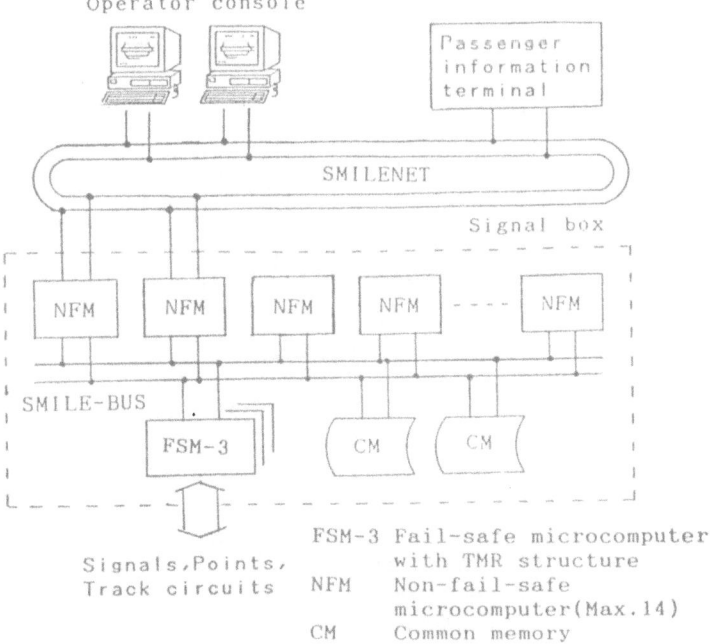

Figure 4. Configuration of SMILE

named FSM-3(fail-safe microcomputer), adopts bus-level comparison among three off-the-shelf computers. On each computer's bus a majority-voting-restorer (MVR) is inserted, and the data bus, address bus and several control signals are voted on bit by bit (Fig.5) [5].

Six FSC's are provided which compare input-output pairs of three MVR's. When an error is detected, based on the results of FSC, a relay circuit (MCC) controls the power switches of the corresponding path in output-voting-circuit (OVC). That is, the control output from a faulty computer will be prohibited from voting, even when it takes part in internal bus level voting. Then the system operates in dual mode. If another disagreement occurs between two buses of normal computers, the relay circuit kills the whole system and switches control outputs to safe-side.

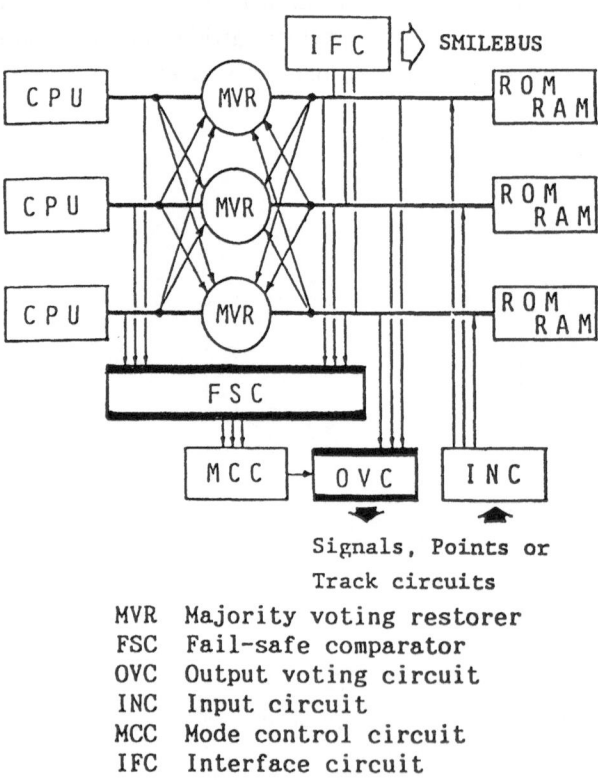

Signals, Points or
Track circuits

MVR	Majority voting restorer
FSC	Fail-safe comparator
OVC	Output voting circuit
INC	Input circuit
MCC	Mode control circuit
IFC	Interface circuit

Figure 5. Fail-safe microcomputer (FSM-3) with TMR structure

OVC receives the outputs from three computers through triplicated output ports and votes them. Any errors in circuits after the output port cannot be detected by FSC, therefore, OVC must have fail-safe characteristics [1].

Protection against a transient error is provided by software with the aid of hardware. Cyclic redundancy code checkers (CRCC's) are provided as shown in Fig.3. Serial data which are the same as those given to the shift register in FSC are supplied to CRCC which calculates the residue of these serial data using a predetermined generator polynomial in every comparison cycle. Three computers read the contents of six CRCC's every Ts (Ts=300ms). If any disagreement among buses is detected,the three computers initialize all the shift registers in FSC's to keep the TMR operation,supposing the disagreement has been caused by a transient error between previous Ts. If another disagreement between the other two buses has occurred during this period, computers never initialize the shift registers. Accordingly the whole system is switched to shutdown. If another disagreement occurs at the same buses during the following Ts, computers judge it to be a permanent error and never initialize the shift registers. Then a failed computer is detached from the system on a hardware basis.

The clock also adopts TMR structure (Fig.6) [Naka87]. This TMR clock circuit with a DPLL and a voter issues a clock pulse with a duty ratio of 50%, using a phase comparator consisting of a D-type flip-flop and a delay element.

Besides, SMILE has such non-fail-safe functions as automatic route-setting, information supply to passengers and so forth. To improve the flexibility and fail-soft feature, we adopted a system structure with functionally divided multi-microcomputers. Computers except FSM-3's have a single or duplex (hot standby) structure depending on their function. Each computer is connected to a duplicated system bus, through which data exchange and mutual diagnosis are done on a software basis [1]. FSM-3, the core of SMILE, identifies a faulty computer based on the results of diagnosis and commands the duplexed computers to switch from master to slave.

5.1.2. Duplex structure

Next comes a hot standby structure,which allows an interruption of processing during the master-to-slave switchover. That is not to say that both master and slave computers (FSM-2) must each have fail-safe features. This structure is

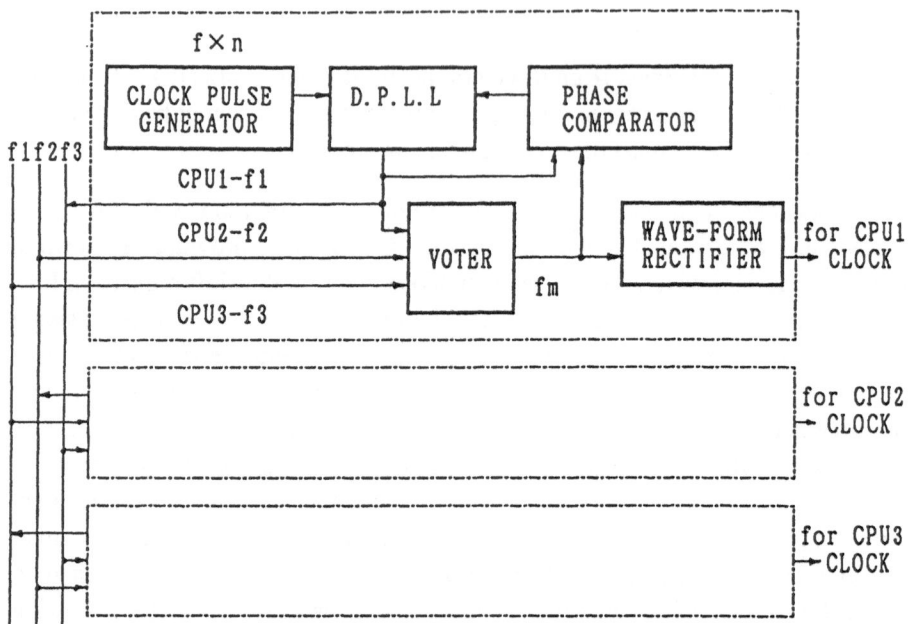

Figure 6. Concept of TMR clock

adopted in μ-SMILE (Fig.7), a simpler type of SMILE for a small station. A computer in the slave mode always receives from another in the master mode such important data on safety as sections being blocked for wayside maintenance work, and keeps them ready for future switchover. These data are transmitted between the two fail-safe computers via interface computers.

This structure reduces the amount of hardware by more than 30% and is considerably more economical than the one with TMR.

5.1.3. Single structure

A single-structure fail-safe computer system costs the least but any errors cause the system to shut down. Following a transient error, however, it is desirable that the system be restartable. The following measures are applicable:

(1) To provide an additional counter circuit with a preset value, and to reset the system on hardware basis until the permitted count is exhausted, whether the error is permanent or not.

(2) To test the system on a software basis in case a disagreement between the two computers has been detected, and to initialize the system, provided the error is judged not to be a permanent one.

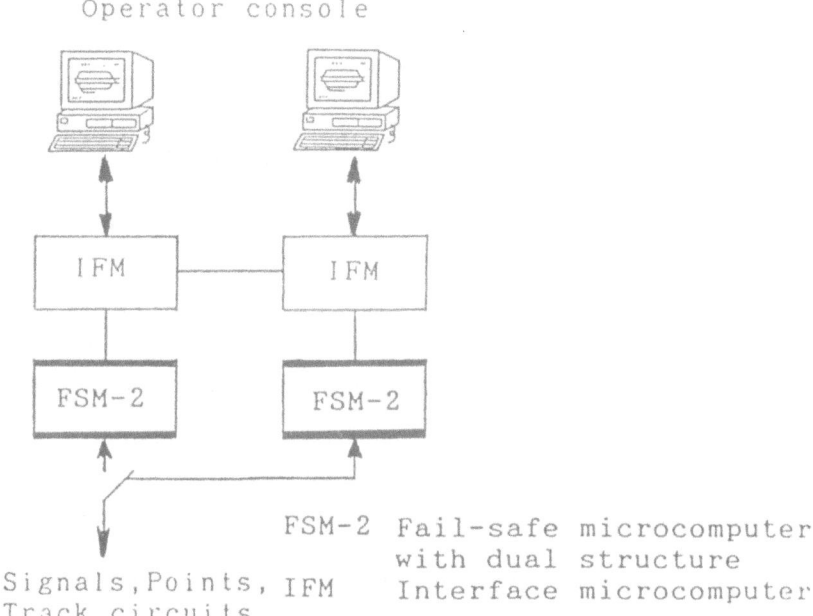

Figure 7. Configuration of μ-SMILE

After inner data of the system are switched to safe-side state,system recovery is executed. Restart operation like this is used in a solid-state track relay (train-position detector) and an automatic train stop system ATS in JR. In such systems as an electronic token system and a level crossing alarm controller, however, important data on safety like the number of trains coming into the control area during system shutdown are not saved. Accordingly an automatic system restart is inhibited to prevent a dangerous situation.

5.2. VDU safety and reliability

Wrong information on VDU (video display unit) can cause an operator to make a wrong judgment. A display which is not updated or whose color is not correct is a typical case. In JR a color mark of RGB (red, green, blue) is displayed periodically at a corner of VDU, to alert the operator of a VDU failure.

When VDU is applied to a man-machine device involving safety like a cab signal display, an incorrect picture must be reliably detected by the operator in the following manner.

Vital information is generated from a fail-safe microcomputer in the form of a dot pattern, and is periodically passed to the screen without editing by a VDU controller, whereby errors in VDU and tansmission can be detected as disarrangements of the picture.

6. Software approach to safety and fault-tolerance

6.1. Reliable software

Software reliability can be attained by structured design and software testing. Our major methods are as follows:

(1) A program is divided into a number of simple modules (for example, about 100-150 steps in the case of Assembler language).

(2) Each module is linked in a single-thread and activated by a periodical timer. Multiprocessing is avoided if possible because of its complexity.

(3) Dynamic and overload tests are done by using a simulator. An automatic data test system and a check-list generation system for functional tests are available in several systems.

6.2. Software structure for fail-safety

Our concepts of methods to protect against hardware errors and also errors in the software itself, if possible are:

(1) Safe-side state and danger-side state are explicitly assigned to input, inner state and output data. For example, 0 means safe-side and 1 means danger-side.

(2) Asymmetrical processing is used. The procedure for giving danger-side state must clear many checking points, while safe-side is given directly.

(3) Periodical processing was adopted, so that it may instantly correct the dangerous output at the worst.

(4) Integrity check of analog input data is done based on limit check, variation check and so on. Abnormal data are changed to safe-side state.

(5) Control output not allowed to remain incorrect even for an instant is issued in time-redundant procedure. For example, a point-machine is controlled in a sequence of unlock, switch operation and lock at different process-timings.

(6) When a serious error is detected, the software reverses internal states and outputs to safe-side.

6.3. Error detection method

As described in chap.2, we have not basically adopted software diversity. Accordingly, if software detects an error of common mode in redundant hardware or one in software itself affecting safety, it does not correct the error but necessarily kills the whole system. In signalling systems the most important thing is the error detection mechanism. Our methods are as follows:

(1) To detect a hardware error by software.

(2) To detect an error of software based on hardware support.

(3) To detect an error by software itself.

As for (1), there are various methods, one of which has been mentioned in Sec. 4.1.2. An error should be detected before it can lead to a double error. In JR we try to detect such latent errors in important circuits as soon as possible:

a. To detect a fault in ROM/RAM areas seldom accessed. In a system with bus level comparison it can be done only by reading ROM areas and by read/write of RAM areas during idle time.

b. To detect a fault in exceptional handling circuits like a memory-protect-interruption, and in input/output circuits which are kept at the same state of value for a considerably long time. Method to detect the former is to diagnose the healthiness of the circuit by error injection. Method to detect the latter is to attach an electrical switch to the circuits and to diagnose dynamically by switching on and off at an instant without disturbing the system operation.

As for (2) and (3), our major methods are as follows:

(1) Processing dropped in permanent loop or runaway is mainly detected by a watch-dog timer (WDT). In a fail-safe computer the WDT itself must have a fail-safe feature. WDT (Fig.8) operates based on both the power supply given by periodical output from software and the oscillation signal from FSC.

(2) Check points of acceptance test for illegal processing are inserted into application programs. Integrity of running sequence of modules linked in a single-thread is also checked.

(3) Diversity methods are partially adopted. SMILE has a double-link-file structure [4].

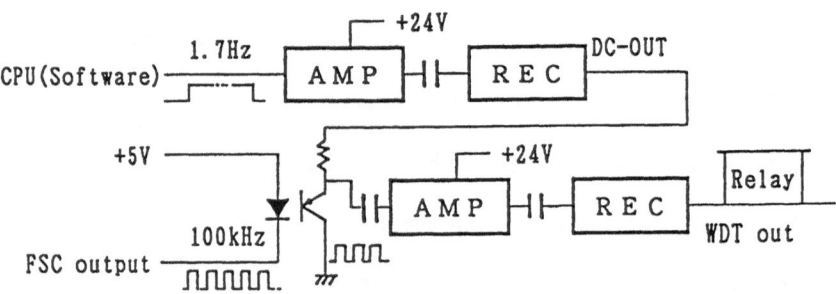

Figure 8. Structure of fail-safe watchdog timer (WDT)

7. System safety management

Referring to Military-Standard-882A we have made a rather simple manual of system safety management. We have also developed guidelines on fail-safe and fault-tolerant technology required for signalling systems. It is composed of 96 requirements for all phases of system design, production, test, evaluation and operation. The outline of our management procedure is as follows:

(1) Safety analysis and verification are done for the circuits or mechanism which is assigned by preliminary safety analysis. FME(C)A, fault simulation on a personal computer and experimental system are usually utilized.

(2) Prototype unit is produced and tests on safety, function, overloaded operation are done. Field testing for comparing the outputs of the prototype unit with those of a conventional system in practical use is also done during a few months.

(3) Design review of a system for practical use is done, to see whether it meets both the requirement specification and the guideline or not.

(4) Quantitative evaluation of a system for practical use is done based on Markov-process-model. For example, SMILE has been predicted to have a mean time to the first dangerous error; 6.1 x 10^{11}(h) and unavailability; 1.0 x 10^{-7} [2]. Both values exclude software errors.

(5) Same kinds of tests as those of prototype and tests against environmental interference are done with a system for practical use. Fig. 9 shows software errors detected in the total SMILE system and in FSM-3 during test stage before practical service. The design errors are mainly due to somewhat ambiguous requirements from railway users. (6) Safety and fault-tolerance techniques which have been newly adopted are recorded to be utilized in other system developments. Each technique is written on a sheet, the items of which are a technical classification such as fail-safe, fail-soft, redundancy and so on; a hazard level; an outline of the technique; and dangerous effects in case of non-application of the technique.

8. Field data on systems in actual use

Since the solid-state interlocking system SMILE was put into service in March, 1985, several systems have been operating in JR. At present there are about 650 fail-safe computers.

Structural characteristics,and field data on safety and availability of 4 kinds of representative systems are shown in Tables 4 and 5. No endangering failure has occurred yet in any of the systems. The rather long downtime for each system is mainly due to time-loss before a maintenance man's arrival.

8.1. Solid-state interlocking system SMILE

SMILE (Safe Multiprocessor System for Inter-locking Equipment) is the first and the largest electronic signalling system ever employed in JR. As shown in Fig.4, it is composed of a fail-safe computer with TMR structure (FSM-3) and many non-fail-safe computers. FSM-3 is the core of SMILE and required to operate without a stop, whether transient or permanent, therefore, TMR structure is used. The volume of application software is rather large and so bus level comparison is used. The mission of this system is to control signals and point-machines and ensure safe train operations at large stations. The interlocking method is logical interlock between control states of signals, point-machines and train running position. Incorrect inputs from an operator and other computer systems can be rejected based on this interlock mechanism. In spite of TMR structure, safe-side system failures occurred 5 times, but during the initial 4 months after the system commenced operation.

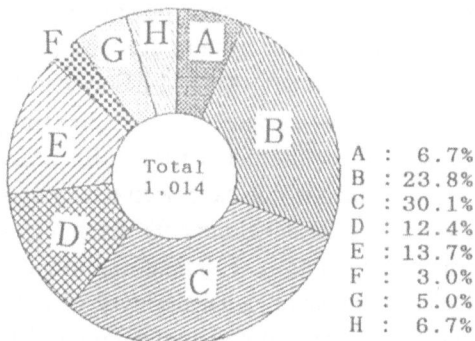

A : 6.7%
B : 23.8%
C : 30.1%
D : 12.4%
E : 13.7%
F : 3.0%
G : 5.0%
H : 6.7%

(a) Software errors in total SMILE system

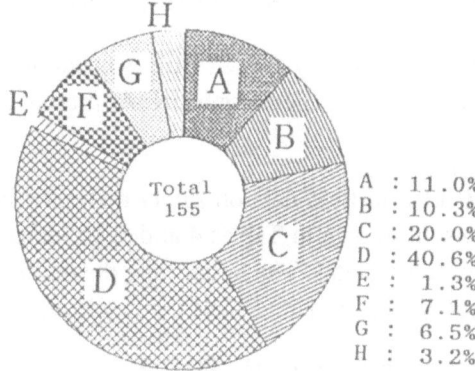

A : 11.0%
B : 10.3%
C : 20.0%
D : 40.6%
E : 1.3%
F : 7.1%
G : 6.5%
H : 3.2%

(b) Software errors in SMILE FSM-3

A : in Requirement specification
B : in design specification
C : in coding
D : in data
E : in software interface
F : in hardware interface
G : in debugging work
H : unknown

Figure 9. Software errors in SMILE during test stage

	SMILE	μ-SMILE	Electronic blocking	Level crossing controller
Redundancy of fail-safe microcomputer	TMR (Intel-8085A)	Dual (Intel-8085A)	Dual (Intel-8031)	Dual (Intel-8085A)
Comparison between fail-safe microcomputers	by hardware (at bus level)	by hardware (at bus level)	by software (at task level)	by hardware (at bus level)
System redundancy	Multiprocessor	Hot-standby	Single	Single
Printed boards per processing unit	102	18	13	9
Size of program (except operator console)	117 ksteps (Assembler)	50 ksteps (Assembler)	23 ksteps (Assembler)	14 ksteps (Assembler)
Size of program in fail-safe microcomputer	27 ksteps (Assembler)	27 ksteps (Assembler)	14 ksteps (Assembler)	8 ksteps (Assembler)

Table 4. Structural characteristics of signalling systems

	SMILE	μ-SMILE	Electronic blocking	Level crossing controller
Numbers of operating sets	17	37	260	124
Dangerous failure	0	0	0	0
Numbers of system down	5	2	65	22
Cause of system down				
1. during initial error period				
Design	3	0	0	0
Production	0	1	19	0
Hardware	0	0	0	0
Software	1	1	0	4
Operation	0	0	6	0
Construction	1	0	2	2
Noise or surge	0	0	7	4
2. during random error period				
Hardware	0	0	2	1
Software	0	0	0	1
Direct lightning strike	0	0	4	0
Noise or surge	0	0	25	10
Mean system down time	55 min (0 min)	40 min (0 min)	70 min (65 min)	45 min (40 min)
Mean time between failures	7.4 year (-)	15.9 year (-)	9.1 year (17.7 year)	6.3 year (9.9 year)
Unavailability	1.4×10^{-5} (0)	4.7×10^{-6} (0)	1.4×10^{-5} (6.9×10^{-6})	1.3×10^{-5} (7.6×10^{-6})
Number of switchovers	11	3	-	-

( ) shows the figure in a random error period.

Table 5. Safety and reliability of signalling systems

The causes of system failures were as follows:

(1) Three were due to unstable contact in one type of connectors. These are errors in design (selection of connector) and production.

(2) One was due to a program bug in exceptional handling of memory protect interruption.

(3) Another was a double error of VDU terminals, the cause of which was a transient voltage drop in the power supply line because of a defect in the rectifier due to a construction error.

After correcting these errors, SMILE has been operating stably until now, though it has suffered partial interruption of operation from errors in components and automatic route-setting programs.

8.2. Solid-state interlocking system μ-SMILE

As shown in Fig.7, μ-SMILE with duplex structure in a small station application is much simpler than SMILE. This system permits an instant interruption of its operation during master-to-slave switchover, because the control area is much smaller than in SMILE. This fail-safe computer with dual structure (FSM-2) of bus level comparison was adopted for the same reason as in SMILE. The first unit was put into service in February,1987. This system suffered 2 shutdowns during the initial error period of 2 months.

The causes were are as follows:

(1) One was due to a fluctuation of 5V (d.c.) for the CPU, which was traced to both a misarrangement of wiring and a transient circuit short.

(2) The other was due to a bug in an additional program peculiar to one of the stations.

After correcting these errors, μ-SMILE has also been operating stably until now, though it has had a few errors in parts of the system.

8.3. Electronic token system

The purpose of this system is to ensure safe train operation on low-traffic single track lines [Sasa86]. A fail-safe computer based on software comparison, which is installed at each station permits only one train to start from the station, after exchanging information on safety over a transmission line between the adjacent stations. When a request for starting a train is issued from a train driver to the fail-safe computer at a station by radio,the computer gives a "proceed-signal aspect", after verifying safety. The first unit commenced operation in November,1986. This system adopts a single structure without backup from a viewpoint of low cost. Major troubles which occurred during an initial error period of 2 months were mainly caused by a production error in a radio transmitter/receiver and by a human error in train schedule-data generation. A computer failure at a station occurred 65 times, the main cause of which was noise or surges. A direct lightning strike occurred 4 times on a cable laid around a station. Countermeasures have been taken to protect against noise and surges.

8.4. Level crossing alarm controller

This system directly controls gates, flashing lights and speakers at level crossings, as a train approaches [6]. The first unit commenced operation in February,1987.

It also adopts a single structure from the view-point of low cost and has had 22 system failures. The main causes were due to noise or surges from outside and input timing errors in the software. Noise or surge problems in power supply lines and signal lines were corrected.

9. Concluding remarks

Several kinds of microcomputer-controlled signalling systems are operating reliably in JR. They are estimated to be superior to the conventional systems using electromagnetic relays,in terms of low-cost, small-size, high-speed, easy operation and maintenance, and so on. As for safety and reliability, there have been several problems, but there have been no dangerous failures at all.

To introduce these systems the following problems had to be solved.

(1) To be cost effective several kinds of redundant structures were developed.

(2) To satisfy railway engineers and users that microcomputer-controlled signalling systems are safe. For this purpose several versions of prototype units were produced and tested during several years.

Results obtained through development and practical use of these signalling systems are as follows:

(1) Fail-safe and fault-tolerant technologies have been perfected and can now be commonly used throughout Japan. A fail-safe computer with bus level comparison is preferred to one with software level comparison in JR, because of software complexity.

(2) Systems with redundant structures which are more complex than single structures have had some problems during an initial error period, but thereafter they have proved to be very reliable.

(3) Fault-tolerance against noise and surges is very important in railway signalling systems.

(4) Systematic safety management is useful for developing a new system. Especially, data on types and effects of errors caused in practical use are being utilized in the design of new systems.

References

[1] K.Akita et al, "Computerized Interlocking System for Railway Signalling Control: SMILE", *IEEE Trans. on Industry Applications*, IA-32, 4, p. 826 (1985).

[2] K.Akita et al,"Development of Computerized Interlocking SMILE", Railway Technical Research Report, No.1361, p.82, in Japanese (1987).

[3] A.Avizienis, "The N-Version Approach to Fault-Tolerant Software", *IEEE Trans. on Software Engineering,* SE-11, 12, p.1491 (1985).

[4] I.Okumura et al,"The Software Structure of a Fail-Safe and Fault-Tolerant Computer for Railway Signalling Device", *Proc. of FTCS-11*, p.56 (1981).

[5] K.Kawakubo et al, "The Architecture of a Fail-Safe and Fault-Tolerant Computer for Railway Signalling Device", *Proc. of FTCS-10*, p.372(1980).

[6] T.Kumagai,"Microelectronic Alarm and Protection Control for Level Crossing Equipment", *Japanese Railway Engineering*, 103, p.18 (1987).

[7] T.Sasaki,"Development of Electronic Blocking System", ibid., 99, p.16 (1986).

[8] H.Nakamura, K.Akita, "Consideration of Synchronism for Tight-coupled Multiprocessor System", *T. IEE Japan*, 108-C, 2, p.135, in Japanese (1988).

[9] D.Nordenfors, A.Sjoberg, "Computer-controlled Electronic Interlocking System, ERILOCK850", *Ericson Review*, No.1 (1986).

DEPENDABILITY OF

DIGITAL COMPUTERS

ON BOARD AIRPLANES

Pascal TRAVERSE
AEROSPATIALE
316, route de Bayonne, 31060 Toulouse - France

Abstract

This paper deals with digital computers on board airplanes. Only the systems unde Aerospatiale's responsibility are dealt with. The airplanes involved are the Airbus and the ATR By studying the evolution of these aircraft families, it can be seen that digital computin; systems are becoming increasingly numerous and used in increasingly critical functions Aerospatiale has therefore had to adapt its design and certification methods to suit. A lesson tc be drawn from this experience is that a problem cannot be solved unless a global approach is used covering all issues.

This paper first deals with the key elements of our approach: the control and monitoring computers and our system design process. As this approach is extensibely used, several practical examples are explained. The safety related research topics are briefly described.

1. Introduction

Aerospatiale's Aircraft Division is a member of the Airbus and the ATR consortiums. In each of these consortiums, Aerospatiale is responsible for almost all the digital computers with stringent safety requirements. Around 300 aircraft equiped with such computers are at present in service. The last model being the Airbus A320 which is also the first airliner to be equiped with digital fly-by-wire controls on all axes and the first of a new generation of aircraft [16]. It is clear that the computers used in a flight control system must have stringent safety requirements. The auto flight ("auto pilot") and engine control systems have similar requirements, this also applies to the flight instruments, and other systems.

In spite of the diversity of the systems studied, the solutions retained exhibit three characteristics:
- the same basic problems are found for each system,
- an overall approach, neglecting none of the possible sources of undependability is absolutely necessary,
- a first set of basic precautions is used to meet the safety requirements. Further precautions are added to these "primary" precautions in order to increase the safety margin.

This paper does not claim to given an exhaustive description of all precautions taken. It first deals with the key elements of our approach: the control and monitoring computers and our system design process. Then, the A320 flight controls system is described, along with our experience in safety related computers. The last section describes the safety-related research topics.

2. Control and monitoring computers

Computers which have high safety requirements are functionally broken down into a **control** channel and a **monitoring** channel. The control channel ensures the function assigned to the computer (for example, to slave a control surface). The monitoring section ensures that the control channel operates correctly.

This functional arrangement can be achieved in different ways (see paragraph 4). However duplex computers are mainly used. These computers are structured around two computing channels (figure 1), each one compares its results with those of the other. Each channel comprises one or more processors, their associated memory, input/output circuits and a power supply unit. When results of one of the two channels deviates perceptibly, the channel(s) which detected the error interrupt(s) the links between the computer and the external systems.The system is designed so that in this case, the computer outputs are in a dependable status. Error detection is essentially made by comparing the difference between the control and monitoring commands and a pre-established threshold. This arrangement serves therefore to detect the consequences of a failure of one of the computer components and to avoid the resulting error propagating beyond the computer. This type of detection is generally completed by monitoring the program good operation via its sequencing (chaining of tasks and time). This monitoring is achieved by exchanges of information between the processors (in the case of a biprocessor "control" channel), or by means of a "super watchdog". (This watchdog is thus qualified to mark its difference with respect to the most widely used watchdogs which only monitor the capacity of the processor to transmit a given signal at a fixed time interval). Acceptance tests are also performed to check the validity of certain data.

This arrangement can be impaired by an error produced both in the control section and in the monitoring section. The software may comprise a first common point. If the software is the same and contains faults, these faults will probably produce errors both in the control and monitoring sections, not necessarily detected by a comparison of the results. The basic method for dealing with this problem is to use great care when writing the software. As laid down by the regulations (and therefore with utmost severity), these software packages meet the most stringent civil aviation standards (software level 1, [15]) and this is sufficient. Also, they are subjected to a considerable number of tests (§ 3).

An additional precaution is to use different software for control than the one used for monitoring. The terms used for this are many but we shall call it "dissimilarity".

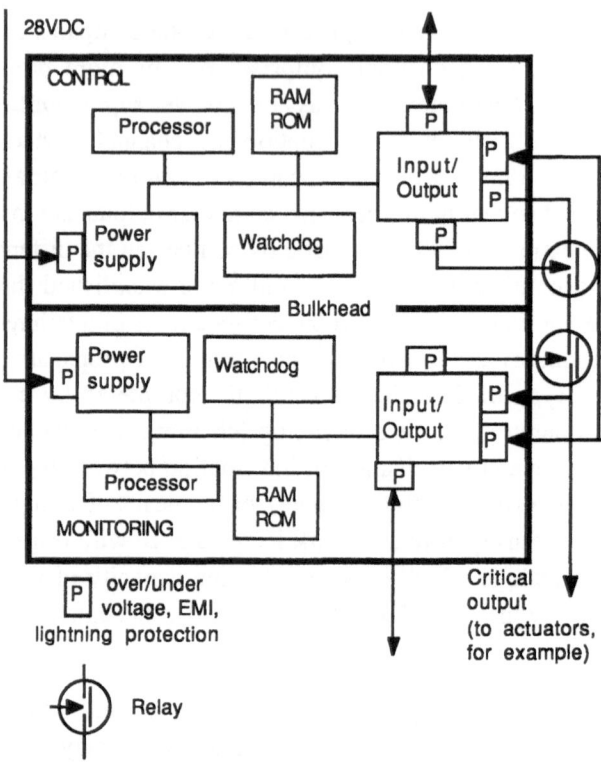

Figure 1. Control and monitoring duplex computer

The aim is to avoid a given fault from being present in the two software packages. The principle used to obtain dissimilar software packages is to have two different software production lines (see figure 2). Thus, a failure of one of the elements of a channel (human element, for example a programmer, or a machine such as a compiler) should not have consequences on the two software packages. Moreover, programming rules are used which tend to increase the dissimilarity, in particular, when a point appears to be complex. The means used to create dissimilar software packages can therefore be placed into four categories [17] :

- inherent differences due to the fact that the control section is different from the monitoring section (structurally and functionally)
- use of two different software production lines
- rules intended to amplify the dissimilarity (complex points of specification)
- others.

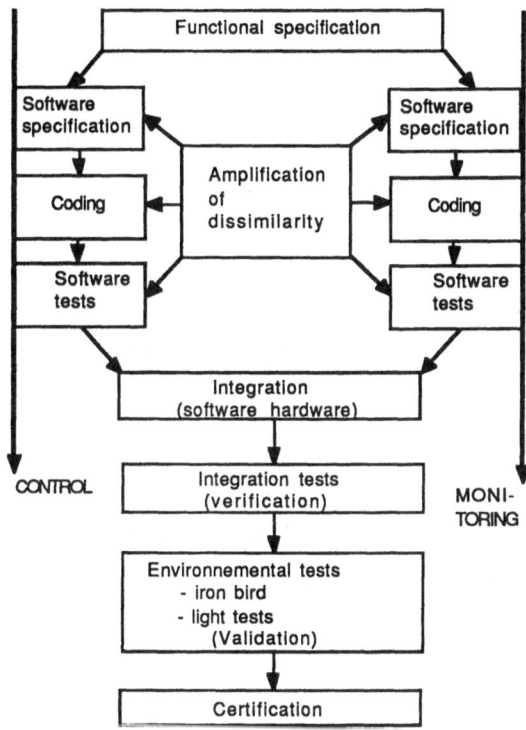

Figure 2. Digital equipment development - dissimilarity

The environment of the computer may be another common point. The control and monitoring channels of the computer have the same electrical source, the aircraft 28VDC network. This electrical source must be converted and regulated within both the control and monitoring channels. These power supplies are in duplicate, each of the units being associated with a computing channel.

The common failure modes of the power supply system are thus easily detectable. The most probable case is the loss of power to the computer. The design retained places this in a dependable status. The computer is also protected against possible over/under voltages and against electromagnetic interference and the indirect effects of lightning.

These protections cover all the agressions that the aircraft is liable to meet. This protection is ensured by filtering all the sensitive computer incoming and outgoing wires. The cables are also protected (shielding, twisting).

An additional protection consists is not strictly synchronizing the control and monitoring channels, introducing a physical separation between the two channels and the designing of system in such a way that the channels have different inputs. The objective is that if the computer is disturbed in spite of its protections, the control and monitoring channels will be affected in different ways thus giving different outputs enabling the disturbance to be detected and rendered passive.

Certain failures may remain masked for a long time after their creation. Typically, this is the neutralization of a monitoring channel detected only when this channel fails. Tests are made periodically to make the probability of occurrence of an underisable event sufficiently low. Typically, a computer performs its self-tests and tests its peripherals during aircraft energization, therefore at least once a day. The objective is to be exhaustive for the most dangerous failures. In-line tests are also performed (for example, a computer can permanently perform a checksum of its ROM memory).

The results are compared in the two channels, in accordance with figure 3. This comparison is made by analogic circuits or by software depending on the computers. The difference between the results of the control and monitoring channels is compared with a threshold ("Threshold", figure 3). An error is detected if the difference between the channels is greater than the permitted threshold. This error must be confirmed before the computer is disconnected. This confirmation consists in checking that error lasts longer than a specified time (T). The detection parameters (Threshold, T) must be sufficiently wide to avoid unwanted disconnections caused by:
- sensor accuracy tolerances,
- computing channel asynchronism,
- software dissimilarity.

However, these parameters must be sufficiently "tight" so that the non-detected errors will be tolerated by the computer environment. For example, for a flight control computer, limited transient control surface runaways are possible and it is therefore necessary to assess their effect on:
- aircraft structure (permissible loads)
- flight qualities.

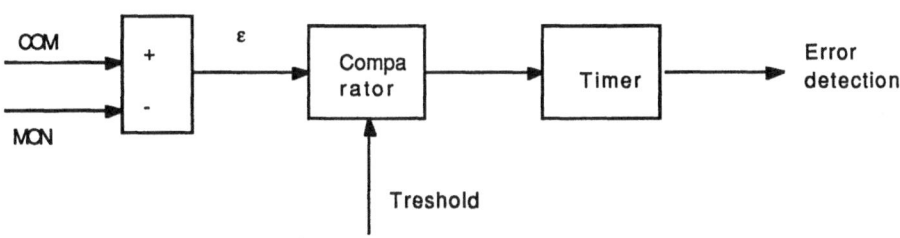

Figure 3. Error detection principle

Generally, these duplex computers are redundant, one being active, the others being "hot" spares, more or less strictly synchronized. As soon as the active computer stops operating, one of the replacements almost instantaneously changes to active mode without transients.

Typically, duplex computers are designed so that they permanently transmit signals of a "I am healthy" type and so that these signals will be interrupted at the same time as the "functional" outputs (to an actuator, for example) following detection of a failure.

The term "fail silent" is sometimes used for this kind of computers.

3. Design process

3.1. Dependability assurance

Fixing objectives

A preliminary safety analysis is made for each aircraft function at the very beginning of the system design phase. This analysis consists mainly in defining the safety objectives enabling the system architecture to be defined. These objectives are then used to design the systems themselves, in particular, the redundancy level, the signals to be monitored and also the level of the effort to be placed in the system software packages. They will guide the Airworthiness Authorities in the quality checks that they perform on the design and production of systems.

The objective definition procedure is as follows. First, all the functions and subfunction are listed and reviewed one after the other. For a given function, the consequences on the aircraft of all possible failures of this function are assessed and classified. These assessments and classifications are also made for all function failure combinations. This classification is based on the airworthiness rules relevant to civil aircraft [8]. The consequences of a failure may be either minor, major, critical or catastrophic.

The criticity of an item of equipment is given by the most serious function failure in which it can participate. For digital equipment, a software level is deduced from this criticity. Each software level corresponds to an effort in the development process. These (3) levels and the associated effort are defined by DO178A [15]. In fact, apart from the lowest level, the difference in the effort concerns the documentation to be supplied and the extent of the tests to be performed. Also, with respect to the criticity of an item of equipment, the Airworthiness Authorities are more or less involved in the monitoring of the equipment qualification process delegating a part to the aircraft manufacturer and/or equipment manufacturer.

Check

The upholding of the safety objectives is initially checked during the Preliminary Safety Analysis. However, all information is not available during this phase ; all the details of the system are not defined, the failure rates are only approximative and the elementary failure modes are not known. The complete justification of a system will be made during the aircraft certification when system design is fixed.

The approach followed is both bottom-up and top-down. On the one hand, for each objective defined during the preliminary analysis, the elementary failures of each item of equipment are analysed to find those which can, alone or in combination, lead to the studied failure mode. On the other hand, all elementary failures must be associated with a system objective. In general, there is no perfect correspondence and the elementary failures which remain apart are studied separately.

These safety analyses are made by a person who participates in the system design, in collaboration with a safety engineer independent from the system design team. These two persons are assisted by specialists from the equipment manufacturer for the analysis of the elementary failure mode lists (FMEA).

The check for the good dependability level reached by a software package is made, naturally throughout the design process. Periodically, audits and inspections are performed on the documentation associated with the software (typically, 6 to 10 reviews). The three parties receiving this process (Airworthiness Authorities, aircraft manufacturer, equipment manufacturer producing the software), participate in these reviews (this check process is described in [2]).

3.2. Validation of functional specifications

The digital equipment under Aerospatiale's responsibility are covered by a functional specification. This specification is a crucial point from a safety point of view. In particular, for most control and monitoring computers, it is very difficult, even impossible, to completely conceive the control software specification independently from the monitoring software specification. It is possible to differenciate the substance of these two specifications (for example, by using a positive logic for one and a negative logic for the other), and this can help in limiting errors common to both control and monitoring during the coding of the software packages. However, this cannot, to all intents and purposes, prevent a general design error at highest level from being found in the two specifications. This problem is not new and is not linked to digital systems. The designers were confronted with this for example during the manufacture of the Concorde automatic landing system. Tools and methods have therefore be used for this problem for quite a long time.

Before writing the detailed functional specifications of an item of equipment, the functions of the equipment are submitted to a functional breakdown. Then, a chapter of the detailed specification is allocated to each equipment subfunction.

The detailed specification is written in a graphic language designated "SAO" (Spécification Assistée par Ordinateur = Computer Aided Specification). The atoms of this language are logic ("AND", "OR", etc) and automatic (filter, integrator, etc.) symbols. The interest of this language is that it has a formal definition for each of its symbols and strict rules concerning their links. The specification is comprised of written sheets of these symbols. These sheets are grouped logically so that each group corresponds to a subfunction of the high level specification. The subfunction and the group of sheets have the same inputs and outputs.

An automatic coherence check is made on sheet acquisition. It checks that the links between the symbols obey certain rules. Also, these sheets are under the control of a configuration management system. A key element of the detailed specification validation is that it is executable.

This enables validation at two levels:

- a sheet, or group of sheets, can be executed after acquisition. Entries to the sheet can be generated, either via a keyboard or, for some, by a more appropriate element (thus, a sidestick is connected to the sheet acquisition computer to bette validate the part of the flight control specification relevant to control laws),

- more globally, the specifications of several items of equipment can be interconnected and simulated together. This serves for validation of the interface between these items of equipment and therefore the system.

A walk through the specification is also practiced by different people (including members of the design team, test engineers, a safety engineer).

This validation is made before the software is produced and does not completely take into account the real time aspects of the system and the actual characteristics of the equipment. This point is dealt with on the test benches [4] which integrate the comple computers, hardware and software. There are two types of benches, the partial benches which test only one item of equipment at a time and the general benches ("iron bird"). A general bench simulates as near as possible a real aircraft. The cockpit, the electrical and hydraulical power, the equipment (computers, actuators) and their integration. Generally, inertial and air data sensors are completely simulated (position of the aircraft in space, speed of the aircraft, etc.), the interfaces are validated on the partial benches. The final validation of the specifications is made during the flight tests (A320, 4 aircrafts which accumulated around 900 flight hours were used with 5000 parameters continuously monitored).

3.3. Software validation

This paragraph dealt with the A320 flight controls systems, on its level 1 software. It is noteworthy that validation in this case is greatly helped by the structure of the software. "Polling" is used, with a periodic reading of the inputs. This means that the clock only generates an interrupt, and that the sequencing of events is deterministic. More, the flight control laws are robust. If the input vector is out of a predetermined subset of the input domain, then the control law is downgraded to a simple law using as input only the side-sticks and surfaces positions.

No software failure rate, or measure of the same type is used. Software validation is based on testing and audits. The result of each step of the development process is tested against its specification. For example, for a given input, "equivalence classes" of values are defined. The test must cover the input space, based on these classes. All the branches of the software have also to be activated (structure). These equivalence classes and a possible supplementary test efforts are agreed between system designers, software designers, quality insurance specialists from the aircraft manufacturer and from the software manufacturer, and the Airworthiness Authorities.

After more than one year in operational use, no error has been found in the flight controls system software.

4. Fly-by-wire controls - A320

The term "Fly-by-wire" is ambiguous. It is, indeed, difficult to imagine a flight control system without mechanical parts (in particular, the actuators and the control elements). The term "wire" underlines the fact that the pilot's orders are transmitted to the actuators in electrical form. Also, fly-by-wire systems can process the pilot's orders in different ways. Thus, distinction must be made between the A320 where the flight controls are an integral part of the aircraft automatic flight controls functions and aircraft where the flight controls only copy the pilots orders (Concorde, A310, etc.). For the first named, there are no longer any biunivocal relations between the flight controls and the position of the control surface.

The A320 flight controls have been described in other document [5, 6, 7, 20]. We shall only deal with them here from a dependability point of view. The flight control system is comprised of controls (side-stick, speedbrake control lever, etc.), computers, aircraft position sensors (inertial and air data systems, accelerometers) and actuators. The computer control the actuators in accordance with references generated from requests made by the pilots via the sticks and in accordance with aircraft return signals.

Distinction can be made between three sets of functions:

- interface with the crew (acquisition and monitoring of controls, artificial feel, information relevant to the position of control surfaces and the status of the system)

- functions associated with the flight control laws (management of air data and inertial information, computation of laws, in particular load factor control, damping of dutch roll, turn coordination, high angle-of-attack protection, etc.),

- control of the aircraft in the three axes (roll, pitch, yaw) and lift control.

One of the advantages that fly-by-wire controls bring to aircraft safety is the protections which are an integral part of the flight control laws. Thus, in normal flight control, the structure is protected (load factor, speed). A third protection, designated high angle-of-attack, stops the aircraft from stalling. These protections lighten the pilot's workload in particular during evasive action concerning an obstacle (near-miss) or windshear. These protections provide increased safety. Thus, a pilot engaged in avoiding another aircraft can concentrate on the path to be followed without worrying about the structural limits of the aircraft or a possible stall. Windshear generally occurs at low altitudes. The safe reaction is tricky to perform as, above all, the aircraft must not be stalled. The fact of having automatic control on the aircraft angle of attack, coupled with an automatic increase in engine speed at high angles of attack, gives the A320 a significant increase in survival probability in windshear conditions. To fully appreciate the interest of such a protection remember that during the last five years windshear was responsible for 2/3 rds of the persons killed in aircraft accidents in the United States [16].

Now that the advantages of the fly-by-wire systems are clear, this system must be made sufficiently safe. The first type of failure not be taken into account is failure of the system equipment.

The computers are of the control and monitoring type (§ 2) and this makes runaway of the control surfaces by a computer Extremely Improbable. A computer failure will therefore lead to shutdown of the computer. The actuators are monitored by the computers via both the computer monitoring channel and the control channel. Both channels can make the actuator passive. The various sensors (on the sticks, actuators, inertial systems, etc.) comprise another source of runaway. Each sensor is at least duplicated so that all information used is consolidated by comparison between at least two different sources of information.

As the system is protected against runaways it must be built so that it is sufficiently available and therefore redundant. Electrical power is normally supplied by two AC generators each one driven by a different engine. Batteries and a Ram Air Turbine are also available. In case of shutdown of the two engines, this RAT is automatically extended. It pressurizes a hydraulic system which drives a third electrical generator. The batteries and an auxiliary generator are also available. The computers are not connected to a single source of energy but to at least two. The aircraft has three hydraulic systems one of which is sufficient to control the aircraft. Two of the systems are each pressurized by one engine the third one is pressurized by an electric pump or by the RAT. The computers and actuators are also redundant. This is illustrated for the elevator control (pitch axis) of the A320 (figure 4). Four control and monitoring computers are used (ELAC: ELevator and Aileron Computer, SEC: Spoiler and Elevator Computer), one alone is sufficient to control the aircraft. In normal operation, one of the computers (ELAC2) controls the elevators, the other ones control the other control surfaces. If ELAC2 or one of the actuators that it controls fails, ELAC1 takes over. In accordance with same failure mode, ELAC1 can hand over control to SEC2. Likewise, elevator control can be switched from one SEC to the other in accordance with the number of control surfaces that one of these computers can control.

Note that three computers are sufficient to meet the safety objectives. The additional computer is fully justified by operational requirements: it would be desirable to tolerate a dispatch with one computer down.

The flight control system went through a very stringent design and production process and we can reasonably estimate that it ensures a largely sufficient safety level. An additional protection has however been taken, which consists in using two different types of computers: the ELACs manufactured by THOMSON-CSF, organized around microprocessors 68010, and the SECs the hardware of which is based on the 80186 and built in cooperation by SFENA/AEROSPATIALE. We have therefore two different design and production teams using different microprocessors (and associated circuits).

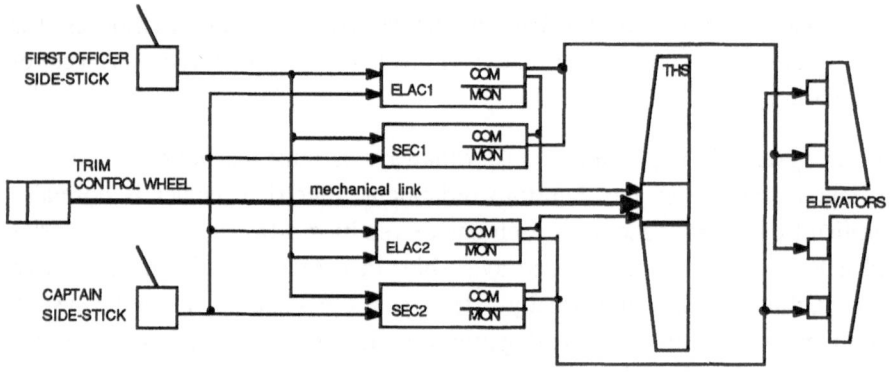

Figure 4. A320 - Pitch control

The electrical installation, in particular the multiple electrical links, also comprise a common point risk.

This is avoided by a thorough segregation: in normal operation, the two electrical power systems exist without common points. The inter-computer links are limited, the links used for monitoring are not routed with those used for control. The destruction of a part of the aircraft is also taken into account: the computers are placed at three different locations, certain links to the actuators are routed underfloor, some in the ceiling and others in the cargo compartments.

With all these precautions taken, it may appear surprising that mechanical back-up has been conserved on the trimmable horizontal stabilize (figure 4) which also enables control in pitch axis and mechanical standby on rudder permitting direct control in yaw axis and indirect control in roll axis. But these mechanical linkages are not expensive and undeniably have a psychological effect.

5. Experience

Our experience in safety-critical computers started with analog automatic control system, with automatic landing capabilities. This can be traced back to Caravelle (early 60'), Concorde (1st flight in 1969) and Airbus A300 (70'). Next steps are (see [17] for descriptions):

- digital automatic control system on A300 FFCC (certification in 1982),

- analog full authority flight controls system on Concorde (with a mechanical back-up that has not yet been used in operation),

- digital flight controls system on secondary surfaces on A310 (certified in 1983).

Our total experience is symbolized by several figures given in table 1.

These are several million flight hours ($> 10^7$) accumulated by the computers. The safety-critical computers on these aircraft are fully under Aerospatiale's responsibility, except for the fuel management computers (CGCC) and the flap and slat control computers (SFCC [11].

SITUATION END OF 1988	A 300 300-600	A 310	A 320	ATR 42
Number of aircraft in service	16	170	12	120
Accrued flight hours	20000	1100000	10000	300000

Table 1. Accrued experience

Up to the present day, the onboard computers have upheld the fixed objectives. However, note that several software design errors were detected in service. These anomalies affected functions which did not concern aircraft safety (and therefore on which the test effort was less extensive) and none required urgent corrective action.

Software engineering is under constant progress. Under equal conditions (errors recorded during the integration and flight and ground test phases, and for computers of similar complexity), the number of residual software errors has been considerably reduced from the A310 to the A320.

The use of dissimilar software is also satisfactory. No common error to a control software and its homologue monitoring software has been reported in service and no error has been found common to the ELACs and the SECs and this even during the debug tests. These results cannot necessarily be generalized. On the one hand, the raw materiel is excellent, it may be that common errors will appear with software of a lesser quality.

On the other hand, note that the control/monitoring structure of the computers is well adapted to suit the dissimilarity, all other structures are more difficult to achieve economically as it is, for example, easier to find two algorithms for a duplex computer than three for a triplex computer.

6. Prospective

This chapter presents two research activities associated with safety and distributed systems. The first concerns the safety assessment of a system, the second future flight control.

6.1. System safety evaluation

Stochastic Petri nets are a first research direction [3 - 18]. Another one is the use of capacity functions. For example, to define the manœuvrability of the aircraft in roll in relation to the degraded status of the flight control system. This manœuvrability can be approximated by the following function which measures the available rool rate as a linear function of the available control surfaces:

$$\Sigma \text{ (roll rate of control surface S)}$$

$$S \in \{ \text{ available surfaces } \}$$

By defining an acceptability threshold ((roll rate > X) ⇒ (sufficient manœuvrability) it is possible to divide the degradation states of the system into success or failure states and thus to calculate the probability of failure of the system with respects to the roll manœuvrability objective. This study has led to a data processing tool which is at present used in the definition of the new Airbus aircraft (A340, A330) and which will also be used to justify this system for certification. This tool (called VERIFCDVE) takes at input an arrangement of computers, actuators, hydraulic and electrical power sources and also specific events such as simultaneous shutdown of all engines and therefore a large number of power sources.

The availability of a control surface is dependent on the availability of certain of these resources. The tool automatically creates the failure combinations and assesses the availability of the surfaces and therefore a manœuvrability function in roll. It compares these results to the objectives. These objectives are relevant both to manœuvrability (availability of elevators, available rool rate, etc.) and reliability (a manœuvrability objective must be taken for all failure combinations where probability is greater than given reliability objective). The tool then lists the failure combinations which do not attain the objectives (if there are any), and gives the non-satisfaction probability for each manœuvrability objective. The tool also takes into account the MEL possibilities (for example, take off with one computer failed). This tool showed itself particularly useful in the A340 flight control system architecture design phase. Many architectures were considered during this phase and each one needs to be checked. This check can be made without tools but the work is fastidious (typically, thousands of failure combinations are to be taken into account to make a very fine check), and errors may occur. The tool enables more architectures to be studied and therefore a better quality final product to be obtained (in termes of number of feedback loops per computer, therefore in terms of computing power and bulk), and globally to gain time.

A new tool is at present being produced using the expert system techniques. It has a dual objective.

1) to be able to refine the system representation (taking into account of reconfiguration logics defined in the functional specifications, refining of capacity function) and

2) to be able to assess systems other than the flight controls (flight instrument system for example).

6.2. Future flight control system

Unstable civil aircraft can be envisaged even though the economical advantages have not yet been fully demonstrated.

The main effect on the flight controls is that the aircraft would not be controllable without an artificial electronic stabilization system, capable of reacting rapidly to the smallest aircraft movement. Because of this, it would not be possible to use a mechanical transmission backup system such as those used on the A320 or the A340. Two solutions can be considered. The first one is to do without the mechanical backup system and not replace it (the fligth control system of the A320 or the A340 meet the safety objectives without taking the mechanical backup system into account). Another solution is to use a standby system structured around specific computers ; these computers would be completely independent from the main flight control system, including technologywise. A project of this type has been studied at Aerospatiale and is now in the laboratory prototype assessment phase. This standby system includes Opto-Electronic Processing Modules (OEPM). This equipment only communicates with the exterior by means of fiber optics both for the main flight control interaction or maintenance functions and for sensing the pilot's orders and flight control positions. All that can be integrated into a single item of equipment is done so, i.e.: the actuator, the computer and the aircraft return inertial sensor. The actuator is hydraulically powered and a micro-generation system creates the electricity for supplying the computer and its sensors. This micro-generation is also integrated into the OEPM (a more detailed description is given in [13]).

In the future, the actuator slave control electronics may be decentralized from the central computers to the actuators. The system would therefore comprise central computers in relation with various sensors (these computers would generate the control surfaces deflections), and "smart" actuators, including slave control electronics, the whole linked by a sufficiently safe communication network. Various studies have been undertaken, both in the industry [9, 10, 18] and in the university circles [1, 12, 19]. The system thus designed can be qualified as distributed in the sense that all subscribers to the communications network are self-dependent, that no element is indispensable and the whole must cooperate to fully perform the task entrusted to the system.

This project is vast and covers various aspects such as the actuator technology. In particular we shall mention the studies concerning:

- the global architecture of such a system (structure, number and type of central computers, slave control electronics, structure of the communication network, integration of an optical back-up system),

- the communication buses to be used (type of access, flow rate, protocol),

- synchronization (time and data) of the central computers, and validation method (see [18]).

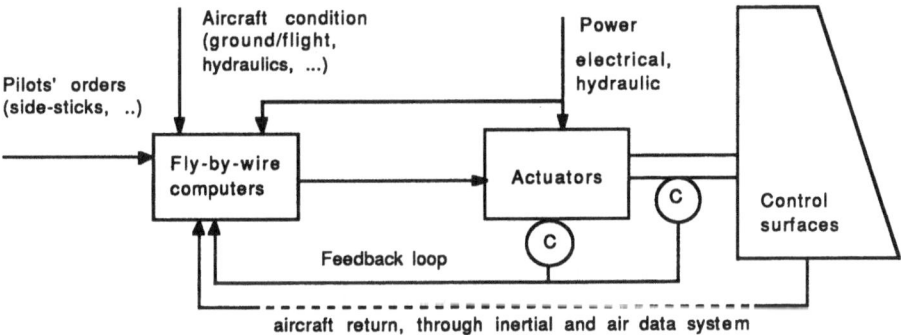

Figure 5. Principle of fly-by-wire controls

References

[1]   A. Avizienis, P. Gunningberg, J.P.J. Kelly, L. Strigini, P. Traverse, K.S. Tso, U. Voges, "The UCLA DEDIX System: A distributed testbed for multiple-version software", *Proc. 15th International Symposium on Fault-Tolerant Computing, FTCS-15*, Ann Arbor, Michigan, June 1985.

[2]   L. Barbaste, J.P. Desmons, "Assurance qualité du logiciel et la certification des aéronefs, expérience A320", *Proc. 1er Séminaire EOQC sur la Qualité des Logiciels*, Brussels, April 1988.

[3]   K. Barkaoui, G. Florin, C. Fraize, B. Lemaire, S. Natkin, "Reliability analysis of non repairable systems using stochastic Petri nets", Proc. 18th International Symposium on Fault-Tolerant Computing, FTCS-18, Tokyo, June 1988.

[4]   D. Chatrenet, "Simulateurs A320 d'Aérospatiale : leur contribution à la conception, au développement et à la certification", INFAUTOM 89, Toulouse, March 1989.

[5]   S.G. Corps, "A320 Flight controls", Proc. 29th Symposium of the Society of Experimental Test Pilots, September 1985.

[6]   M. Durandeau, J. Troyes, "Les commandes de vol des avions de transport", Cercles des Officiers Mécaniciens et Ingénieurs Navigants - COMINAC, Roissy en France, January 1986.

[7]   J. Farineau, "Lateral electric flight controls laws of the A320 based upon eigenstructure assignment technique", *Proc. AIAA Guidance, Navigation and Control Conference,* Boston, August 1989.

[8]   Federal Aviation Authority (FAA) and Joint Airworthiness Authorities (JAA), "Federal Aviation Regulation part 25 and Joint Airworthiness Requirements part 25".

[9]   D.P. Glutch, M.J. Paul, "Fault-tolerance in distributed digital fly-by-wire flight control systems", *Proc. 7th Digital Avionics Systems Conference - DASC,* Fort Worth, Texas, October 1986.

[10]   J. Grossin, P. Traverse, "Système de commande vol pour aéronef", French Patent n° 88.03.343, March 1988.

[11]   A.D. Hills, "Digital fly-by-wire experience", *AGARD Lectures Series n° 143,* 1985.

[12]   C. Hourtolle, "Conception de logiciels sûrs de fonctionnement : analyse de la sécurité des logiciels ; mécanismes de décision pour la programmation N-version", Thesis for the doctorate of the Institut National Polytechnique de Toulouse, n° 122, October 1987.

[13]   J.P. Laborie, P. Desjean, J.P. Domergue, P. Palandjian, "Système pour la commande aérodynamique mobile d'un aéronef", French Patent n° 86.01.576, February 1986.

[14]   H. Lansdorf, "Terminal weather", *Flight International,* May 23rd, 1987.

[15]   Radio Technical Commission for Aeronautics (RTCA) and EURopean Organization for Civil Aviation Electronics (EUROCAE), "Software considerations in airborne systems and equipment certification", n° DO178A and ED12A, March 1985.

[16]   C.R. Spitzer, "All-digital jets are taking off", *IEEE Spectrum,* September 1986.

[17]   P. Traverse, "Sûreté des systèmes informatiques embarqués à bord d'avions", *Proc. 3ème Colloque ANAE International sur la Sécurité Aérienne et Spatiale,* Toulouse, September 1988.

[18]   P. Traverse, "Evolution dans les applications civiles", *Proc. AGARD Conference on Fault Tolerant Design Concepts for Highly Integrated Flight Critical Guidance and Control Systems,* CP-456, Toulouse, October 1989.

[19]   J.M. Wensley, L. Lamport, J. Goldberg, M.W. Green, K.N. Levitt, P.M. Melliar-Smith, "SIFT: The design and analysis of a fault-tolerant computer for aircraft control", *Proceedings of the IEEE,* vol. 66, n° 10, October 1978.

[20]   B. Ziegler, M. Durandeau, "Flight control system on modern civil aircraft", International Council of the Aeronautical Sciences - ICAS84, Toulouse, September 1984.

LONGLIFE DEPENDABLE

COMPUTERS FOR SPACECRAFTS

Tadashi TAKANO * - Takahiro YAMADA *
Hatsuhiko KATO ** - Toshiyuki TANAKA ** - Hirokazu IHARA **
Nobuyasu KANEKAWA *** - Hideo MAEJIMA ***

* Institute of Space and Astronautical Sciences
3-1-1 Yoshinodai, Saga mihara-Shi, Kanagawa 229 - Japan

** Space Systems Division, HITACHI Ltd.
216 Totsukamachi, Totsukaku, Yo kohama-Shi 244 - Japan

*** HITACHI Research Laboratory, HITACHI Ltd.
4026 Kujimachi, Hitachi-Shi, Ibaragi 319-12 - Japan

Abstract

Computers for spacecraft are being developed utilizing fault-tolerant configurations. To attain high dependability mostly using components available for commercial use, many fault-tolerance concepts are systematically combined. Besides adopting the conventional approaches such as spatial diversity based on the majority voting, the usage of ECC, watch dog timers and current limiter, we introduce some new concepts such as softwarevoting with inter-cell communication, time diversity based on the output feedback and stepwise negotiating voting. Most of the results were implemented on the flight model of On-Board Computer (here in after OBC) to be loaded on MUSES-A (Mu Space Experimental Satellite) which will be launched onto lunar swing-by orbit early in 1990. The design approach to construct a fault-tolerant computer system mostly using components in commercial use will be verified on MUSES-A as one of its mission apparatuses for experiment.

1. Introduction

Electronic equipments used in space are required to maintain their functional integrities under severe environmental conditions especially the incidence of cosmic rays. Particularly, the computers loaded on spacecraft must have high dependability since they are required notonly to survive but also to perform their functions all through their mission time without maintenance.

The conventional approach has been to use radiation-hardened components such as CPUs or memories which are specifically developed and manufactured for space or nuclear engineering fields. However, these components are of low integration density consisting only of staticflip-flops. Low integration density causes the increasein chip count and as a result, an increase in weight, size and power consumption. The radiation-hardened com-ponents have another shortcoming, specifically, cost. The maintenance of a dedicated manufacturing process fora product lifecycle of long duration and low production rate is an economic burden. On the other hand, as the applications of computers in space increase in scale andcomplexity, the usage of radiation-hardened components becomes prohibitive because of limitations in weight, size, power consumption and cost.

The usage of commercially available components can be an attractive solution for these problems. Those components are of high integration density consisting ofdynamic flip-flops and fine circuit lines. They are in a constant production under a tight quality control, which eliminates the latency of accidental degradation of product. Commercially available components also havean economic advantage of being produced in large quantity and their costs coming down very steeply along the learning curve of semiconductor process. Accordingly, they do not have the shortcomings of radiation-hardened components stated above.

However, we have to consider the problems inherent in the commercially available components before using them in space. They do not have sufficient durability to be simply replaced for the radiation-hardened components in space use. Therefore, to obtain a highly dependable computer system composed of non-radiation- hardened components, a systematic approach must be implemented.

The princial mission of MUSES-A OBC is to implement and verify this approach. The configuration of MUSES-A satellite is shown in Fig. 1. The satellite system is controlled by the data processing unit. The OBC is classified as one of the mission apparatuses which are not engaged in satellite system control. In another words, the failure in OBC will not lead the whole satellite into uncontrollable status. Only for the evaluation of the approach, OBC performs its experimental function of packet telemetry.

In order to construct a dependable computer system using components which are vulnerable to cosmic rays, we have to utilize techniques for selection of components and system configuration to maintain fault-tolerance and fault-avoidance.

Significant contributions to the area of dependable computing include those by JPL, SRI, Draper Laboratory and others. [1,2,3,5,12,15]. We evaluated those achievementswith respect to their applicability under the specific requirements for our development.

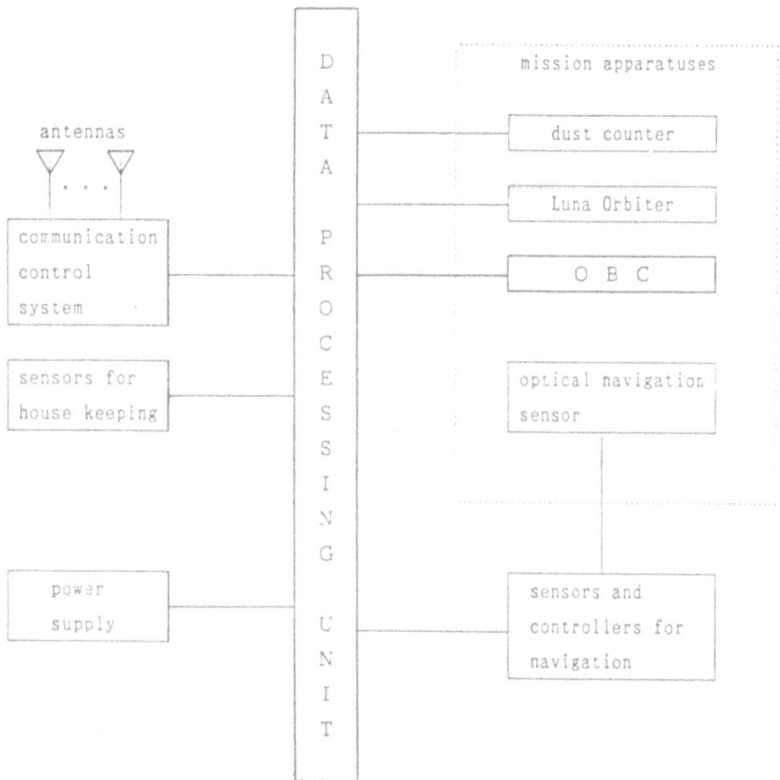

Figure 1. Configuration diagram of MUSES-A

2. Requirements on MUSES-A OBC

As the result of our requirement specification for MUSES-A OBC, we pointed out the following items:

(1) Reliability

The survival probability of MUSES-A OBC is required to be more than 99% at the end of its 1 year mission time. MUSES-A OBC is required to perform its function of packet telemetry throughout its mission time even after some of its subsystems or components fail.

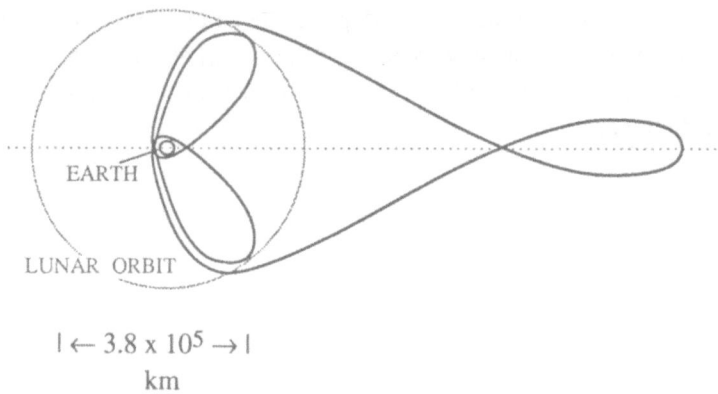

$$| \leftarrow 3.8 \times 10^5 \rightarrow |$$
km

Figure 2. MUSES-A orbit

(2) Radiation

The orbit of MUSES-A is shown in Fig.2. The satellite travels along a perilunar orbit and traverses the ionospheric zone where radiation is significant one day out of its orbital period of 14 days. The influence of cosmic rays on the satellite consist of the three phenomena [9,11]:

(a) Degradation due to the total dose

The influence of cosmic ray dosing accumlates in the semiconductor device resulting in its gradual de gradation. The dosing rate varies depending on the distance from the earth, with the total dose amount estimated from the orbit and mission time of a satellite. Based upon the result of measurement, we estimated the dosing rate in the ionospheric zone to be less than 1,000rad/year. The dosing rate will be decreased by a factor of 10 by encapsulating the device with a 2mm thick aluminum case and 0.5gr/cm^2 potting material.

(b) Latch-up of CMOS ICs

The CMOS ICs have stray PNPN junctions, which can be incidentally turned on by cosmic rays causing a flow of large intensity current. This phenomenon is called a latch-up and can lead to the complete failure of the components unless immediately recovered by shutting off the power of the ICs.

(c) Soft errors of a flip-flop

The state of the CMOS flip-flop is inverted by the incidence of a charged particle. The maximum rate of its occurrence is $10^{-6} \sim 10^{-4}$ times/day·bit.

(3) Shock and vibration

The equipments loaded on the satellite must with stand the shock and vibration imposed on them at the time of launching. The testing conditions for the shock and vibration are specified to be 25G and 20~2,000 Hz.

(4) Temperature range

The temperature inside the satellite will be conditioned within a range of -30°C ~ 60 . This is a very severe condition for the commercially available semi-conductor devices constituting the circuitry of the OBC.

(5) Weight and size

The weight of MUSES-A OBC is restricted to be within 3kg. The size is specified to be within 260mm x 234mm x 110mm. These conditions coincide with weight and size limits on the satellite. If the system was designed using only radiation-hardened components, the system would require 4000 chips resulting in a combined weight and size 5 times that of the requirements.

(6) Power consumption

Due to the shortage in power supply of the satellite, the power consumption of a computer should be minimized. In the case of MUSES-A OBC, power consumption is required to be within 4W. If bipolar ICs or discrete CMOS ICs of low integration density are used, the power consumption would be 40W, 10 times that of the requirements or more.

3. Approaches to meet the requirements

3.1. Overall approach to system design

As stated above, it is the mission of MUSES-A OBC to construct and test a computer system using mostly commercially available components. In order to attain this mission, a systematic approach is required, evaluating and combining all the feasible technologies. The individual requirements shown in the previous chapter are interrelated with one another and the solutions for them also correlate and trade off with one another. We examined the relationships among the requirements and solutions to obtain the approaches described in the following sections. In combination with conventional fault avoidance techniques, we adopted the fault-tolerance approaches applied in the field of on-line control to fulfill the overall dependability requirements.

3.2. Components and packaging

In order to reduce the weight, size and power consumption, we employed CMOS devices, which attain higher integration density and lower power consumption than devices from other processes.

CMOS devices have another advantage of withstanding the total dose of radiation. However, the ICs with high environmental durability are of low integration density, thus increasing the chip count and weight. Accordingly, the usage of radiation-hardened IC should be minimized.

We tested the non-radiation-hardened CMOS LSIs and found that they can withstand a total dose of more than 5×10^{-3} rads. This value of durability is more than sufficient for the mission time and orbit of MUSES-A.

However, these LSIs are vulnerable to single events such as soft errors and latch-up. Single events can not be avoided by shielding. On the contrary, the reradiation from the shielding material will multiply the charged particles which shower on the components and cause single events. Latch-up can be recovered by a current limiter but we have to prepare some other way to avoid hazards caused by soft errors.

Additional approach to reduce weight, size and power consumption is to use gate arrays. We used two CMOS gate arrays with 25,000 gates on each to encapsulate the circuitry performing the system control functions for the fault tolerance.

Besides the usage of high density components, we considered the usage of the most up-to-date packaging technologies. We took into account the multichip carrier hybrid ICs using the multi-layer ceramic base as a possible candidate. However, we decided to postpone the usage of this technology until the next project because of its immaturity.

For the case of MUSES-A OBC, we used the conventional packaging technologies which have been proven in space. Individual components are mounted on glass epoxy boards and shielded with 2mm thick aluminum case. The spaces between boards are stuffed with potting material to suppress shock and vibration.

3.3. Fault-tolerant configuration

The problem in components and packaging can be avoided only by a fault-tolerant configuration of the computer system. The malfunctions caused by soft errors should be prevented by a multiple computer system which performs a majority voting. By using a fault-tolerant configuration, the usage of radiation-hardened components can be minimized. The fault-tolerance techniques consist of spatial diversity, time diversity and others.

4. The details of dependability improvement

4.1. Spatial diversity

We adopted the traditional concept of spatial diversity, in which identical functional parts of the system are spatially distributed, so that the system can maintain its integrity even after some parts in it fail. Concepts of spatial diversity has been widely applied to antennas used for wireless communication. The spatial diversity approach requires a spatial redundancy resulting in extra hardware. This approach can be considered as one of the techniques of fault tolerance rather than the techniques of fault avoidance. Spatial diversity improves the reliability of data processing under hazardous environments but not the reliability of hardware. The individual techniques of spatial diversity will be classified into different levels and stated here together with their application to MUSES-A OBC.

(1) Intra-component

Level As the spatial diversity at the intra-component level, there are techniques such as memory parity check and error correcting codes. The recent progress in wafer- scale integration brought on-chip multiprocessors into reality. ECC is commonly used to detect and correct errors in RAM. 6 bits ECC can detect and correct 1 bit error in a data of 16 bit width and detect 2bits error in RAM as is the case of MUSES-A OBC. We adopted this because we had experiences and proven results of ECC in space applications. If 1 bit error is detected during a 16 bits data access, the data will be automatically corrected and stored at the same location by dedicated circuit implemented on the gate array. We implemented a residual-error-free mechanism by software so that the RAM area is scanned by a read operation to correct any 1 bit data error and to prevent it from growinginto a multi-bit error. This operation will be performed as a background task of the lowest priority activatedonly when the CPU is not occupied by other tasks.

(2) Inter-component

Level Reliability of a data processing system can be improved by introducing parallelism. A multiprocessor system composed of processors tightly coupled and driven by a single clock is an example of spatial diversity at the inter-component level. For instance, the Triad of FTMP (Fault Tolerant Multiprocessors) developed at Draper Laboratory is composed of 3 processors which operate in a bitwise synchronism.

In the case of our OBC, we adopted the technique of redundant storage of data in dispersed locations on RAM chips. One set of data will serve as the backup of one another in the event one location is hit by cosmic rays and the data is lost. This method is effective even for the multi-bit errors which ECC can not negotiate. The readout of the multiple data from RAM, the comparison and correcting are executed in the same task as the RAM scanning stated previously.

(3) System level

Spatial diversity can also be conceived at the system level. A loosely coupled multiprocessor system with software synchronization is an example of spatial diversity incorporated at the system level. SIFT (Software Implemented Fault Tolerance) of SRI and COMTRAC (COMputeraided TRAffic Control) of Hitachi perform synchronization of elements and comparison of result by software [6].

We adopted the spatial diversity approach based upon our experience in COMTRAC, in which the subsystems are loosely coupled with each other and the synchronization and the comparison of results are performed by software.

We adopted the multiprocessor configuration incorporating the concept of "autonomous decentralized system", patterned after biological systems [7]. The system consists of several units called BPUs (Basic Processing Units), which operate with autonomy like cells in a biological system. Even if a BPU fails, the rest of the BPUs can maintain the functional integrity of the system by controlling themselves and cooperating with one another. The BPUs are loosely coupled as they are driven by individual clocks.

Besides using the hardware voters, the BPUs perform "software voting" based upon the processing results exchanged through intercell communication channels. There is no upper limit to the number of BPUs for the generic OBC. Three BPUs are used in the case of MUSES-A OBC. Three is the minimum number to constitute a fault- tolerant configuration, since majority voting requires at least 3 BPUs. The dual BPU system is fault-detectible but not fault-tolerant being unable to identify the faulty BPU.

Synchronization among the BPUs is performed by the software using the output data themselves as synchronizing signal which are exchanged among BPUs with a certainperiod. For the case of MUSES-A OBC, this period is approximately 1 second. The output data are edited into units called "packets" and their transmission requires 300 to 400ms/ unit. The BPUs perform their functions independently so the timing of output from different BPUs will slightly differ from one another. Usually, the difference in timing is not serious because the 1 secondtime frame leaves sufficient margin for a data duration of 400ms. If by some chance, a data unit from one BPU is retarded beyond the time margin, software voting willbe continued by only 2 BPUs. If the same BPU outputs retarded data in two successive periods, the output from this BPU will be considered different from the others.

An individual BPU consists of a 68000 CPU, a DMA controller, ROMs, RAMs and gate arrays. The BPUs are physically dispersed to prevent the errors of the same context from occurring to more than one BPU due to a single incidence of cosmic rays.

4.2. Time diversity

The problem inherent in the voting system is the malfunctioning of the voter. To guarantee the normal functioning of the system, the voter should not be faulty because the reliability of the output data depends on the reliability of the voter. This reasoning can be quantitatively expressed as:

$$R = (3R_B^2 - 2R_B^3) R_V$$

where R, R_B and R_V are respectively the reliabilities of the system, individual BPU and the voter. Here the reliability means the probability in which the hardware survives after a period of time. Differentiating R with respect to R_B , we obtain:

$$\frac{\partial R}{\partial R_B} = 6R_B (1 - R_B) R_V .$$

The partial derivative is almost 0 for $R_B = 1$, while R affects R by linear order. This statement indicates that the reliability of the voter has a significant in fluence on the reliability of the total system. If we have triple redundancy for the voter to improve its re liability, we obtain a relation:

$$R = (3R_B^2 - 2R_B^3) (3R_V^2 - 2R_V^3)$$

In this case, R_V does not affect R when $R_V = 1$ since:

$$\frac{\partial R}{\partial R_V} = 6 (3R_B^2 - 2R_B^3) R_V (1 - R_V) .$$

But we need a selector to diagnose the voters and to select one which is judged to be normal. Now the elimination of single point failure of the selector poses itself as a problem. One solution for this is the usage of ICs of MIL-Class S grade to construct the selector preventing permanent errors [9]. Together with this, to eliminate transient errors, we can extend redundancy concept to the selector operation and multiply it. Here, in order to prevent the recursive argument to select one from multiple selectors, we use only one selector and multiply its operation in time domain. Thus we came up with the idea of output feedback. Fig. 3 shows the exemplary configuration of an OBC. Outputs from the voters are transmitted to the selector, whose output is transmitted to the BPUs through feedback lines for acknowledging the soundness of the system. In this configuration the selector can be of a simpler constructionthan the voter. This approach of maintaining the data reliability requires extra processing time, which is regarded as a diversity in time. A combination of BPU and voter is called a cell. In the example of Fig.3, the number of cells is 4.

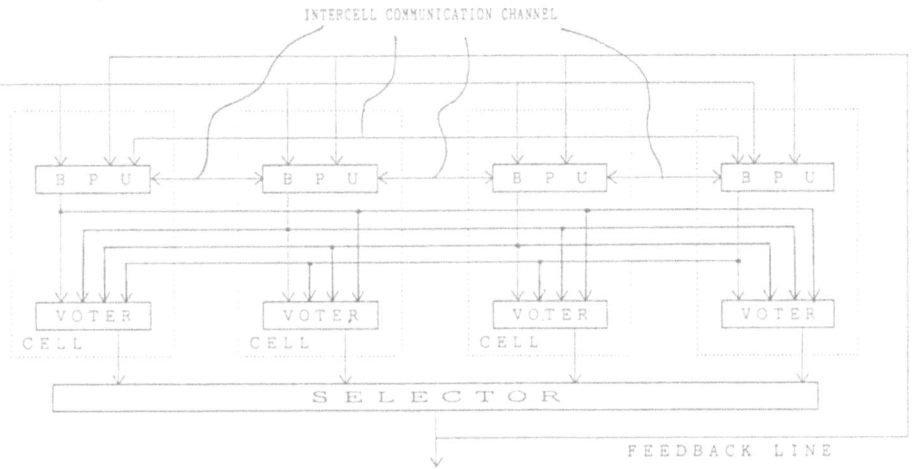

Figure 3. Exemplary configuration of OBC

4.3. Stepwise negotiating voting [8]

We created the concept of "stepwise negotiating voting", in which the level of required reliability can be adapted to the situation. This concept is patterned after a situation in which a group of persons are attending a meeting. As opposed to the "Byzantine Generals problem", the members of the meeting try to adapt themselves to other opinions based upon the degree of their selfconfidence [10]. Here, all the BPUs participate the voting except those in cold stand-by status. The BPUs can be classified int o following four classes depending upon the degree of their reliability of data processing.

Class A

(The BPU is judged to be fault-free as the result of self-diagnosing) and (the output data of the BPU agrees with the output data of other BPUs). Here "fault-free" refers to a status in which the BPU is currently not in a faulty operation.

Class B
The output data of the BPU agrees with the output data of other BPUs.

Class C
The BPU is judged to be fault-free as the result of self-diagnosing.

Class D
Otherwise. The BPU is judged to be faulty as the result of self-diagnosing and the output data of the BPU does not agree with those of other BPUs. This BPU is in a faulty status.

The reliability of Class A BPU is the highest with respect to data processing. Class D system is of the lowest reliability and the selector outputs a fault-secure signal, indicating that the output data from the system should be disregarded.

4.4. Other methods to improve dependability

Besides the above techniques for the improvement of dependability, we considered the following:

(1) Fault-detecting and correcting hardware

The local faults due to noise or radiations should be detected and corrected immediately after their occurrence to prevent them from growing into a more serious breakdown of the system. The CPUs have features to detect errors which occur internally, such as address error, bus error, illegal instruction and spurious interrupts. The dedicated hardware such as watch dog timers detecting the overrun of the software and the current limiter detecting and correcting the latch-up will serveto the improvement of fault-tolerance. These features of hardwares ar e implemented and fully utilized for MUSES-A OBC.

(2) Fault-avoidance by software engineering

We also considered ways in which to eliminate the liability of the initial error due to the low software quality. We are using HI68K, the ITRON (ITRON: Industrial TRON, TRON is an acronym of The Realtime Operating System Nucleus) specification realtime operating system, which will be contained in the ROM and installed into the system [13]. The software was designed using the PAD (Problem Analysis Diagram), a tree-structured diagram, to maintain the design quality and to improve reliability [14]. The software development was carried out using the development tools including C language compiler and in circuit emulater.

5. Diagnosing algorithm

A diagnosing algorithm is necessary to detect errors, to discriminate the faulty BPU or voter and to select a reliable output data when some failure occurs in them. Here, the selector can be considered fault-free since it consists of ICs of MIL-Class S grade and of very simple construction. It can also be reasoned that the probability in which a faulty BPU is judged to be normal is far lower than the probability of other way around. As shown in Fig.3, the output from the BPUs are distributed to all the voters. The voters diagnose the BPUs according to the results of software voting and self-diagnosing based upon the comparison of output data with the feedback data. At the same time, the voters are diagnosed by the BPUs. Here, only three BPUs will be in operation regardless of the total number of BPUs in the system. The others are in cold stand-by status if present. The faulty BPUs or voter are temporarily replaced with other ones which are in normal status. The diagnosing is executed as follows:

(1) If no pair of BPU outputs coincide as a result of software voting performed by BPUs, all the 3 BPUs are judged to be faulty.

(2) When (1) does not apply, and if the feedback output from the voter does not match any BPUs, the voter currently selected is judged to be faulty.

(3) When neither (1) nor (2) apply, a pair of BPUs which do not match the feedback output from the voter are judged to be faulty. This judgement is exempt when this pair of BPUs match after the software voting and the voter is judged to be faulty.

(4) When a BPU does not match the feedback output or excluded by the software voting, the BPU is judged to be faulty.

The diagnosing algorithm to be performed by each BPU and other cicuitry stated above are summarized in Table 1. Here, A, B, and C indicate the BPUs and V indicates the voter which is currently in use. The equalities and inequalities in the leftmost column show the matching (or unmatching) of the corresponding BPUs as theresult of software voting. The equalities and inequalities in the upper rows show matching (or unmatching) of the feedback data as a result of self-diagnosis.

A, B, C and V in the entries of the table are the components which are judged to be faulty for the combination of the results of software voting and self-diagnosis.

Feedback Results — A	=	=	=	=	≠	≠	≠	≠
Software Voting — B	=	≠	=	≠	=	≠	=	≠
Results by BPUs — C	=	=	≠	≠	=	=	≠	≠
$A = B = C$	normal	B	C	B, C	A	A, B	A, C	V
$A = B \neq C$	C	B, C	C	B, C	A, C	V	A, C	V
$A \neq B = C$	A	A, B	A, C	V	A	A, B	A, C	V
$A = C \neq B$	B	B	B, C	B, C	A, B	A, B	V	V
$A \neq B \neq C \neq A$	A, B, C	A, B, C	A, B, C	A, B, C	A, B, C	A, B, C	A, B, C	A, B, C

Table 1. Diagnosing algorithm

In the case of MUSES-A OBC, the system consists of only three BPUs. Accordingly, there can be no BPUs in stand-by status. The results of diagnosis are used to select the most reliable output data and to select voters.

The faulty BPU is temporarily isolated from the other BPUs and continues its processing but its outputs are disregarded. The software voting is continued by the other 2 BPUs. The other 2 BPUs occasionally resume communication with the isolated BPU and inspect its output by software voting. If the isolated BPU is judged to be normal it comes back to the triple system.

MUSES-A satellite will be supported by a ground support equipment which monitors and controls it. The status of the OBC will be transmitted to the ground by the data processing unit and communication control system and an appropriate command will be transmitted to the OBC by an operator through the ground support equipment. In the most serious case such as the one in whichall the BPUs and voters fail, the system will be restarted by the command from the ground.

Fig.4 shows the overall configuration of fault-tolerance functions of MUSES-A OBC.

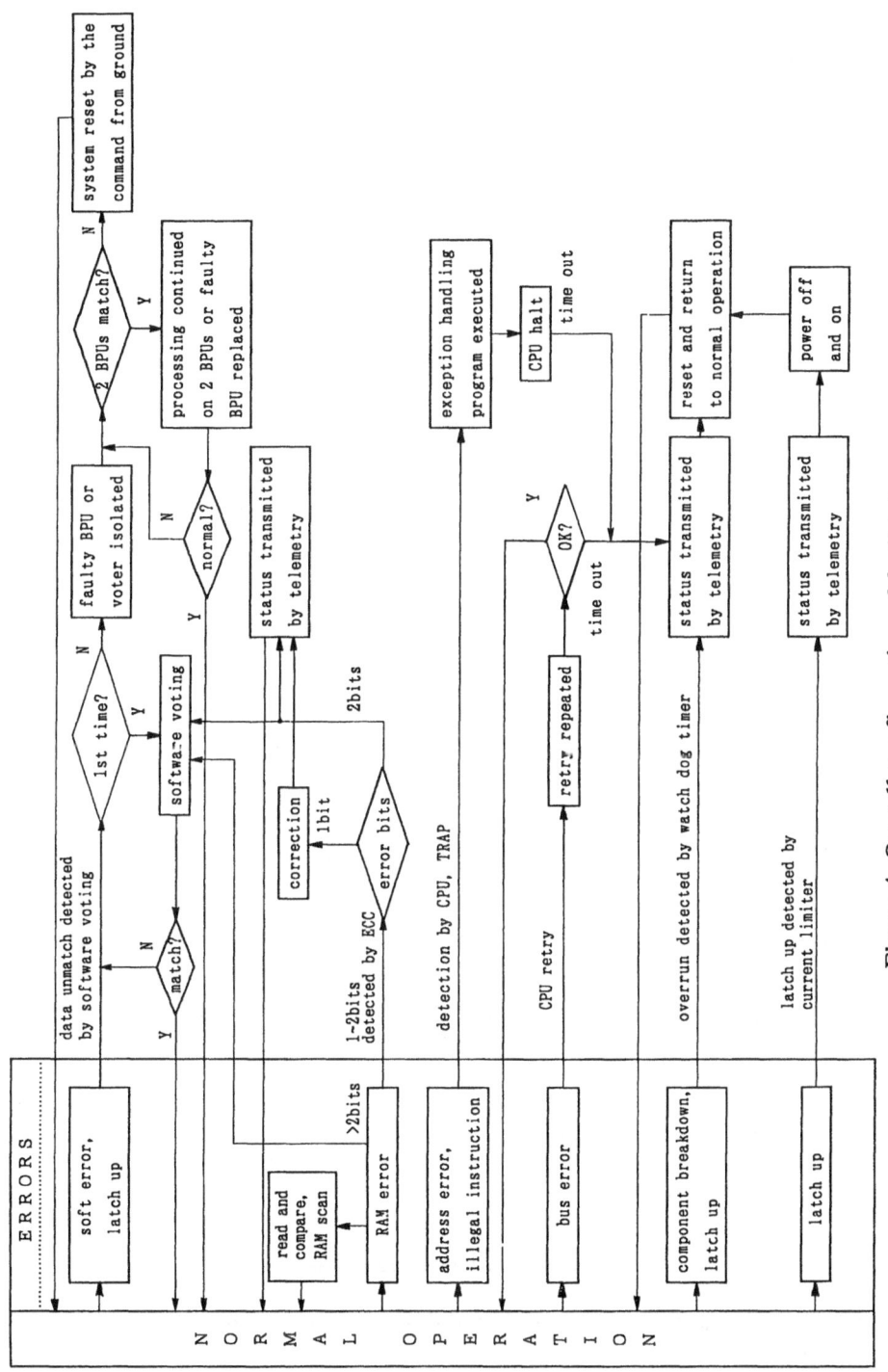

Figure 4. Overall configuration of fault-tolerance

6. Evaluation of reliability [14]

Here we make a quantitative consideration of reliability improvement due to the output feedback. We compare the reliabilities of the case with and without the output feedback with respect to the hardware reliability and data reliability.

(1) Hardware reliability

Hardware reliability is referred to as the reliability in the conventional sense and expressed as

$$R_h (\lambda t) = \exp (- \lambda t)$$

where λ is the failure rate. The hardware reliability is not improved by the output feedback.

(2) Data reliability

Data reliability is defined to be the probability in which the current output data is correct. Let us consider a situation in which a specific set of K subsystems out of N are participating the voting operation. For each of $_NC_K$ combinations of K BPUs, the probability in which the wrong data from K subsystems coincide and judged to be normal as the result of self-diagnosis is estimated to be

$$R_d(K/N) = (1 - P_\varepsilon^K \cdot P_{\varepsilon d}^K \cdot P_{a\varepsilon}^{K-1})$$

where P_ε is the probability of occurrence of transient fault, $P_{\varepsilon d}$ is the failure rate of self-diagnosis and $P_{a\varepsilon}$ is the probability in which the wrong data coincide. K can assume values 2, 3,...thru N, among which the case K=2 is the most significant since the cases K>2 are infinitesimal of higher order.

(3) Improvement of reliability

The data reliability with the output feedback can be estimated as

$$R_d (K/N, feedback) =$$

$$(1 - P_\varepsilon^K \cdot P_{\varepsilon d}^K \cdot P_{a\varepsilon}^{K-1} ((P_{a\varepsilon} + P_{\varepsilon rb} - P_{a\varepsilon} \cdot P_{\varepsilon rb})^K)$$

where $P_{\varepsilon rb}$ is the failure rate of data due to the fault in the output feedback mechanism. Since usually

$$P_{a\varepsilon} + P_{\varepsilon rb} - P_{a\varepsilon} \cdot P_{\varepsilon rb} < 1$$

holds,

$$R_d (K/N) < R_d (K/N, \text{feedback})$$

i.e. data reliability is improved by the output feed- back. Specifically assuming $P_\varepsilon = P_{\varepsilon d} = P_{a\varepsilon} = P_{\varepsilon rb} = 0.1$ and K=2, we obtain $R_d (2/N) = 0.999999609$ and $R_d (2/N, \text{feedback}) = 0.999999995$, indicating that the data reliability is improved by a factor of 100.

7. Example of implementation

As an initial target, we applied our concepts of dependability to MUSES-A OBC. The designing, implementation and testing were already completed. The OBC has been connected to the other equipments to be loaded on the MUSES-A satellite and is functioning without trouble.The fault tolerance function proved its robustness by detecting and/or correcting the faults which were artificially implanted into the system.

The source program size was 6179 lines of code in Clanguage and 4348 lines of code in assembly language both including comments. The target program size was 61.6 kilobytes/BPU excluding 32 kilobytes of HI68K real-time operating system.

8. Possible extensions

Our concepts can be extended in several ways for additional improvement of reliability. Here, we discuss the extensions in three ways, i.e. horizontal and hierarchical spatial diversities and multiple time diversity.

(1) Horizontal spatial diversity

In order to enhance the concept of spatial diversity, we can introduce a system in which more than 3 BPUs are normally in operation. For example, let 5 BPUs be in operation. There can be several more BPUs in stand-by status. The results of software voting can be classified into following 5 classes:

Class A
   All the outputs from 5 BPUs match
1 case.

Class B

The outputs from 4 BPUs match but 1 BPU shows a different output
$$_5C_1 = 5 \text{ cases.}$$

Class C

The outputs from 3 BPUs match. This class can be splitted into two subclasses according to the relationships between the outputs of the other 2 BPUs.

Subclass C1: The outputs from the 2 BPUs do not match
$$_5C_3 = 10 \text{ cases.}$$

Subclass C2: The outputs from the 2 BPUs match
$$_5C_3 = 10 \text{ cases.}$$

Class D

The outputs from 2 BPUs match. This class can be splitted into 3 distinct subclasses according to the relationships among the outputs of the other 3 BPUs.

Subclass D1:   The outputs from 3 BPUs match. This subclass is identical to subclass C2 and the number of cases is counted in subclass C2.

Subclass D2:   The outputs from 2 BPUs out of 3 match.

$$\frac{_5C_2 \times {}_3C_2}{2!} = 15 \text{ cases}$$

Subclass D3:   The outputs from 3 BPUs do not match
$$_5C_1 = 10 \text{ cases.}$$

Class E

The outputs from all BPUs do not match in any combination
$$1 \text{ case.}$$

The total number of cases is 52. For each of these cases, the results of self-diagnosis can come out in $2^5 = 32$ different cases. As a result, there can be $52 \times 32 = 1664$ combinations of cases. The diagnosing algorithm shown in Chapter 6 can be extended to this configuration with 5 BPUs. If the resuit of software voting falls into class E, all the BPUs are judged to be faulty. If the feedback output from the voter does not match any BPUs, the voter is judged to be faulty. The BPUs whose outputs do not match those from other BPUs are judged to be faulty.

(2) Hierarchical spatial diversity

The fault-tolerant configuration can be hierarchically organized by recursive application of the spatial diversity concept. A fault-tolerant system is configur ed whose subsystems are also fault-tolerant multiprocessors instead of BPUs. The diagnosing algorithm is also recursively applied. Recursive application of the identical diagnosing algorithm to different hierarchies unifies the arguments an d standardizes the implementation, requiring no modification depending upon the number of BPUs.

(3) Multiple time diversity

Extensions to the above concepts can be conceived not only in spatial domain but also in time domain. Self-diagnosis based upon the feedback data can be repetitively applied to ensure a correct result. Keeping a record of matching and unmatching outputs from each BPU, more accurate diagnosis can be done. For instance, retrying will eliminate erroneous diagnosis due to transient noise. A BPU whose output does not match with others repetitively can be considered faulty and replaced.

9. Conclusions

The concepts of dependable computer were established based upon the fault-tolerant configuration for the purpose of verifying the approach which makes use of commercially available components to build computers for space. It has been expected that the usage of these components of high integration density can be an effective solution for reducing the weight, size, power consumption and cost of computer systems.

Besides utilizing some of the methods which have been conventionally taken for space use, we developed new concepts such as time diversity and stepwise negotiating voting based upon the software synchronization. By these concepts, the level of required reliability can be flexibly adapted to the situations and as a result, the reliability of data processing is improved.

We developed a diagnosing algorithm to support these concepts and designed a software to execute it. The concepts and the algorithm were applied to implement the On-Board Computer to be loaded on the MUSES-A satellite and tested. The results of testing show that the algorithm has a robu stness to negotiate the conceivable errors in the computer system.

The result of implementation and testing indicate that the decrease in weight, size and power consumption more than compensates for the increase in hardwares required by the redundant configuration to execute fault- tolerant functions.

The overall approach to build computers based upon these concepts will be subject to its final verification as one of the mission of MUSES-A satellite to be launched early in 1990.

Acknowledgements

The authors wish to show their gratitude to Professor Algirdas Avizienis of UCLA, to Professor Yoshihiro Tohma of Tokyo Institute of Technology and to Professor Yoshiaki Koga of National Defence Academy for their technical advices starting from the early stage of conceptual development.

The gratitude is also directed to the members of IFIP WG 10.4 for general discussions.

References

[1]   A. Avizienis, "Fault-Tolerance:The Survival Attribute of Digital Systems", *Proceedings of the IEEE*, Vol.66, No.10, pp.1109-1125, 1978.

[2]   A. Avizienis, "The Dependability Problem: Introduction and Verification of Fault Tolerance for a Very Complex System", *FJCC Proc.*, pp.89-93, 1987.

[3]   A. Avizienis et al., "The STAR (Self-Testing and Repairing) Computer: An Investigation on the Theory and Practice of Fault-Tolerant Computer Design", *IEEE Trans. Comp.*, Vol. C-20, No.11, pp.1312-1321, 1971.

[4]   Y. Futamura et al., "Problem Analysis Diagram (PAD)", *Japan Annual Review in Electronics, Computers and Telecommunications*, Vol.12, 1984.

[5]   A.L. Hopkins Jr. et al., "FTMP-A Highly Reliable Fault-Tolerant Multiprocessor for Aircraft", *Proceedings of the IEEE*, Vol.66, No.10, pp.1221-1239, 1978.

[6]   H. Ihara et al., "Fault-Tolerant Computer System with Three Symmetric Computers", *Proceedings of the IEEE*, Vol.66, No.10, pp.1160-1177, 1978.

[7]   H. Ihara al., "Autonomous Decentralized Computer Control Systems", *Computer*, Vol.17, No.8, pp.57-66, 1984.

[8]   N. Kanekawa et al., "Dependable Onboard Computer System with a New Method-Stepwise Negotiating Voting", *Proc. 19th FTCS*, 1989.

[9] S.E. Kerneset al. (Ed.), "Special Section on Space Radiation Effects on Microelectronics", *Proceedings of the IEEE*, Vol. 76, No.11, 1988.

[10] L. Lamport et al., "The Byzantine Generals Problem", *ACM Trans. on Prog. Lang. and Sys.*, Vol.4, No.3, pp.382-401, 1982.

[11] E.L. Petersen, "Single Event Upsets in Space: Basic Concepts, Tutorial Short Course, IEEE 1983 Nuclear and Space Radiation Effects Conference.

[12] D.A. Rennels, "Architectures for Fault-Tolerant Spacecraft Computers", *Proceedings of the IEEE*, Vol.66, No.10, pp.1255 -1268, 1978.

[13] K. Sakamura, *ITRON: An Overview, Tron Project 1987*, Springer-Verlag, pp.75-82, 1987.

[14] T. Takano et al., "Fault-Tolerant Onboard Computers", *Proc. 16th ISTS*, pp.1097-1100, 1988.

[15] J.H. Wensley et al., "SIFT: Design and Analysis of a Fault-Tolerant Computer for Aircraft Control", *Proceedings of the IEEE*, Vol.66, No.10, pp.1240-1255, 1978.

Experimental Evaluation

Chair: J.F. Meyer (University of Michigan, Ann Arbor, Michigan, USA)

A FAULT BEHAVIOR MODEL FOR AN AVIONIC MICROPROCESSOR: A CASE STUDY

Gwan S. CHOI, Ravi K. IYER, Resve SALEH
Computer Systems Group - Coordinated Science Laboratory
University of Illinois at Urbana-Champaign
1101 West Springfield Avenue - Urbana, Illinois 61801 - USA

Victor CARRENO
National Aeronautics and Space Administration
Langley Research Center - Hampton, Virginia 23665 - 5225 - USA

Abstract

This paper describes an experimental analysis of the impact of transient faults on a microprocessor-based jet-engine controller, used in the Boeing 747 and 757 aircrafts. A hierarchical simulation environment based on SPLICE which allows the injection of transients during run-time and, the tracing of their impact is described. Results show that given a transient fault, there is approximately an 80% chance that there is no impact on the chip. If no latch-errors occur within 8 clock cycles, no significant damage is likely to happen. Thus, the overall impact of a transient is well contained. An empirical model is also derived to identify and isolate the critical fault propagation paths, the module most sensitive to fault propagation and, the module with the highest potential of causing external pin-errors.

Keywords: Simulation, transient faults, fault injection, error propagation, empirical models.

1. Introduction

This paper describes an experimental analysis of the impact of transient faults in a microprocessor-based jet-engine controller used in the Boeing 747 and 757 aircrafts. A hierarchical simulation environment based on SPLICE [17] for the run-time injection of transients and for the tracing of their impact is described. The probability that a transient results in latch, pin or functional errors is determined. Given a transient fault, there is approximately a 80 percent chance that there is no impact on the chip. The probability of a latch error is over 20 percent while that of a pin error is approximately 12 percent.

An empirical model to depict the process of error explosion and degeneration in the target system is derived. The model shows that, if no latch-errors occur within 8 clock cycles, no significant damage is likely to happen. Thus, the overall impact of a transient is well contained. A state transition model is derived from the measured data to describe the error propagation characteristics within the chip and, to quantify the impact of transients on the external environment. The model is used to identify and isolate the critical fault propagation paths, the module most sensitive to fault propagation and the module with the highest potential of causing external pin-errors.

2. Related research

Several researchers have investigated the impact of transients in computer systems. An early study of failures in digital systems reported in [2] showed that nearly 90 percent of failures were transient in nature. Recent studies using failure data from IBM mainframes reported in [10,86] also showed that nearly 85 percentp of major system errors were transient in nature. Furthermore, a strong relationship was found between the occurrence of transients and the level of system activity.

Device-level analysis of the mechanisms of transient upset has been in progress for quite some time. The hazards of transient upset in dynamic RAM's was first reported in [15] where the behavior of alpha-particle induced soft errors was explored. An approximate analytical model for a current transient was developed in [16].

A series of experiments, aimed at error analysis through the physical insertion of faults, were conducted by several investigators at the NASA AIRLAB test bed facility. An experiment to study fault latency distributions through hardware fault injections is described in [19]. An investigation of fault propagation in microprocessors is discussed in [14] and [12]. The analysis quantified the dependency of the measured error propagation on the location of the fault and the type of instruction/micro-instruction activity. In [3], new experiments to study fault and error latencies under varying workload conditions are discussed. Experience gathered from these studies shows that the data generated can provide considerable insight into error manifestation. More recently in [1], physical fault injection was used to validate a computerized interlocking system for the French railways. A new approach referred to as "accelerated fault injection" has recently been proposed in [4] and implemented on a large IBM multiprocessor system.

At the microprocessor level, studies have primarily focused on vulnerability assessment and on evaluating the efficiency of error detection methods. An assessment of different transient error test methods is presented in [13]. In [7], a detailed analysis of the vulnerability of the Z80 microprocessor based on ion bombardment testing is described. An approach which involves the development of a state transition matrix to describe the response to transient faults is described in [9]. In [20], transient faults which result in steady-state failures are analyzed and detection methods are discussed. In [6] and [18], techniques to determine the efficiency of error-detection mechanisms are described.

An important question not addressed in the above studies is the propagation of transients from the device-level through the microprocessor functional units to the pins. Apart from furthering the knowledge of transient fault propagation in microprocessors, this information is crucial for further defining the vulnerability of microprocessors to transients. In [8], a preliminary experiment to quantify the impact of transients from the device to the pin-level was described. Transients with charge-levels of 0.5, 1, 2, 3 and 4 picoCoulombs were injected. Logic upsets and first-order latch and pin errors were measured and analyzed via analysis of variance methods.

The above results point toward the need for more complete analysis of fault propagation characteristics. The type of functional-errors which can result from the injected transients need to be determined. Such errors can result in serious system malfunction, especially in avionic systems. In order to isolate the critical paths in the circuit, the fault propagation between the functional units and to the external pins must be quantified. In particular, the mechanisms involved in internal propagation of latch-errors (i.e. transient fault latency) and their effect at the pin-level needs to be investigated and modeled.

3. Target system

The target system for our study is a microprocessor used for real-time control of jet-engine functions. The system is currently used in commercial aircraft, including the BOEING-747 and the 757. The control system architecture thus contains microprocessors, memory units, I/O gate array chips, communication channels, frequency samplers, A/D converters and D/A converters. In this experiment we simulate the microprocessor and its associated memory with a focus on the impact of transient errors.

The 16-bit HS1602 microprocessor (Figure 1), which is the heart of the controller consists of six major functional units. The arithmetic and logic unit (ALU), which contains six registers, can perform double precision arithmetic operations. The control unit, which is responsible for issuing signals to control the operations of the ALU, is made up of combinational logic and several registers. The decoder unit decodes I/O signals, the multiplexer unit provides the discrete lines and buses and, the countdown unit is used to drive chip-wide clock signals. The watchdog unit provides protection against fault by resetting the processor in the event of parity error or when the application software is timed out by the software sanity timer. Also the signal to synchronize the dual system is provided by this unit. The chip runs at 6 Mhz and is implemented in a 3 micron technology CMOS gate-array made of 2688 blocks of 4 N-channel and 4 P-channel transistors.

4. The experimental environment: FOCUS

4.1. Simulation environment

In order to perform fast and accurate analysis, a mixed-mode transient fault simulator [8] based on SPLICE [17] was used. The modified simulator provides a fault injection and analysis environment which uses the SPLICE1 relaxation algorithm for circuit analysis. A transient fault injection is implemented as a run-time modification of the circuit whereby a current source is added to the target-node[1], thus altering the voltage level of the node over the time interval of the injected current waveform. The method allows both single and multiple transient injections. Since the injected current source is specified in a mathematical functional form, the transients can be of varying shapes and duration. Details of the implementation are given in [8].

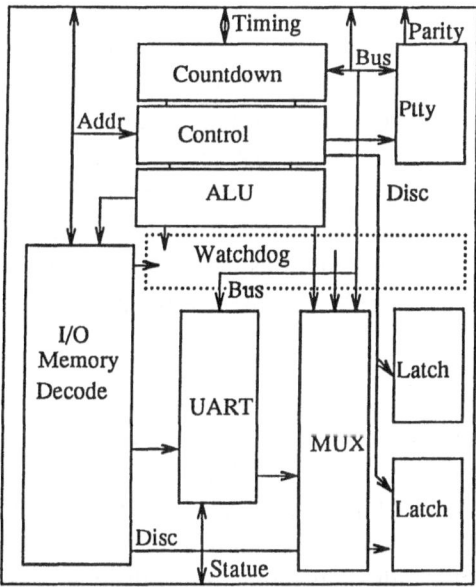

Figure 1. Data flow diagram of HS 1602

For a comprehensive study of fault propagation in the microprocessor a tracing facility was also developed to monitor all of the internal nodes (over 4000) in the HS1602. The tracing facility is capable of monitoring each node for all processed events. The trace data for each event consists of the time of the event, the hierarchical node name and the new and previous voltage levels (for electrical nodes) or the new and previous logic levels and their strengths (for logic nodes).

A graphical analysis facility was developed (on a color SUN Workstation) to visualize the error activity in different functional units of the processor and, the

---

[1] A node is defined as a point in a conductive interconnection between electrical and/or logical elements.

fault propagation on the major interconnects and at the external pins. Details of the automated environment "FOCUS" are given in [5].

4.2. The experiment

In our experiment, the entire HS1602, was simulated along with its associated memory modules. In the simulations, the gates around the region of fault injection were simulated at the electrical-level and the rest of the processor was simulated at the logical-level. The memory modules, which were not subject to fault injection, were simulated at the functional level. The actual design parameters of HS1602 and the capacitances extracted from the circuit layout were used in the simulations. The initialization phase of the microprocessor (about 72 clock cycles), which consists of a watchdog test, a parity test, an instruction set test and a ROM sum test and ensures that all of the functional units are exercised, was simulated. The simulation includes the processor accessing one external ROM for instructions and another external ROM for the initialization parameters. Arithmetic processing and address generation is also performed.

Transients with charge-levels in the range 0.5 to 9.0 picoCoulombs [2] were injected at seven nodes in each of the six major functional units. The nodes were selected to increase the chance of a latch-error (e.g., a transient was injected to the fan-in node of a latch). Each charge-level was injected at five different time-points during the execution of the application code sequence. The specific waveforms used in the fault simulations follow the double-exponential function proposed in [16]:

$$I(t) = \zeta \left[e^{-t/\alpha} - e^{-t/\beta} \right]$$

where ζ is the approximate maximum current, α is the collection time-constant for junction and, β is the ion track establishment time-constant.

The error-data for the analysis were generated by comparing each faulted simulation with a fault-free simulation. An error was assumed to occur if the injected transient caused the node voltage to vary beyond a defined logic threshold. For each simulation, the recorded data included the time of fault

---

[2] The charge-levels chosen represent transient response of various heavy ions including 100 MeV [56] Fe ions, which are commonly found in the cosmic environment. These levels were chosen so as to ensure that no permanent errors occur. Charge-levels approximately greater than 10 picoCoulombs are known to cause permanent latch-ups (device failure) in.

occurrence, the location of fault, the faulted value, and the fault-free value. Each fault event was also classified as either a timing error (premature or late firing) or a value error.

The error-data were then processed by a series of programs that collected statistics on the fault injections which resulted in a voltage transient large enough to result in latch and pin errors and errors at the interconnections of the functional units. Statistics on errors resulting in a functional alteration of the processor functions were also collected. The collected statistics were classified by the charge-level and by the location. In total, over 2100 fault injections/simulations were performed.

5. Impact of transients

Table 1 summarizes the overall impact of transients in the range 0.5 to 9.0 picoCoulombs. In the table, a first-order error is defined as one which occurs during the first clock cycle following a transient fault injection; second and higher order errors are those that occur during the second and subsequent clock cycles [3]. The second column shows the number of fault injections which result in errors. The third column shows the total number of resultant errors. For example, out of 2100 fault injections, one or more first-order latch-errors occurred in 470 cases (22.4 percent) and, a total of 2149 latch-errors were observed.

TRANSIENT FAULT SEVERITY			
Type	Occurrences	Count	Percentage
Injected transients	2100		100%
First-order latch errors	470	2149	22.4%
Second and higher order latch errors	120	1829	5.7%
First-order pin errors	255	1168	12.1%
Second and higher order pin errors	90	839	4.3%
Functional errors	193	747	9.2%

Table 1

---

[3] Transients modeled in the experiment last no longer than one clock cycle. This is typical of effects of cosmic rays and the like.

A number of issues relating to the fault sensitivity of the chip are highlighted by this data: Over 20 percent of the injections result in latch-errors. Given that a transient results in a latch-error, the chance of multiple errors is high (an average of 4 latch-errors per transient). The existence of such multiple latch-errors is potentially a serious problem since, these errors can subsequently propagate to the pins and lead to multiple failures.

In addition, even though only 25 percent (120 out of 470) of the latch-errors propagated past the first clock cycle (i.e. the first-order), each such propagation can result, on the average, in about 15 latch-errors, thus further intensifying the propagation problem. An effect of second and higher order latch-errors is an increase in the probability of functional-errors (erroneous control signals or data, which result in an alteration of the microprocessor functions). Chance of having functional-errors is almost a 10 percent Over one-third of the total number of functional-errors were due to transients in the ALU unit. Further analysis of the error data showed that a significant number of functional-errors, resulting from transients in the ALU, were due to first-order effects. This is because, transients that latch directly on to the ALU registers result in an immediate alteration of address or data information. Functional-errors caused by second and higher order effects of transients were more dispersed among different functional units. A relationship between the second and higher order latch-errors and functional-errors is discussed further in the section 5.1.

From Table 1, the percentage of first-order pin-error occurrences is significant (over 10 percent). Given a pin-error, the chance of recurrence during the subsequent clock cycles is relatively high (90/255) and each propagation can result, on the average, in approximately 9 pin-errors (in comparison we have approximately 4 pin-errors resulting from the first-order propagation).

5.1. Charge-level analysis

This section quantifies the impact on the chip of the charge-level in a transient. Statistical analysis of the the error data was performed to determine the effect of different charge-levels, in the injected transients, on the severity of latch, pin and functional errors. Figure 2 shows the frequency of latch, pin and functional errors as a function of the charge-level. First, we see that beyond 7 picoCoulombs, the number of error occurrences remain relatively constant, i.e., additional charge does not result in an increase in the error probability. This is because, at this charge-level, essentially all the latches in the propagation path have been affected (i.e. hold erroneous values).

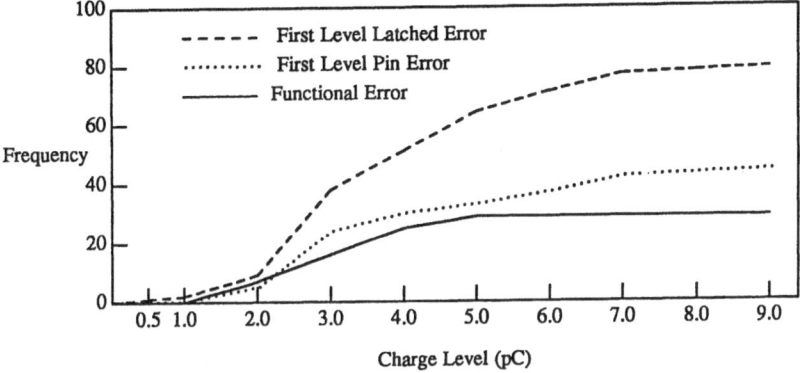

Figure 2. Error frequency vs. charge-level

For latch and pin errors we see a charge threshold of 2 picoCoulombs, at which there is a sharp increase in error activity. Over 95 percent of the latch-errors occurred at charge-levels greater than 2 picoCoulombs and 100 percent of the pin-errors occurred for charges at or above 2 picoCoulombs. For functional-errors, however, the threshold is not so well defined. This is most likely due to the fact that, functional-errors can also result from second and higher order latch-errors (in addition to being caused by the first-order effect of a transient). The higher order effects, of course, are not charge dependent, hence a charge threshold does not occur. Figure 3 shows the frequency of second and higher order latch-errors and the functional upsets. Note that the frequency of the second and higher order latch-errors also lacks the distinctive charge-threshold.

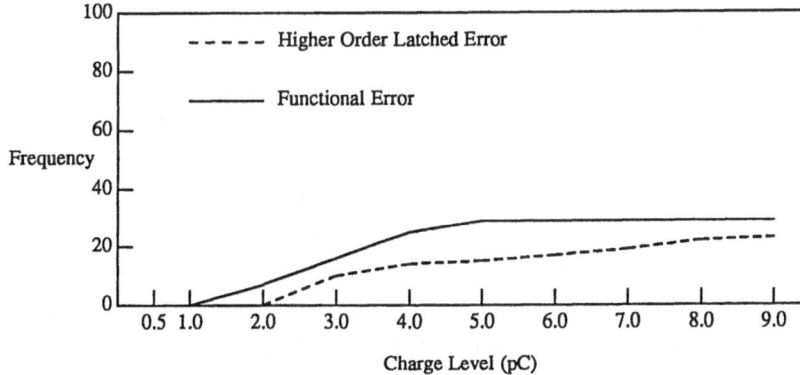

Figure 3. Comparison of functional errors and higher order latch-errors

Figure 4 shows, for each functional unit, the first-order latch and pin error distributions by the charge-level in injected transients. For charge-levels above the threshold, the ALU and the watchdog units have the highest latch-error probability. The watchdog unit has high latch-error occurrences; pin-errors occur only for charges above 6 picoCoulombs. The reason is that, although an error can quite easily get latched in the numerous feed-back paths in the watchdog, it does not always propagate to the external pins. The Decoder unit shows a relatively low pin-error propagation probability. The chance of transients below the threshold being latched is generally small, except for the control unit where the possibility of having latch-errors is high even at 2 and 3 picoCoulombs. The relatively small capacitive loading of the feedback paths to the latches in the control circuit explains this low charge sensitivity. As shown in Figure 4, the Multiplexer does not have any latch or pin errors. This is because the electrical nodes in the Multiplexer unit have high capacitances due to the large number of fanouts.

6. Error propagation and state transition model

Fault propagation usually occurs because errors can get latched and then migrate to different sections of the chip. A latch-error can stay latent and undetected until it migrates to the pins at a later time. The additional internal propagation between latches can increase the probability of generating functional upsets. Thus a characterization of the latch-to-latch fault propagation patterns is important. A latch-error can either get re-latched, propagate out to the I/O pins and/or disappear in each clock cycle.

6.1. Error latency

To characterize the latency of transient faults in the circuit, the expected time (in clock cycles) for an injected transient to migrate to the pins was calculated. The expected error latency was defined as the mean value of the interval between the time of fault injection and, the time at which a resultant pin-error occurred. Table 2 shows the expected error latency for transients in different functional units. The expected error latency for transients in the control unit is the highest. This is because the majority of the pin-errors, due to transients in the control unit, resulted from latch-errors. Note that the mean latency for pin-errors in the countdown and the decoder units is less than a clock cycle. All pin-errors in these

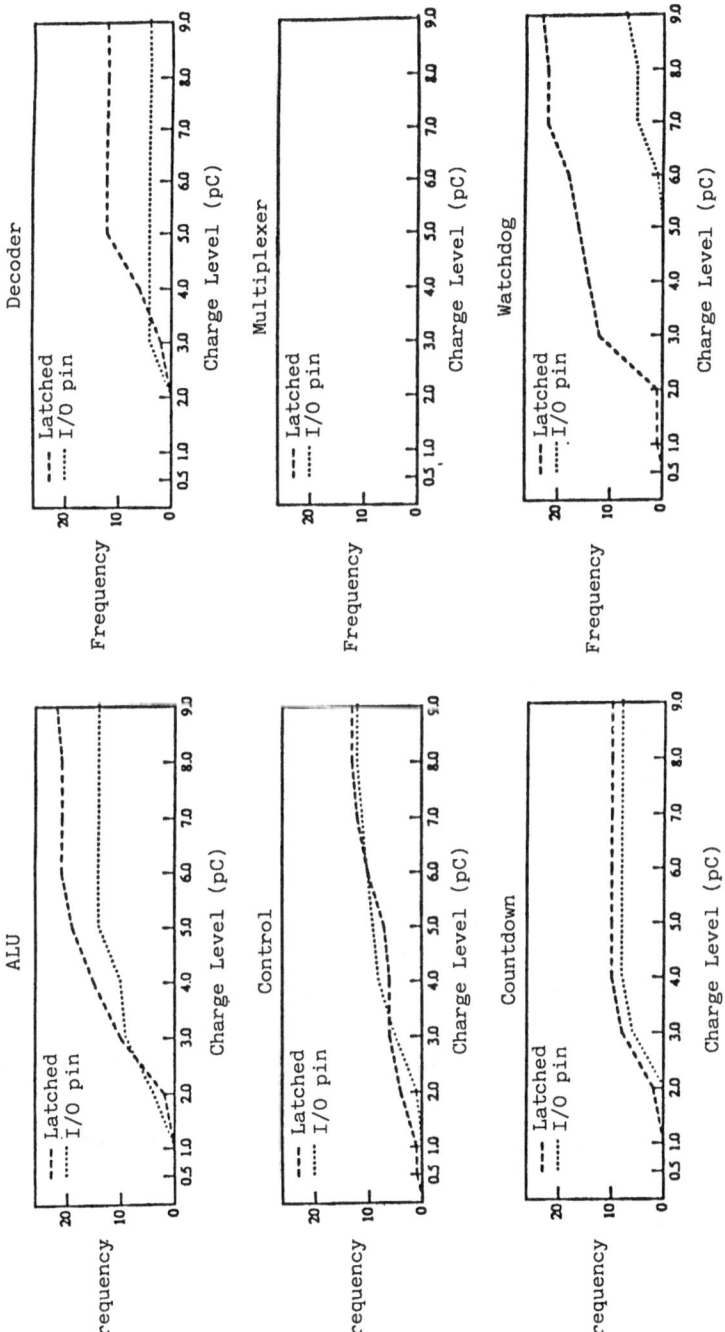

Figure 4. Error frequency by charge level for each functional unit

units resulted directly from the injected transients (i.e. no pin-errors in these units were do latch errors). Thus, the mean latency for the pin-errors in thecountdown and the decoder units is simply the signal propagation delay from the injected location to the external pins. No pin-errors were observed for transients in the multiplexer unit.

MEAN ERROR LATENCY	
Functional unit	Clock cycle
ALU	1.43
Countdown	0.23
Control	4.67
Decoder	0.14
Multiplexer	-
Watchdog	1.94

Table 2.

6.2. Error propagation model

The propagation of the latch-errors in time (in clock cycles) for the control unit is illustrated in Figure 5. In this figure, the x-axis represents the clock cycles from fault-injection time and the y-axis represents the total latch-error count for each clock cycle. It can be seen that, given a certain number of latch-errors in the first clock cycle, this number degenerates until the fourth clock cycle. At approximately the fifth clock cycle, the number of errors rapidly multiply. Thus, despite the fact that only on few occasions do the latch-errors last until the fifth clock cycle, when they do, the number of errors is large. This is because at this time period, the error signal enters a unit with a large number of latches and high fan-out, e.g., the ALU registers. After the sixth cycle, the number of errors degenerate significantly until they finally disappear after the eighth cycle. Thus, the impact of latch-errors last at the most up to 8 clock cycles from the time of injection.

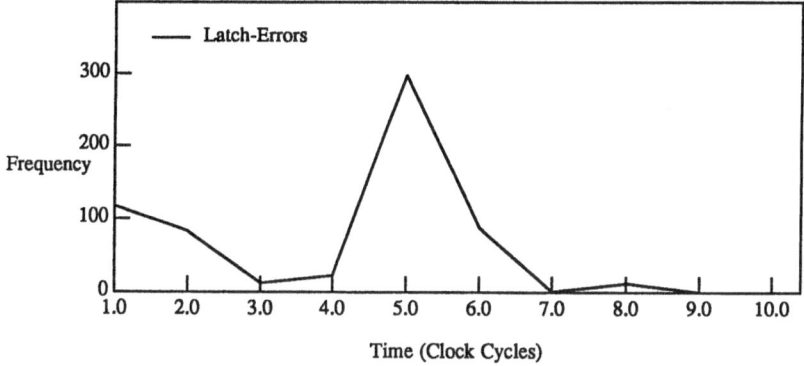

Figure 5. Latch-error occurrences vs. time

We define the clock cycles in which the number of latch-errors increases in comparison with previous cycle as "error explosion" cycles. Clock cycles where the number of errors decreases in comparison with previous cycle are defined as "error degeneration" cycles. In Figure 5 an error explosion occurs in the fifth clock cycle. The chance of functional-errors and pin-errors is likely to be maximal during this period of error explosion. In the sixth clock cycle (a degeneration cycle), the number of latch-errors decreases to about one-third of the number in the fifth clock cycle.

A model to depict the process of error explosion and degeneration for the overall system is shown in Figure 6. The model is derived from the measured data to quantify the dynamics of the error propagation in the system. As seen from the model, an injected fault either gets latched (represented by "latch-error" state) or has no impact on the circuit (represented by "fault-free state"). The "explosion" state represents the situation where the number of latch-errors in the current clock cycle is greater than that in the previous cycle. And the "degeneration" state represents the scenario where the number of latch-errors in the current clock cycle is smaller than that in the previous cycle. The value assigned to a state is the average number of latch-errors in that state.

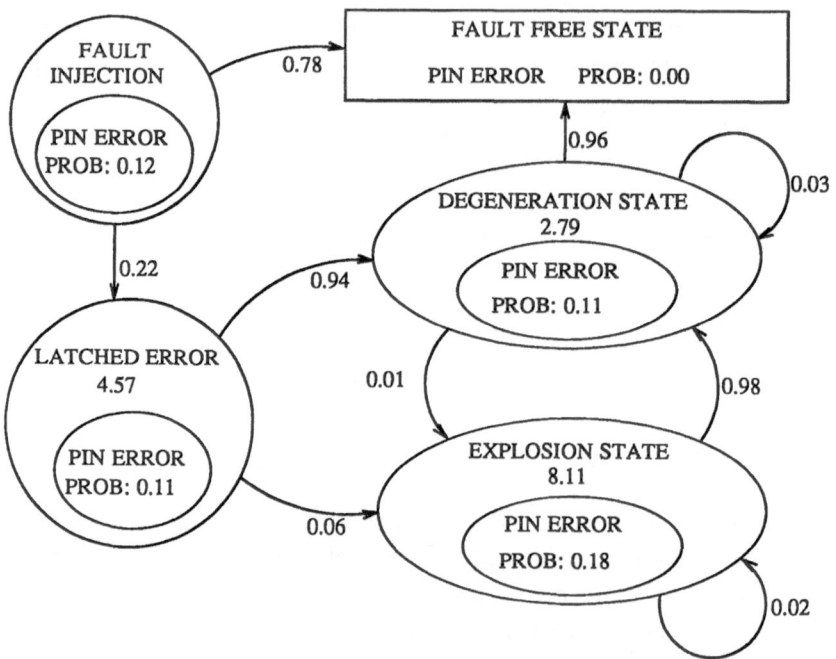

Figure 6. Error explosion/degeneration model

Given a transient fault, there is approximately a 80 percent chance of having no impact on the chip. Although the probability of a latch-error resulting in an error explosion is small (0.06), when it does occur, the average number of latches holding an erroneous values is large (8.11), i.e., although the explosion event rarely occurs, it is potentially disastrous. The probability of latch-errors, in the explosion state, causing pin-errors is higher than that for the degeneration state (0.18 compared to 0.11). This is clearly so because, with the larger number of latch-errors, the probability of error propagation to the pins is increased. After an explosion, there is a 98 percent chance of the latch-errors degenerating and becoming fault-free.

In summary, the probability of sustained explosion is very low with 0.02 probability, i.e., the chance of uncontrolled propagation is small thus the overall impact of a transient is well contained. Further, if no latch-error occurs within 8 clock cycles, no significant damage is likely to happen. Thus, limited roll-back recovery techniques which can keep track of the machine state for up to 8 clock cycles may be very successful.

6.3. State transition model

The foregoing section presented an analysis, of the forward propagation in time, of an injected transient in particular of its potential to cause latch and pin errors. This section examines the question: given a latch-error in a unit, where did it come from? The question of the internal module-to-module latch-error propagation is addressed. It will be seen that the result of this analysis is useful in identifying several critical aspects of the system. Some examples include the identification of the critical error propagation paths, the determination of the module most sensitive to fault propagation and the module with the highest potential for causing external pin-errors.

The latches in each functional units, the external pins and interconnects between the functional units were monitored to determine error propagation among the functional units and to the external pins. Figure 7 shows a state transition diagram, based on the measured data, to quantify the inter-module latch-error propagations. In the figure, each state (except the fault-free state) represents an error condition in the specified unit, e.g., the state ALU represent a latch error condition in the ALU. The numerical value shown with each state is the average number of latches holding erroneous values (e.g., given a latch-error occurrence in the ALU, on the average 9.89 latches are affected). Given a latch-error in a specified unit, the model shows the probabilities that each of the other functional units are the likely error sources. Thus, the model is of the inverse Markov type [4]. For example, in the figure, given an external pin-error, the probability of the ALU being the error source is 0.16; the probability of the control unit being the error source is 0.27; the multiplexer unit being the source is 0.02 [5].

---

[4] A normal forward transition Markov model to describe latch-error propagation can become quite complicated since a latch-error can propagate out to the multiple locations at once, i.e. a latch-error can propagate to both the external pins and other latches in the circuit. Thus, an inverse Markov model is used to describe transition from the error source to error for different states.

[5] Recall the fact that no transients injected in the multiplexer unit have resulted in a pin-error. Although no pin-error has occurred due to the injected transients in the multiplexer unit, two percent of all pin-errors have resulted from the propgagation of the latch-errors (result of the transients in the other units) in the multiplexer unit.

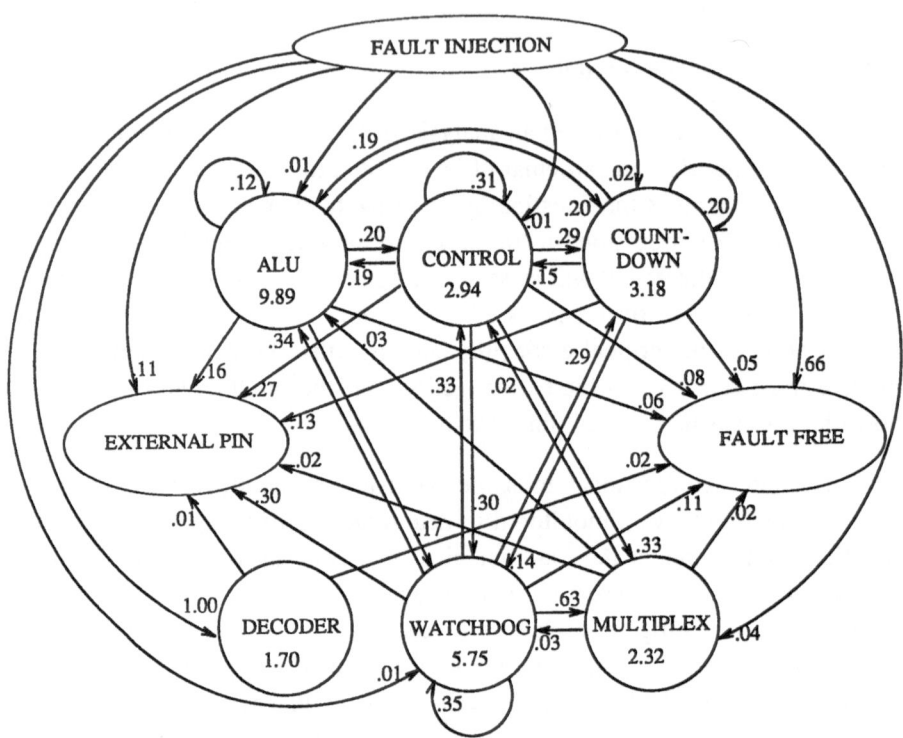

Figure 7. State transition model for error propagation

It can be seen from the model that latch-errors in the decoder unit do not propagate to other functional unit. All latch-errors that occurred in this unit are due to the direct effect of the injected transient. This is because the latches in the decoder unit are well isolated from the inputs of the other units. The probability of a latch error in the decoder unit propagating to the pins is small (0.01). Thus, the decoder is not a critical unit from a fault propagation point of view.

The model addresses several issues raised at the beginning of this section. The critical fault propagation path in the system is between the control and the watchdog units. Given a latch-error in the control unit, the probability that it propagated via the watchdog unit is 0.33. Conversely, the probability of the control unit being the source for a latch-error in the watchdog unit is also high (0.30). In examining the other units it is seen that although the one-way propagation probability is high in some cases (e.g., 0.63 from the watchdog unit to the multiplexer), none has a higher two-way propagation probability. Therefore, all other factors being equal, the best way of reducing inter-module error propagations is to protect the interconnections between the watchdog and the control units. Since a significant number of functional-errors result from the second and higher order latch-errors, the system level impact of providing this protection is expected to be a decrease in the probability of functional-errors.

The model also shows that the module with the highest potential to cause external pin-errors is the watchdog unit. Thirty percent of all pin-errors were due to the latch-errors in the watchdog unit. Hence, to reduce the number of pin-error occurrences, the outputs of the watchdog unit should be protected. The module most sensitive to fault propagation is seen to be the ALU unit. Of all functional units, an error occurrence in the ALU is likely to lead to the largest number of latch-errors (9.89). Applying internal retry to ALU operations may be a successful way of reducing the number of latch-errors.

Finally, it is seen that the probability that an injected transient directly causes pin-errors is low. More than 90 percent of the pin-errors are due to second and higher order propagation from latch-errors. Similarly the probability that an injected transients directly causing a latch-error is also low. Notice that less than 5 percent of the latch-errors are due to the direct propagation from the injected transients. The fact that over 95 percent of latch-error occurrences are due to propagations from other latch-errors, makes fault propagation a critical issue from a reliability perspective.

7. Conclusion

In this paper, the effect of transients in a microprocessor-based jet-engine controller was investigated. A design automation environment to allow the run-time injection of transients was described. The results show that, given a transient fault, there is approximately a 80 percent chance that it has no impact (latch or pin errors) on the chip. Only 25 percent of the latch-errors propagated past one clock cycle although, each such propagation resulted, on the average, in about 15 latch and 9 pin errors. More than 40 percent of the pin-errors were due to multi-step latch-error propagations. Approximately 10 percent of the transients result in the errors in the microprocessor's control functions.

The probability of a latch-error resulting in an error explosion was very small (0.06). However, when it did occur, the average number of latches holding erroneous values was large (8.11), i.e., although the explosion event rarely occurs, it is potentially disastrous. The probability of sustained error explosion was also very low (0.02), i.e., the chance of uncontrolled error propagation was small. Thus, the overall impact of a transient was well contained; if no latch-error occurred within 8 clock cycles, there was no significant damage to the microprocessor functions. Thus, limited roll-back recovery techniques which can keep track of the machine state for up to 8 clock cycles may be very successful. A state transition model, to describe the error propagation within the chip and, to quantify the impact of transients on the external environment was proposed. The model was used to identify and isolate the critical fault propagation paths, the module most sensitive to fault propagation and the module with the highest potential of causing external pin-errors.

Acknowledgments

This work was supported by National Aeronautics and Space Administration under NASA grant NAG-1-602. The authors thank the researchers at NASA AIRLAB, for many useful discussions. In particular we thank Celeste Belcastro for providing insight into the EEC131 controller. Thanks are also due to our co-workers in the Computer Systems Group and to Dr. J. Arlat of LAAS for their comments of an earlier draft of this paper.

References

[1]   J. Arlat, Y. Crouzet and J.C. Laprie, "Fault-Injection for Dependability Validation", LAAS Research Report no. 88-363, December 1988.

[2]   H. Ball and F. Hardy, "Effects and Detection of Intermittent Failures in Digital Systems", *1969 FJCC, AFIPS Conference Proceedings*, vol. 35, pp. 329-335.

[3]   R. Chillarege and R.K. Iyer, "Measurement-Based Analysis of Error Latency", *IEEE Transactions on Computers*, vol. C-36, pp. 529-537, May 1987.

[4]   R. Chillarege, "Understanding Large System Failures - A Fault Injection Experiment", *Digest, FTCS-19, The Nineteenth International Symposium on Fault Tolerant Computing*, pp. 356-363, 1989.

[5]   G. Choi and R.K. Iyer, "An Automated Environment for Dependable VLSI Design", Technical Report, Computer Systems Group, University of Illinois, Urbana, 1989.

[6]   B. Courtois, "Some Results about the Efficiency of Simple Mechanisms for the Detection of Microcomputer Malfunctions", *Digest, FTCS-9, The Ninth International Symposium on Fault Tolerant Computing*, pp. 71-74, 1979.

[7]   J. Cusick, R. Koga, W.A. Kolasinski and C. King, "SEU Vulnerability of the Zilog Z-80 and NSC-800 Microprocessors", *IEEE Transactions on Nuclear Science*, vol. NS-32, pp. 4206-4211, December 1985.

[8]   P. Duba and R.K. Iyer, "Transient Fault Behavior in a Microprocessor A Case Study", *1988 ICCD, Proceedings*, October 1988.

[9]   R.E. Glaser and G.M. Masson, "Transient Upsets in Microprocessor Controllers", *Digest, FTCS-11, The Eleventh International Symposium on Fault Tolerant Computing*, pp. 165-167, 1981.

[10] R. K. Iyer and D. J. Rossetti, "A Statistical Load Dependency Model For CPU Errors at SLAC", *Digest, FTCS-12, The Twelveth International Symposium on Fault Tolerant Computing*, pp. 363-372, 1982.

[11] R.K. Iyer and D.J. Rossetti, "A Measurement-Based Model for Workload Dependence of CPU Errors", *IEEE Transactions on Computers*, vol. C-35, pp. 511-519, June 1986.

[12] S.Kim and R.K. Iyer, "Impact of Device Level Faults in a Digital Avionic Processor", *AIAA/IEEE 8th Digital Avionics Systems Conference*, pp. 428-436, Oct. 1988.

[13] R.Koga, W.A. Kolasinski and M.T. Marra, "Techniques of Microprocessor Testing and SEU-Rate Prediction," *IEEE Transactions on Nuclear Science*, vol. NS-32, pp. 4219-4224, December 1985.

[14] D. Lomelino and R.K. Iyer, "Error Propagation in a Digital Avionic Processor: a Simulation-Based Study", *Proc. Real Time Systems Symposium*, pp. 218-225, Dec. 1986.

[15] T.C. May and M.H. Woods, "Alpha-Particle-Induced Soft Errors in Dynamic Memories", *IEEE Transactions on Electron Devices*, vol. ED-26, pp. 2-9, January 1979.

[16] G.C. Messenger, "Collection of Charge on Junction Nodes from Ion Tracks", *IEEE Transactions on Nuclear Science*, vol. NS-29, pp. 2024-2031, December 1982.

[17] R.A. Saleh, "Nonlinear Relaxation Algorithms for Circuit Simulation", Memorandom No. UCB/ERL M87/21, Electronics Research Laboratory University of California, Berkeley , 1987.

[18] M.E. Schmid, R.L. Trapp, A.E. Davidoff and G.M. Masson, "Upset Exposure by Means of Abstraction Verification", *Digest, FTCS-12, The Eleventh International Symposium on Fault Tolerant Computing*, pp. 237-244, 1982.

[19] K.G. Shin and Y.H. Lee, "Measurements of Fault Latency: Methodology and Experimental Results", Technical Report CRL-TR-45-84, Computing Research Laboratory University of Michigan, Ann Arbor, 1984.

[20] J. Sosnowski, "Evaluation of Transient Hazards in Microprocessor Controllers", *Digest, FTCS-16, The Sixteenth International Symposium on Fault Tolerant Computing*, pp. 364-369, 1986.

USE OF HEAVY-ION RADIATION FROM ^{252}CALIFORNIUM FOR FAULT INJECTION EXPERIMENTS

Johan KARLSSON, Ulf GUNNEFLO, Jan TORIN
Department of Computer Engineering
Chalmers University of Technology
S-412 96 Göteborg - Sweden

Abstract

The use of heavy-ion radiation from ^{252}Californium as a fault injection method for experimental validation of dependable computing systems is described and discussed. Heavy ions from ^{252}Cf can cause transient faults as they pass through depletion regions in integrated circuits. Irradiation of a circuit must be performed in a vacuum. A design of a portable miniature vacuum chamber which contains one integrated circuit and a ^{252}Cf source is presented. The vacuum chamber can be plugged directly into an IC socket of the same type as the irradiated circuit uses. For an effective system validation, the fault injection method used must cause a variety of errors in the irradiated circuit. Results from irradiation of the MC6809E 8-bit microprocessor are presented. They show that errors occurred on all output pins of the microprocessor, and that both single and multiple bit errors were observed. Sensitivity to heavy ion radiation from ^{252}Cf for various IC technologies is discussed briefly.

Keywords: dependability validation, fault injection, transient faults, single event upsets.

This work was supported by the Swedish National Board for Technical Development under contract # 86-3585.

1. Introduction

Computer systems are being used in an increasing number of applications throughout society. They will be used in the future in many critical applications where a malfunction would lead to catastrophic consequences. To avoid this, dependable computing systems must be used. They are designed either to tolerate faults or to be shut down in a safe way when faults occur. Such systems will be common in the manufacturing industry, for automobiles and aircraft, etc. As the number of critical applications increases, the demand for safety certification and validation of dependable computing systems will grow. An important step in dependability validation of a dependable computing system is to inject faults into hardware in order to validate the error handling mechanisms in the system.

Dependability validation, according to the definition by Laprie [1], can be divided into two different activities, error removal and error forecasting. In error removal, fault injection can be used for verification of fault tolerant mechanisms. Discovery of a design error in a hardware error detection mechanism is an example of this. Fault injection can also be used in error forecasting, e.g. fault coverage can be estimated and later used in predictions of system reliability. Fault injection techniques, which have been used for validation and similiar purposes, include various methods that introduce errors on the pins of ICs [2]-[10], the use of thermal stress to provoke IC failures [11], and exposure of ICs to heavy ions, alpha particles and neutrons by the use of particle accelerators [12]-[13].

In safety certification of systems it is necessary to standardize validation procedures. For fault injection it would be desirable to have a universal method for injection of faults, which could be used for many types of systems and circuits. A universal fault injection method must also produce errors which in one of several possible ways are well spread and hard to detect in order to reveal, as effectively as possible, any deficiencies present in the system under validation. On the other hand, it will not, perhaps, be possible to find one comprehensive fault injection method. Instead, a combination of several fault injection methods may have to be used. In any case, it is important to investigate various fault injection methods to determine their usefulness in experimental validation of dependable computing systems.

In this paper we discuss the method of using heavy-ion radiation from ^{252}Californium for injection of transient faults into integrated circuits, a method which the authors believe could play an important role in experimental validation of dependable computing systems, and which could perhaps become a universal fault injection method for certification of such systems.

Fault injection by heavy-ion radiation from ^{252}Cf has become a popular method of assessing Single Event Upset (SEU) vulnerability of integrated circuits considered for use in space. SEUs are transient faults caused by heavy ions, which in space are found in cosmic rays. The method was originally developed by research groups at UKAEA, AERE Harwell, the UK, and ESA-ESTEC, The Netherlands [14]. Heavy ions emitted from ^{252}Cf are highly ionizing particles capable of creating transients when they pass through a depletion region in an integrated circuit (IC). Since the heavy ions are attenuated in air, irradiation of circuits must be performed in vacuum with the lid removed from the circuit package. We present here a design of a miniature vacuum chamber which contains the irradiated circuit and the ^{252}Cf source. The vacuum chamber is equipped with electrical feed-throughs for the pin connections and can be plugged directly into an IC socket of the same type as the irradiated circuit uses. In this way it is very easy to inject faults into a circuit which is part of a large system.

Single event upset tests have shown that many types of circuits fabricated by various technologies are susceptible to heavy ion radiation, and that susceptibility tends to increase with the level of integration. The major strength of the ^{252}Cf method is the capacity to inject faults at internal locations in the ICs, faults which can be expected to cause many different kinds of errors. We have used the ^{252}Cf method for fault injection experiments with the MC6809E 8-bit microcomputer. The variation of the error behavior of the MC6809E has been studied and results are presented later in this paper. The effectiveness of various error detection mechanisms for this microprocessor has also been investigated and these results are presented in [15]. More detailed results of the error behavior investigations will appear in [16]. The main purpose of this paper is to introduce the method for those who are involved in experimental validation, and to provide practical hints to those who are interested in using the method. More research is needed, both to investigate the sensitivity to heavy ion radiation of different IC technologies and to assess the usefulness of the method for verification and evaluation of dependable computing systems.

In the next section of this paper we present the design of the portable miniature vacuum chamber. In Section 3 we discuss the fault mechanism, the nuclear characteristics of ^{252}Cf, and the susceptibility to heavy ion radiation of circuits and IC technologies. Advantages and potential problems of using ^{252}Cf for fault injection are discussed in Section 4. In Section 5 results are presented from initial fault injection experiments with an MC6809E 8-bit NMOS microprocessor. Finally, Section 6 is devoted to conclusions.

2. Miniature vacuum chamber

The design of a miniature vacuum chamber for irradiation of an IC is shown in Figure 1. The vacuum chamber consists of two parts. One part consists of a short tube of stainless steel 25 mm long and 60 mm in diameter. The tube is sealed at one end by a stainless steel plate welded onto it. At the other end of the tube there is fixed flange. Inside the tube, a fixture for the Californium source is mounted, as shown in the figure. A commercially available 37 kBq (1 µCi) ^{252}Cf source is used. It is produced by high temperature vacuum evaporation onto a stainless steel disc. The backing tray has a diameter of 25 mm, a thickness of 0.5 mm, and the active diameter is 4 mm.

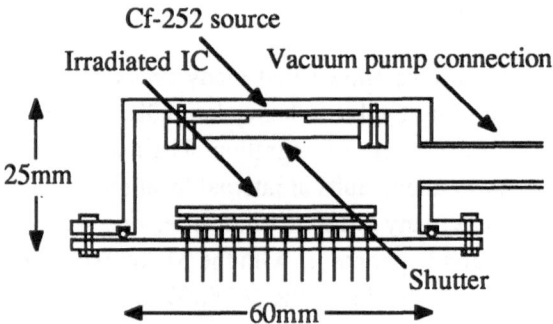

Figure 1. Miniature vacuum chamber, crossectional view.

Beneath the source is mounted an electrically maneuvered shutter, which is used to shut radiation on and off. There is an air outlet on the side of the tube to which a vacuum pump is connected. A vacuum valve (not shown in the figure) can be connected to the air outlet to allow the vacuum chamber to be disconnected from the vacuum pump.

The other part of the vacuum chamber is a plate on which a socket for the irradiated circuit is mounted. The socket pins extend through the plate to the outside of the vacuum chamber. The socket plate is fastened to the tube flange by screws. An o-ring rubber sealing is inserted into the flange to prevent leakage. The socket plate can be exchanged for different kinds of sockets. The distance between the ^{252}Cf source and the irradiated circuit can be extended by inserting pieces of tubes of appropriate lengths between the socket plate and the other part of the vacuum chamber.

To prevent radioactive contamination, the vacuum chamber must be assembled in a radiologically controlled area under a vented hood. Once the air has been pumped out and the vacuum pump has been disconnected, the miniature vacuum chamber can be handled freely, without special arrangements for radiation protection.

3. The fault mechanism and susceptible technologies

Heavy ions from ^{252}Cf are highly ionizing particles capable of causing Single Event Upsets in integrated circuits. A Single Event Upset (SEU) is a change of the level of a logic signal in a circuit caused by the passage of a single ionizing particle. When a heavy ion from ^{252}Cf hits a semiconductor it creates electron hole pairs. If the heavy-ion passes through a depletion region in a circuit, the high electric field present there will cause the charge to be collected. This causes a current pulse which in turn may result in a voltage pulse large enough to reverse the logic state of a signal line. The change of the logic signal may lead to a change of the logic state of a memory element, a bit flip. Such a bit flip is called a soft error because no hardware is damaged and the error can be recovered from by reloading the memory element with the correct binary value. A detailed description of the SEU fault mechanism is given by Messenger and Ash [17].

^{252}Californium decays both by alpha particle emission and by spontaneous fission. For fault injection, the fission fragments are of primary interest. The charge generation rate for the alpha particles is too small to affect most circuits presently in use. The heavy ions created by the spontaneous fission have considerably higher charge generation rates However, as device dimensions decrease, sensitivity to alpha particles can be expected to increase. ^{252}Californium also emits neutrons which activates exposed material resulting in alpha emission. This effect is, however, negligible for small sources such as the one used in the miniature vacuum chamber described in the previous section. The nuclear characteristics of ^{252}Cf are summarized in Table 1. Each fission results in two heavy fragments of the original nuclide. The fission process is asymmetric and results in emission of one heavy isotope and one light isotope.

Table 1. Nuclear characteristics of ^{252}Californium

Alpha decay:	96.91% 5.97-6.12 MeV
Spontaneous fission decay:	3.09%
Effective half life:	2.639 years
Mean fission fragment masses:	106.2 amu* 142.2 amu
Mean fission fragment energies:	102.5 MeV 78.7 MeV
Alpha particle LET** in silicon:	~1.6 MeV mg-1 cm2
Fission fragment LET in silicon:	41-45 MeV mg-1cm2
Penetration depth for fission fragments in silicon:	~15 mm

37 kBq 252Cf source:

     3.6×10^4 alpha particles/s

     1.1×10^3 fission particles/s

     ~10^3 neutrons/s

*   amu = atomic mass units
** LET = Linear Energy Transfer. This figure is proportional to the charge
      generation rate.

The heavy-ion radiation sensitivity for a specific integrated circuit (IC) depends on the IC technology (NMOS, CMOS, Bipolar), device dimensions in the circuit, logic design, etc. It is generally hard to predict the heavy-ion sensitivity for an IC without testing it, mainly because the charge collection mechanisms in semiconductors are very complicated and not well understood. In a sensitive IC there are usually many sensitive volumes in which a passing heavy ion can cause SEUs. For example, the drain of an MOS transistor in the turned off state can constitute such a sensitive volume. Transistors used in modern chip designs are of approximately the same size, except for interfaces to external pins. Hence, a spread of the sensitive volumes within the IC can be expected. During operation, the sensitive volumes in a circuit will change as the transistors change between the on and off state. Therefore, an analysis of how SEUs effect the function of a circuit becomes a very complex task. Testing is thus the only practical way to assess SEU susceptibility.

The SEU susceptibility of existing ICs and fabrication technologies is not known in detail. However, much information can be obtained from results of SEU susceptibility tests of ICs intended for use in space. Results from SEU testing of more than 200 different integrated circuits fabricated by various technologies have been complied by Nichols et al. [18],[19]. The tests have been performed either by the use of ^{252}Cf or by the use of heavy ions generated in particle accelerators, (a much more expensive method for SEU testing than using ^{252}Cf). Most of the ICs tested are SSI or MSI devices, but some LSI and VLSI devices have also been tested.

For SSI and MSI circuits, clear trends in the SEU susceptibility can be identified for different IC technologies. The results show that CMOS circuits are less susceptible to SEUs than NMOS and Bipolar circuits. Many MSI and SSI circuits did not show any upsets in the tests, but in most cases these circuits used less common IC technologies such as CMOS fabricated on sapphire or epitaxial silicon. These technologies are preferred in space applications because they are known to be less sensitive to SEUs. For LSI and VLSI circuits, on the other hand, clear differences between IC technologies have not been observed. All such circuits tested were susceptible to SEUs, except for those that use special radiation-hard designs. This suggests that SEU susceptibility increases with packing density, which is also supported by other experiments. For example, VLSI circuits such as the INMOS T414 Transputer and the Motorola MC68020, both 32-bit microcomputers, have been reported to be sensitive to heavy-ion radiation from ^{252}Cf [20],[21].

Very few of the SEU tests available describe in detail the error behavior of the circuits tested. Such information could be used to estimate the spread of the sensitive volumes within the chip. More testing of circuits must be performed, both to assess sensitivity to heavy ions and to estimate the spread of the sensitive volumes.

4. Advantages and problems

In validation of systems that use ICs which are sensitive to ^{252}Cf heavy ions, and which have the sensitive volumes well spread in the chip, some of the advantages of the ^{252}Cf method are:

- Faults can be internally injected into the ICs;
- Faults are spread evenly in the chip;

- Large variation of error behavior seen on the output terminals of the IC;

- There is inherent randomness; and

- The fault injection equipment is low cost.

Injection of faults internally into the ICs becomes more important as integration level increases. It is feasible to experimentally validate a fault tolerant system consisting of SSI, MSI and LSI circuits by insertion of faults on the interconnections of the the circuits, but this would not be possible for a system using just a few VLSI circuits. Other fault injection methods that do cause internal faults have many shortcomings. Interfering with the power supply to a circuit [7],[22], may cause simultaneous faults in different parts of a circuit. Also, a small number of devices inside the IC could be more sensitive than the rest, and therefore account for the majority of the faults. Using a laser beam [23] is feasible only for circuits in which the transistors are not covered by metal layers.

An even spread of faults can be expected for circuits with the sensitive volumes well spread within the chip, provided that the flux of ions is constant over the area of the circuit being tested. For circuits without internal fault tolerance mechanisms, one can expect a large variation of the error behavior seen on the output terminals, in terms of error location, number of bits affected and error duration, if the origin of the error is only one single ion. The reason for this is that a single ion can generate both multiple and single bit errors, and that these errors are generated at various locations in the chip. Single bit errors should dominate but multiple bit errors will not be uncommon. Multiple bit errors can be generated directly or indirectly. A transient that affects a clock signal connected to many memory elements may directly cause a multiple bit error. An indirect multiple bit error can, for example, occur as a result of an arithmetic or logic operation. If there is a single bit error in the input data to the operation then the result of the operation may contain a multiple bit error. The assumptions about the error behavior being made here, are partially based on the results of the experiments presented in section 5 of this paper.

An attractive property of the ^{252}Cf method is the inherent randomness due to the fact that radioactive decay can be modeled as a Poisson process. One can then also expect the error occurrences to be Poisson distributed, even though this has to be verified by experiments separately for each circuit type.

The cost of the fault injection equipment is low for the ^{252}Cf method compared with, for example, the equipment needed for using a laser beam, which is at least one order of magnitude more expensive [23]. The cost of using a particle accelerator is even higher.

Besides the question of radiation sensitivity of different ICs, there are also other problems associated with the use of ^{252}Cf for experimental validation of dependable computing systems. We will discuss some of the problems here and propose solutions.

One major drawback to using ^{252}Cf fault injection is the problem of latch-ups, which may occur in bulk CMOS circuits when they are exposed to heavy-ion radiation. A latch-up is the transformation of the circuit to an anomalous state, in which no response to input signals can be seen. The latch-up is caused by the triggering of a parasitic structure in the device, typically an npnp or a pnpn switch. This could lead to destruction of the circuit due to excessive current flow within the chip. If the circuit is undamaged, it can be restored to its original function by temporarily turning off the power supply to the circuit. Permanent damage can be avoided if a current limiter is connected in series with the power supply line to the circuit. In this way systems containing latch-up sensitive circuits could also be evaluated by ^{252}Cf fault injection, provided that the SEU rate is higher than the latch-up rate [21].

A limitation of the ^{252}Cf method is the difficulty in controlling time and location for the injection of faults. To restrict fault injection to a specific part of a chip, a thin plastic or metal folio could be used to shield those parts of the chip in which no faults are wanted. A slot in the folio would allow the heavy ions to pass and inject faults in a specific area of the chip. The precision of the masking of the radiation depends on how close to the chip the folio could be positioned. If the chip is located in a ceramic package with the lid removed, then the package or the bonding wires may prevent the folio from being placed close to the surface of the chip. Nevertheless, it should be possible to restrict fault injection at least to different functional blocks of the integrated circuit.

If the shielding material could be applied on top of the chip before bonding and packaging, then fault injection could be restricted down to a specific transistor. The thickness required to stop the heavy ion is at most 30-50 mm, so it should be possible to use photolithographic methods and a thick layer of photoresist to produce slots in the shield small enough to expose a single transistor. The ^{252}Cf source must then be positioned close to the slot to achieve a sufficient flux of heavy ions. Otherwise, a ^{252}Cf source with much higher activity must be used.

The point in time at which faults are injected into a circuit is, of course, impossible to control exactly with the ^{252}Cf method, because there is simply no way to control the decay of a radioactive element. On a macroscopic time scale it is, however, possible to turn radiation on and off by means of a shutter mechanism inserted between the source and the irradiated chip. One could use an ordinary camera shutter which should open or close with a response time of less than 10 ms. A shutter mechanism should be very useful in validation of systems. For example, a central processing unit (CPU) can be allowed to execute program code necessary to start-up a system before faults are injected into it. The shutter could also be used to turn off radiation after the detection of an error in order to prevent further faults from being injected.

In many cases it is unnecessary to know the exact time when a fault is injected into a circuit, as long as the first manifestation of the fault on the output terminals of the circuit can be detected. This can be done by using a "golden chip", a reference circuit operated in synchrony with the irradiated circuit and given the same input signals. An error is detected by comparing the outputs from the two circuits. The "golden chip" method was used in the initial experiments described below.

5. Initial experiments

As an initial experiment to assess the usefulness of ^{252}Cf for experimental validation of dependable computing systems, the effects of heavy ion radiation from ^{252}Cf on the MC6809E 8-bit microprocessor have been investigated. The main purpose of these experiments has been to verify that this processor is sensitive to radiation from ^{252}Cf and to investigate the error behaviour. Some data about the behaviour of microprocessors exposed to radiation from heavy ions and neutrons have been reported previously [12],[13]. However, in these experiments, radiation was created by particle accelerators.

In this experiment, a small microcomputer system with two MC6809E CPUs operating in synchrony and using the same clock were used. The crystal frequency was 4 MHz which lead to a bus cycle time of 1µs. One of the CPUs was irradiated and the other was used as a reference. Errors were detected by comparison of the output signals from the two CPUs. The comparison of each signal was made during the whole time period of a bus cycle for which the signal was specified as valid according to the MC6809E bus protocol. At the same moment a mismatch was detected, the state of all pins on both CPUs was immediately clocked into an array of flip-flops. The contents of the flip-flops were used to determine in which signal, or signals, a fault was first manifested. The mismatch also triggered a logic analyzer which recorded the address, data and control signals of the irradiated CPU for 85 bus cycles. The contents of the flip-flops and the data recorded in the logic analyzer were transferred to a VAX 11/750 which was used as a host computer.

The activity of the source used was 37 kBq (1 μCi). At the time of the experiments its activity had been reduced to 18kBq due to its age. The distance from the 252Cf source to the MC6809E chip was 35mm which led to a heavy-ion flux of approximately 7 ions/s cm2 impinging onto the chip. The MC6809E chip used had an area of 0.25 cm2. Errors occurred at a rate of about one per minute.

Table 2. Location and number of bit flips

# of bits	A	D	C	A+D	A+C	D+C	A+D+C	Total	
>9	12	-	-	0	18	0	2	32	(3%)
9	2	-	-	0	1	0	0	3	(<1%)
8	3	0	-	0	14	0	1	18	(2%)
7	12	1	-	0	11	0	2	26	(3%)
6	13	0	0	0	6	0	0	19	(2%)
5	21	2	0	0	10	0	1	34	(4%)
4	18	1	0	0	4	1	1	25	(3%)
3	31	0	13	1	5	2	0	52	(6%)
2	44	4	21	0	1	0	-	70	(7%)
1	464	16	182	-	-	-	-	662	(70%)
Tot	620 (66%)	24 (3%)	216 (23%)	1 (0%)	70 (7%)	3 (0%)	7 (1%)	941*	(>99%)

A = Address bus (16 bits)
D = Data bus (8 bits)
C = Control bus (6 bits)

An error with bit flips in both address and data lines is listed in column A+D. In each column, the total number of bit flips defines the row for the entry.

* One error was excluded as its only effect was to force tri-state outputs to their high impendance state.

Table 3. Classification of errors

ERROR CLASS	No. of Errors	Est. Prob.	Std. Dev.
1. Control Flow Errors			
a) Errors that caused execution to diverge permanently from the correct program.	119	0.59	0.03
b) Errors that caused execution to diverge temporarily from the correct program.	15	0.08	0.02
c) Control flow errors that did not become effective within 85 bus cycles.	35	0.17	0.03
2. Data Errors	22	0.11	0.02
3. Control Signal Errors	70.04	0.01	
4. Other Errors	20.01	0.01	
	$\Sigma 200$		

The software executed in the MC6809E system consisted of a sorting program using the Quicksort algorithm. The program was written in C, and sorted an array of pointers to 50 data records, according to the value of a 16-bit integer variable located at the beginning of each data record. The program started by resetting the pointer array to an unsorted state before the sorting began. These two steps were executed in an infinite loop. When the recording of an error was completed, both code and data for the Quicksort program was reloaded into the MC6809E system. The iteration of the Quicksortprogram was then restarted and continued until the next error occurred.

The contents of the error flip-flops were used to determine the distribution of bit flips on the output pins during the first erroneous bus cycle. The data is based on recordings of 942 errors. Two percent of the errors caused the processor to put the address bus, the data bus and the control bus into a floating state. These errors were therefore not included in the results for these buses since the signal levels were undefined. The location of the erroneous bits on the processor bus and the number of bit flips are shown in Table 2. The majority of the errors occurred either in the address bus alone, 66%, or in the control bus alone, 23%. Only 3%

of the errors occurred in the data bus alone. Most of the errors affected only one bit, although multiple bit errors were not unusual. As much as 49% of the errors were first manifested as one bit errors in the address bus.

Error behaviour of the first 200 of the recorded errors were studied more thoroughly by manual inspection. Each of these errors was classified into one of four major groups; control flow errors, data errors, control signal errors and other errors.

Control flow errors (class 1) are defined as errors that caused the processor to diverge from the correct program, i.e. to read program words in incorrect sequence. This deviation was permanent (class 1.a) if execution did not return to the correct 'track', i.e. if the error caused what is usually referred to as a system crash. The deviation could also be temporary (class 1.b). The third subclass (class 1.c) contains errors that had not caused the program to diverge from the correct program within 85 bus cycles but were predicted to do so later on. These errors changed the stack pointer so that an incorrect return from a subroutine would be made later.

Data errors (class 2) are defined as errors that caused data to become erroneous without affecting the control flow.

Control signal errors (class 3) include errors in control signal lines that did not affect either control flow or data.

Other errors (class 4) consist of errors that resulted in output of incorrect addresses during bus cycles with the address defined as "don't care", thus these errors had no effect on execution.

The result of this classification is shown in Table 3. The majority (59%) of the errors were control flow errors that caused execution to diverge permanently from the correct program.

The SEU susceptibility of existing ICs and fabrication technologies is not known in detail. However, much information can be obtained from results of SEU susceptibility tests of ICs intended for use in space. Results from SEU testing of more than 200 different integrated circuits fabricated by various technologies have been complied by Nichols et al. [18],[19]. The tests have been performed either by the use of ^{252}Cf or by the use of heavy ions generated in particle accelerators, (a much more expensive method for SEU testing than using ^{252}Cf).

Most of the ICs tested are SSI or MSI devices, but some LSI and VLSI devices have also been tested.

For SSI and MSI circuits, clear trends in the SEU susceptibility can be identified for different IC technologies. The results show that CMOS circuits are less susceptible to SEUs than NMOS and Bipolar circuits. Many MSI and SSI circuits did not show any upsets in the tests, but in most cases these circuits used less common IC technologies such as CMOS fabricated on sapphire or epitaxial silicon. These technologies are preferred in space applications because they are known to be less sensitive to SEUs. For LSI and VLSI circuits, on the other hand, clear differences between IC technologies have not been observed. All such circuits tested were susceptible to SEUs, except for those that use special radiation-hard designs. This suggests that SEU susceptibility increases with packing density, which is also supported by other experiments. For example, VLSI circuits such as the INMOS T414 Transputer and the Motorola MC68020, both 32-bit microcomputers, have been reported to be sensitive to heavy-ion radiation from ^{252}Cf [20], [21].

6. Conclusions and indications for future work

We have shown how faults can be internally injected into integrated circuits in an inexpensive and simple way by using a miniature vacuum chamber and a ^{252}Cf radiation source.

In order for the ^{252}Cf method to work it is necessary that the irradiated devices be sensitive to heavy-ion radiation. Tests have shown that this is the case for LSI and VLSI circuits that do not have special radiation-hard designs, and for many SSI and MSI circuits. In order to be suitable for validation of dependable computing systems, the fault injection method should cause a large variety of errors in the irradiated circuit. We have shown that irradiation of the MC6809E microprocessor with heavy ions causes much variation in the error behaviour. However, more research is needed to further assess the sensitivity of heavy ions for different circuits and technologies.

More reserach is also needed to further assess the role of fault injection in validation of dependable computing systems. Different fault injection methods need to be characterized and compared. For example, it could turn out that some may be better in revealing design errors, while others may be more suited for estimation of fault detection coverage. On the basis of such characterization, fault injection procedures for certification of dependable computing systems can be established. To characterize the ^{252}Cf method, further experiments must be done to gain more experience with the method.

References

[1]   J. C. Laprie, "Dependable Computing and Fault-Tolerance: Concepts and Terminology", *Digest of Papers, 15th International Symposium on Fault-Tolerant Computing (FTCS-15)*, pp. 2-11, IEEE, Ann Arbor, MI, USA, June 1985.

[2]   V. Tasar, "Analysis of Fault Detection Coverage of a Self-Test Software Program", *Digest of Papers, 8th International Symposium on Fault-Tolerant Computing (FTCS-8)*, *pp.65-71*, IEEE, Toulouse, France, June 1978.

[3]   R. P. Kurlak and J. R. Chobot, "CPU Coverage Evaluation Using Automatic Fault Injection", *Digest of Avionics*, pp. 294-300, (AIAA), 1981.

[4]   M. E. Schmid, R. L. Trapp, A. E. Davidoff, and G. M. Masson, "Upset Exposure by Means of Abstraction Verification", *Digest of Papers, 12th International Symposium on Fault-Tolerant Computing (FTCS-12)*, pp. 237-244, IEEE, Santa Monica, CA, USA, June 1982.

[5]   Y. Crouzet and B. Decouty, "Measurement of Fault Detection Mechanisms Efficiency: Results", *Digest of Papers, 12th International Symposium on Fault-Tolerant Computing (FTCS-12)*, pp.373-376, IEEE, Santa Monica, CA, USA, June 1982.

[6]   J. H. Lala, "Fault Detection Isolation and Reconfiguration in FTMP: Methods and Experimental Results", *5th AIAA/IEEE Digital Avionics Systems Conference*, pp. 21.3.1 - 21.3.9, 1983.

[7]   A. Damm, "The Effectiveness of Software Error-Detection Mechanisms in Real-Time Operating Systems", *Digest of Papers, 16th International Symposium on Fault -Tolerant Computing (FTCS-16)*, pp. 171-176, IEEE, Vienna, Austria, 1986.

[8]   M. A. Shuette, J. P. Shen, D. P. Siewiorek, and Y. X. Zhu, *"Digest of Papers, 16th International Symposium on Fault-Tolerant Computing (FTCS-16)*, pp. 373-376, IEEE, Vienna, Austria, 1986.

[9]   J. Arlat and Y. Crouzet, "MESSALINE: A Fault Injection Tool for Dependability Validation of Fault-Tolerant Computing Systems", LAAS Report No 86356, LAAS, Toulouse, France, Dec. 1986.

[10]   R. Hummel, "Automated Fault-Injection for Digital Systems", *Proc. Annual Reliability and Maintainability Symposium*, pp. 112-117, LA, USA, Jan. 1988.

[11]   M. L. Cortes and E. J. Mc Cluskey, "An Experiment on Intermittent Failure Mechanisms", CRC Technical Report No. 87-7, Stanford University, Stanford, CA, USA, March 1987.

[12]   C. S. Guenzer, A. B. Campbell and P. Shapiro, "Single Event Upsets in NMOS Microprocessors", *IEEE Trans. Nuclear Science*, vol. NS-28, no. 6, pp. 3955-3958, Dec. 1981.

[13] J. Cusick, R. Koga, W.A. Kolasinski, and C. King, "SEU Vulnerability of the Zilog Z-80 and NSC-800 Microprocessors", *IEEE Trans. Nuclear Science*, vol. NS-32, no. 6, pp. 4189-4194, Dec. 1985.

[14] J. H. Stephen, et al., "Cosmic Ray Simulation Experiments for the Study of Single Event Upsets and Latch-up in CMOS Memories", *IEEE Trans. Nuclear Science*, vol. NS-30, no. 6, Dec. 1983.

[15] U. Gunneflo, J. Karlsson, and J. Torin, "Evaluation of Error Detection Schemes Using Fault Injection by Heavy-ion Radiation", *Digest of Papers, 19th International Symposium on Fault Tolerant Computing (FTCS-19)*, IEEE, Chicago, IL, USA, June 1989.

[16] J. Karlsson, U. Gunneflo, and J. Torin, "The Effects of Heavy-ion Induced Single Event Upsets in the MC6809E Microprocessor", *Proc. 4th International Conference on Fault-Tolerant Computing Systems*, Baden-Baden, W. Germany, Sept. 1989.

[17] G. C. Messenger and M. S. Ash, *The Effects of Radiation on Electronic Systems*, Van Nostrand Reinhold, 1986.

[18] D. K. Nichols, W. E. Price, W. A. Kolasinski, R. Koga, J. C. Pickel, J. T. Blandford jr., and A. E. Waskiewicz", Trends in Parts Susceptibility to Single Event Upset from Heavy Ions", *IEEE Trans. Nuclear Science*, vol. NS-32, no. 6, pp. 4189-4194 Dec. 1985.

[19] D. K. Nichols, L. S. Smith, W. E. Price, R. Koga, and W. A. Kolasinski, "Recent Trends in Parts Susceptibility to Single Event Upset from Heavy Ions", *IEEE Trans. Nuclear Science*, vol. NS-34, no. 6, pp. 1332-1337, Dec. 1987.

[20] J. Thomlinson, L. Adams, and R. Harboe-Sorensen, "The SEU and Total Dose Response of the INMOS Transputer", *IEEE Trans. Nuclear Science*, vol. NS-34, no. 6, pp. 1803-1807, Dec. 1987.

[21] U. Gunneflo and J. Karlsson, "Latch-up and SEU Test of Motorola MC68020 32-bit microprocessor", Technical Report No. 60, Dept. of Computer Engineering, Chalmers University of Technology, Göteborg, Sweden, 1988.

[22] M. L. Cortes, E. J. McCluskey, "Properties of Transient Errors Due to Power Supply Disturbances", *Proc. International Symposium on Circuits and Systems*, IEEE, pp. 1046-1049, May 1986.

[23] S. P. Buchner et al., "Laser Simulation of Single Event Upsets", *IEEE Trans. Nuclear Science*, vol. NS-34, no. 6, pp. 1228-1233, Dec.1987.

Modeling and Evaluation

Chair: A. Goyal (IBM T.J. Watson Research Center, Yorktown Heights, New York, USA)

A UNIFIED APPROACH FOR SPECIFYING MEASURES OF PERFORMANCE, DEPENDABILITY AND PERFORMABILITY

W. H. SANDERS
The University of Arizona
Computer Eng. Research Laboratory
Dept. of Electrical &
Computer Engineering
Tucson, AZ 85721 - USA

J. F. MEYER
The University of Michigan
Computing Research Laboratory
Dept. of Electrical Engineering &
Computer Science
Ann Arbor, MI 48109 - USA

Abstract

Methods for evaluating system performance, dependability, and performability are becoming increasingly more important, particularly in the case of critical applications. Central to the evaluation process is the definition of specific measures of system behavior that are of interest to a user. This paper presents a unified approach to the specification of measures of performance, dependability, and performability. The unification is achieved by 1) using a model class well suited for representation of all three aspects of system behavior, and 2) defining a variable class which allows for the specification of a wide range of measures of system behavior. The resulting approach permits the specification of many non-traditional as well as

This work was supported in part by Office of Naval Research under Contract n° N00014-85-K-0531.

traditional measures of system performance, dependability, and performability in a unified manner. Example instantiations of variables within this class are given and their relationships to variables used in traditional performance and dependability evaluations are illustrated.

Keywords: Performance Evaluation, Dependability Evaluation, Performability Evaluation, Reward Models, Stochastic Petri Nets.

1. Introduction

With growth in the complexity of computing systems and their applications, means of system evaluation are becoming increasingly more complex and difficult. One source of this difficulty is the dependence of what is to be evaluated, as reflected by the measures employed, on the specific nature of the system's application. This is particularly so in the case of critical applications since, here, the measures used must indeed capture what the user perceives as crucial to successful operation. Development of methods for the evaluation of system performance (see [5, 12, 14, 28], for example), dependability (see [13, 2], for example), and performability [17, 18] has thus become an activity of recognized importance. Central to this activity is the definition of specific measures of performance, dependability, and performability that are of interest to a user. Many different measures in these categories have been proposed. Typically, however, the measure definition is linked to a particular class of models for that measure, and different measures require different model classes. This paper presents a unified approach to the specification of measures of performance, dependability, and performability. The unification is achieved by 1) using a model class well suited for representation of all three aspects of system behavior, and 2) defining a variable class which allows for the specification of a wide range of measures of system behavior.

The model class used for system representation is a stochastic extension of Petri nets known as "stochastic activity networks". Stochastic activity networks (SANs) [21, 24] were developed to facilitate unified performance/dependability evaluation and have features which permit the representation of parallelism, timeliness, fault tolerance, and degradable performance [20]. Through the introduction of several new primitives (relative to Petri nets), they allow a model to be specified in a convenient way, as evidenced in applications of SANs to computer networks (e.g., [1, 16, 22]), computer systems (e.g., [23]), and automated manufacturing systems, while providing the formal structure necessary for analytic solutions [25]. When model characteristics preclude analytical evaluation, performability (as well as performance and dependability) can be evaluated via simulation.

However, before methods to do this can be developed, it is necessary to specify the range of measures (i.e. "types" of variables) that may be considered. To some extent, this range is determined by the choice of representation scheme. For example, if queueing networks are employed as the representation scheme, one is typically limited to asking questions regarding server utilizations, queue lengths, waiting times, and service times. If stochastic activity networks are used, the class is larger, due to the lower-level nature of the model primitives.

It is therefore useful to formally categorize measures of system behavior in a manner that suggests methods which may be used to obtain their solution. Previous work done regarding "reward models" [10] (and associated "reward variables") provides an instructive example in this regard. Informally, a reward model consists of a stochastic process and a "reward structure". The reward structure relates possible behaviors of the process to a specified performance variable. Typically, this is done by associating a "reward rate" with each state, the interpretation being that this rate is the rate at which reward accumulates while the process is in the state. The performance variable in this case is then taken to be the reward accumulated over some utilization interval (either finite or infinite). By associating different reward rates to states, one can construct performance variables with many different interpretations.

We take a similar approach in this paper, but develop reward structures that quantify behaviors at the stochastic activity network level, instead of the state level. This approach has several distinct advantages over the state-level approach outlined above. First, the assignment of rewards and interpretation of solutions is more natural, since it is done at the level at which the modeler thinks. Second, since rewards are assigned at the network level, they can be used in the construction procedure (i.e. the procedure by which a stochastic process or simulation program is generated from the network representation and performance variable).

The remainder of this paper is organized as follows. In the next section, the basic definitions and concepts concerning stochastic activity networks are reviewed. Traditional reward models and variables are then reviewed and a general framework for classifying reward variables based on the "type" of their reward structure is given. This framework is then used to generate particular variable types that will be considered. Variables based on a particular type of reward structure which captures information regarding activity completions and numbers of tokens in places are then investigated. Finally, example instantiations of variables of this type are given and their relationships to variables used in traditional performance and dependability evaluations are illustrated.

2. Stochastic activity networks

Stochastic activity networks (SANs) [21, 24] incorporate features of both stochastic Petri nets and queueing models. Structurally, SANs have primitives consisting of *activities, places, input gates, and output gates*. Activities ("transitions" in Petri net terminology) are of two types, *timed* and *instantaneous*. Timed activities represent activities of the modeled system whose durations impact the system's ability to perform. Instantaneous activities, on the other hand, represent system activities which, relative to the performance variable in question, complete in a negligible amount of time. Cases associated with activities permit the realization of two types of spatial uncertainty. Uncertainty about which activities are enabled in a certain state is realized by cases associated with intervening instantaneous activities. Uncertainty about the next state assumed upon completion of a timed activity is realized by cases associated with that activity. Places are as in Petri nets. Gates are introduced to permit greater flexibility in defining enabling and completion rules.

The stochastic nature of the nets is realized by associating an *activity time distribution function* with each timed activity and a *probability distribution* with each set of cases. Generally, both distributions can depend on the global marking of the network. A *reactivation function* [21] is also associated with each timed activity. This function specifies, for each marking, a set of *reactivation markings*. Informally, given that an activity is activated in a specific marking, the activity is *reactivated* whenever any marking in the set of reactivation markings is reached. This provides a mechanism for restarting activities that have been activated, either with the same or different distribution. This decision is made on a per activity basis (based on the reactivation function), and is not a net-wide execution policy.

The execution of stochastic activity networks is discussed in detail in several places, including [25]. Informally, SANs execute in time through completions of activities that result in changes in markings. More specifically, an activity is chosen to *complete* in the current marking based on the relative priority among activities (instantaneous activities have priority over timed activities) and the activity time distributions of *enabled* activities. A case of an activity chosen to complete is then selected based on the probability distribution for that set of cases. These two choices determine uniquely the next marking of the network, which is then obtained by executing the input gates connected to the input of the activity chosen and the output gates connected to the chosen case. This procedure is repeated by considering the activities enabled in the new marking.

Stochastic activity networks can be solved by both analysis and simulation, depending on system characteristics. Informally, SANs can be solved via analytic methods when all activity time distributions are exponential and activities are reactivated often enough to ensure that their rates depend only on the current state. When this is the case, stochastic processes exist that can be used to obtain analytic solutions for a wide class of variables characterizing both activity and marking related behavior. If this is not the case, simulation can be used to evaluate system behavior.

In order to be effectively applied to realistic systems, model construction and solution techniques require machine implementation. Both the complexity of the construction procedures and the typical sizes of resulting base models make this a necessity. To fill this need an extensive software package, called METASAN [1], [26], has been developed specifically for the construction and solution of SAN-based performability models.

METASAN, developed at the Industrial Technology Institute, was written using UNIX tools (C, Yacc, Lex, and Csh) and contains some 37,000 lines of source code. Models consist of two parts: a description of the structure of the net, and a description of the desired performance variables and solution method to be used in the evaluation process. Solution options include analytical techniques as well as both terminating and steady-state simulation.

3. Measure specification

As stated in the introduction, a reward model consists of a stochastic process and a reward structure. The stochastic process represents the dynamics of the system and can be constructed by hand or, automatically, from some network level description. The reward structure is typically a set of one or more functions defined on the states or transitions between states in the process. In all cases

---

[1] METASAN is a registered Trademark of the Industrial Technology Institute.

known to the authors, the interpretation given to each function is either that it is a *rate* at which reward is accumulated or that it is an *impulse* of reward that is obtained at the time of some "event" of the process. These events are typically either entrances to states, exits from states, or transitions between pairs of states. If the interpretation is of the first type we say that the reward is *rate-based*; reward functions with the second interpretation are said to be *impulse-based*. Performance variables can then be written in terms of the reward structure.

As with the reward structure itself, the manner in which this is done varies greatly in the literature. Variables can be written in terms of the state of the process at a particular time, during an interval of time, or during a time-averaged interval of time. In the first case, the variable typically represents the "status" of the modeled system at some time t and is said to be an *instant-of-time* variable. In the second case, the variable typically represents accumulated benefit derived from operating the system for some interval of time and is said to be an *interval-of-time* variable. If the reward accumulated during some interval is divided by the length of the interval, one obtains a variable which represents the (time-averaged) rate at which reward is accumulated during the interval. Variables of this type are called *time-averaged interval-of-time* variables.

An excellent early exposition of a general reward structure and variable class is given by Howard [10]. In [10], Howard postulates a reward structure on semi-Markov processes that consists of both "yield rates" and "bonuses". In the terminology introduced above, the "yield rates" specify rates at which reward is accumulated and the "bonuses" specify impulses of reward that are obtained at state changes. More precisely, *yield rates* are associated with pairs of states, the interpretation being that for a pair of states i and j, $y_{i,j}(\alpha)$ is rate at which reward is accumulated in state i α time units after i was entered when the successor state is j. Furthermore, *bonuses* are associated with state transitions, where $b_{i,j}(\tau)$ is the reward awarded upon exit from i and subsequent entry into j given that the holding time in i was τ time units. The bonuses paid at state transitions depend both on the transition made and the holding time in the state preceding the transition. The generality of this structure is difficult to fully exploit, due to the complexity of the resulting solution. The analysis required is simplified if one considers reward rates that are constant during the occupancy of each state and bonuses that do not depend on the holding time in the previously occupied state.

In this case,

$$y_{i,j}(\alpha) = y_{i,j} \text{ and } b_{i,j}(\tau) = b_{i,j}.$$

Howard then considers the solution for the expected value of an interval-of-time variable written in terms of reward structures of this type.

Further work focused on developing solution methods for reward models and did not make use of reward structure types that were as general as those considered by Howard. In particular, most researchers have limited their attention to a reward structure type with a single function that is rate-based. For acyclic systems, two general approaches have emerged. The first is a time-domain approach. Examples of work that take this approach include Meyer [19] (rate-based time-averaged interval-of-time variable, specific two-processor system), Furchtgott and Meyer [6] (rate-based interval-of-time variable, acyclic nonrecoverable [32] system), and Goyal and Tantawi [9] (rate-based interval-of-time-variable, acyclic nonrecoverable system). The second approach is to use transform techniques. For example, see Donatiello and Iyer [4] (rate-based interval-of-time variable, acyclic system), Iyer et al. [11] (rate-based interval-of-time variable, acyclic system), and Ciciani and Grassi [3] (rate-based interval-of-time variable, acyclic system).

Later work considered systems that were cyclic, as well as more general reward variables. Notable here is the work of Trivedi et *al.* [31] (rate-based instant-of-time and interval-of-time variable, cyclic and acyclic system), Smith et *al.* [29] (rate-based instant-of-time, interval-of-time, and time-averaged interval-of-time variable, cyclic and acyclic system), and de Souza e Silva and Gail [30] (rate and impulse based interval-of-time variable, cyclic and acyclic system).

While each of these efforts extended known solution techniques for reward models, they did little to extend the generality of reward structure types and hence performance variables that could be considered. Except for the work by Howard and de Souza e Silva, little use has been made of impulse rewards. In addition, the utility of these methods has been limited by having all rewards assigned at the state level. While reasonable for state spaces that are small or have a high degree of regularity, it is often difficult to assign meaningful rewards to large numbers of states. We address both these issues by 1) constructing general reward structures at the network level and 2) systematically generating variables from these reward structure types.

The variables that we consider are systematically organized according to reward structure type, category within a reward structure type, and variable type within a category. The manner in which we do this is outlined in Figure 1. As depicted in this figure, categories of variables are distinguished at the highest level by the choice of a reward structure type. By type we mean one or more classes of functions that have a particular interpretation in terms of the networks. For a given reward structure type, variables can be further distinguished by the interval of time that they depend on. Three categories of variables are distinguished at this level, as was discussed earlier in this chapter. The first category, instant-of-time variables, represents the status of the SAN at either a particular time t or in steady state, as shown in Figure 1. Interval-of-time and time-averaged interval-of-time variables will also be considered.

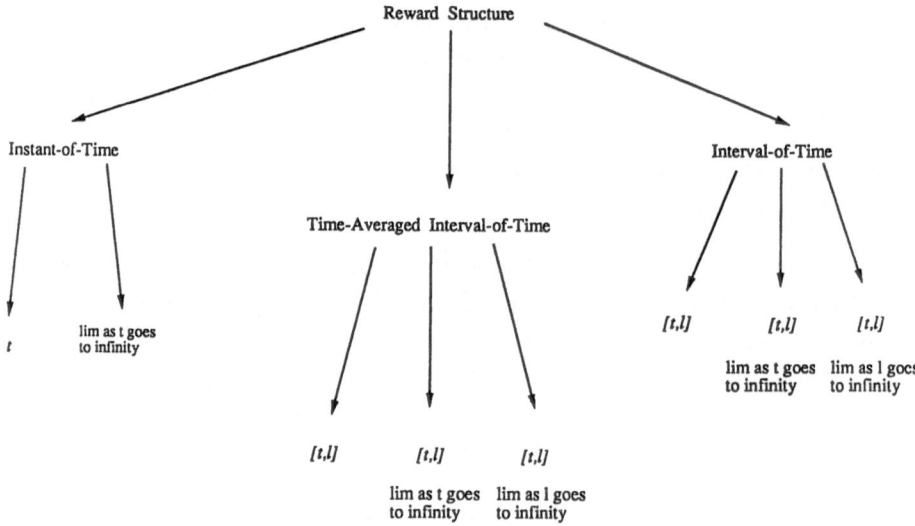

Figure 1. Types of variable considered

Within each of the other two categories, the interval-of-time variables and time-averaged interval-of-time variables, three types of variables are considered. The first type represents the total or time-averaged reward (relative to a particular reward structure) accumulated during some interval $[t, t + l]$. The second type corresponds to an interval of length l as t goes to infinity, and is useful in representing the reward that is accumulated during some interval of finite length in steady-state. The final variable type corresponds to the total or time-averaged reward accumulated during an interval starting at t and of length l as $l \to \infty$. Thus, as can be seen in Figure 1, we consider eight variable types for each reward structure type.

We now consider a reward structure type that quantifies benefits associated with activity completions and particular numbers of tokens in places. By associating impulse rewards with activity completions, as well as reward rates with particular numbers of tokens in places, we greatly extend the measures of performability that can be considered. Variables based on a reward structure of this type can be used to determine many traditional and non-traditional measures of performance, including queueing time, queue length, processor utilization, steady-state and interval availability, reliability, and productivity. In addition, if some high-level measure of "worth" is defined, this can be expressed as a particular reward structure of this type.

3.1. Structure and variable definitions

We define an "activity-marking oriented reward structure" as follows:

Definition 3.1: An *activity-marking oriented reward structure* of a stochastic activity network with places P and activities A is a pair of functions:

$C : A \to I\!R$ where for $a \in A$, $C(a)$ is the reward obtained due to completion of activity a, and

$\mathcal{R} : \mathcal{P}(P, I\!N) \to I\!R$ where for $v \in \mathcal{P}(P, I\!N)$, $\mathcal{R}(v)$ is the rate of reward obtained when for each $(p,n) \in v$, there are n tokens in place p,

where $I\!N$ is the set of natural numbers and $\mathcal{P}(P, I\!N)$ is the set of all partial functions between P and $I\!N$.

Informally, impulse rewards are associated with activity completions (via C) and rates of reward are associated with numbers of tokens in sets of places (via $\mathcal{R}$). An element $v \in \mathcal{P}(P, I\!N)$ is referred to as a *partial marking*. The marking is partial in the sense that natural numbers are assigned some subset of P, namely the domain of the partial function v. This assignment is made in a manner identical to the way a (total) marking assigns natural numbers to all the places in the set P. Although $\mathcal{R}$ has a countably infinite domain, the number of elements v that are of interest to the modeler and, hence, deserving of a non-zero reward assignment will generally be small compared to, say, the number of reachable stable markings of the SAN. Similarly, it will usually be the case that only a fraction of the SAN's activities will have non-zero rewards associated with their completions. We thus use the convention, in practice, that rewards associated with activity completions and partial markings are are taken to be zero if not explicitly assigned otherwise.

Given a SAN with a reward structure of this kind, there are a variety of ways of defining different types of performance (reward) variables, as suggested in the previous section. In particular, we consider two variable types in the instant of time category. The first of these quantifies the behavior of a stochastic activity network at a particular time t. More precisely, if we let V_t denote this variable type then

$$V_t = \sum_{v \in \mathcal{P}(P,I\!N)} \mathcal{R}(v) \cdot I_t^v + \sum_{a \in A} C(a) \cdot I_t^a,$$

where

I_t^v is an indicator random variable representing the event that the SAN is in a marking such that for each $(p,n) \in v$, there are n tokens in p at time t, an

I_t^a is an indicator random variable representing the event that activity a is the activity that completed most recently at time t.

This variable expresses the total reward (according to the reward structure defined above) associated with a SAN's status at an instant of time t. Depending on the instantiation of the reward structure, the variable can represent a variety of things including queue length and component status (e.g. idle, busy, blocked, failed, functioning). In view of our above observations concerning typical reward structures, and since zero values of $\mathcal{R}(v)$ can be ignored in the summation, the number of elements which must be accounted for in this sum is again relatively small.

Depending on the nature of the stochastic activity network in question, I_t^v and I_t^a may converge in distribution for all v and a with non-zero rewards as t approaches ∞. When this happens, the "steady-state" reward obtained at an instant of time can be studied. If we denote the random variable with this steady-state distribution as $V_{t \to \infty}$, its value can be expressed as

$$V_{t \to \infty} = \sum_{v \in \mathcal{P}(P,I\!N)} \mathcal{R}(v) \cdot I_{t \to \infty}^v + \sum_{a \in A} C(a) \cdot I_{t \to \infty}^a,$$

where

$I^v_{t \to \infty}$ is an indicator random variable representing the event that the SAN is in a marking such that for each $(p,n) \in v$, there are n tokens in p in steady-state, and

$I^a_{t \to \infty}$ is an indicator random variable representing the event that activity a is the activity that completed most recently in steady-state.

Variables of the interval and time-averaged-interval categories can also be considered. In these cases, the reward accumulated is related both to the number of times each activity completes and time spent in particular markings during an interval. As was discussed in the previous section, we consider three variable types in each of these categories corresponding to an interval of length l starting at time t $[t,t + l])$, an interval of length l as $t \to \infty$ $([t,t + l], l \to \infty$, and an interval starting at t as $l \to \infty$ $([t,t + l], l \to \infty)$. In the following, variable types of the interval category are denoted by "Y" while variables types of the time-averaged category are denoted by "W", each with the appropriate subscript. In particular, let :

$$Y_{[t,t+l]} = \sum_{v \in \mathcal{P}(P,I\!N)} \mathcal{R}(v) \cdot J^v_{[t,t+l]} + \sum_{a \in A} C(a) \cdot N^a_{[t,t+l]}, \text{ and}$$

$$W_{[t,t+l]} = \frac{Y_{[t,t+l]}}{l}$$

where

$J^v_{[t,t+l]}$ is a random variable representing the total time that the SAN is in a marking such that for each $(p,n) \in v$, there are n tokens in p during $[t,t+l]$, and

$N^a_{[t,t+l]}$ is a random variable representing the number of completions of activity a during $[t,t+l]$.

If $J^v_{[t,t+l]}$ and $N^a_{[t,t+l]}$ converge in distribution as $t \to \infty$ for all v and a that have non-zero reward assignments, the time-averaged reward accumulated and total reward accumulated during some interval of length l in steady-state can be studied. If we denote the random variables with these steady-state distribution as $Y_{[t,t+l],t} \to \infty$ and $W_{[t,t+l],t} \to \infty$ then

$$Y_{[t,t+l],t \to \infty} = \sum_{v \in \mathcal{P}(P,\mathbb{N})} R(v) \cdot J^{v}_{[t,t+l],t \to \infty} + \sum_{a \in A} C(a) \cdot N^{a}_{[t,t+l],t \to \infty},$$

and

$$W_{[t,t+l],t \to \infty} = \frac{Y_{[t,t+l],t \to \infty}}{l}$$

where

$J^{v}_{[t,t+l],t \to \infty}$ is a random variable representing the total time that the SAN is in a marking such that for each $(p,n) \in v$, there are n tokens in p during a interval of length l in steady-state, and

$N^{a}_{[t,t+l],t \to \infty}$ is a random variable representing the number of completions of activity a during an interval of length l in steady-state.

Similarly, if $J^{v}_{[t,t+l]}$ and $N^{a}_{[t,t+l]}$ converge in distribution as $l \to \infty$ for all v and a that have non-zero reward assignments, the total reward and time-averaged reward accumulated during an infinite interval starting at time t can be expressed as

$$Y_{[t,t+l],l \to \infty} = \sum_{v \in \mathcal{P}(P,\mathbb{N})} R(v) \cdot J^{v}_{[t,t+l],l \to \infty} + \sum_{a \in A} C(a) \cdot N^{a}_{[t,t+l],l \to \infty},$$

$$W_{[t,t+l],l \to \infty} = \lim_{l \to \infty} \frac{Y_{[t,t+l]}}{l}$$

where

$J^{v}_{[t,t+l],l \to \infty}$ is a random variable representing the total time that the SAN is in a marking such that for each $(p,n) \in v$, there are n tokens in p during $[t,\infty)$, and

$N^{a}_{[t,t+l],l \to \infty}$ is a random variable representing the number of completions of activity a during $[t,\infty)$.

4. Example variable instantiations

Traditional measures of dependability and performance as well as more general performability measures can be specified easily using the performance variables just discussed and particular instances of the activity-marking oriented reward structure. To illustrate this, we consider a simple multiprocessor system where all processors service tasks from a single degradable buffer. The normal, fault-free operation of the system is as follows. Tasks arrive as a Poisson process with rate α. If the buffer is full, they are rejected. If not, they are placed in the buffer to be served by the first available processor in a FIFO manner. In addition, processing times are independent and exponentially distributed with each processor having a processing rate β.

Faults can occur both due to a failure of a buffer stage and due to a failure of a processor. In each case, the fault may be covered (i.e. the system degrades successfully to a less productive structure state) or it may result in a total loss of processing capability (i.e. total system failure). Additionally, certain processor failures are repairable. Repairs are performed on one processor at a time, with an exponentially distributed repair time with rate ζ. We assume further that faults in both a buffer stage and a processor occur as Poisson processes with rates λ and γ, respectively.

A stochastic activity network representing changes in the structure of the multiprocessor due to faults is given in Figure 2. Since our intent is to illustrate the specification of traditional dependability and performance variables, the model is kept simple. System resources (i.e. processors and buffers) are represented by tokens in places. Place A represents the number of processors queued for repair, place B represents the number of fault-free processors, and place C represents the number of fault-free buffer stages. Activities *processor_failure* and *buffer_failure* represent the occurrence of faults in the processors and buffer stages, respectively. Three types of processor faults are possible, corresponding to the three cases associated with activity *processor_failure*. Case 1 represents the occurrence of a fault that is repairable. Case 2 represents a total system failure, and case 3 represents the occurrence of a non-repairable fault. Cases for *buffer_failure* are similar, except that buffer stages may not be repaired. Here case 1 represents the occurrence of a non-repairable fault and case 2 represents total system failure. Processor repairs are represented by activity *processor_repair*.

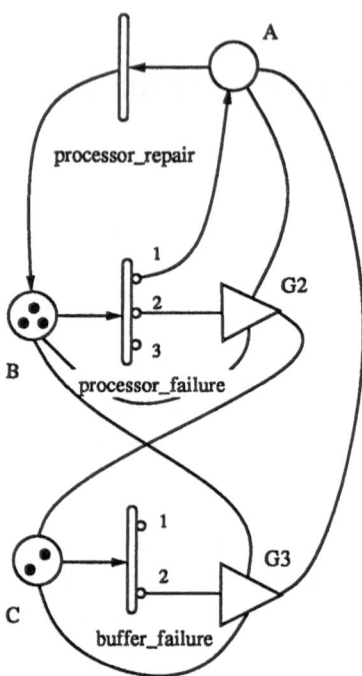

Gate	Enabling Predicate	Function
G2	-	MARK(A)=MARK(B)=MARK(C)=0;
G3	-	MARK(A)=MARK(B)=MARK(C)=0;

Activity	Rate	Probability		
		case 1	case 2	case 3
processor_failure	$\gamma * MARK(B)$	cp1	cp2	cp3
buffer_failure	$\lambda * MARK(C)$	cb1	cb2	-
processor_repair	ζ	-	-	-

Figure 2. Multiprocessor fault model

Before dependability measures can be formulated, a definition of "system failure" must be given. In this regard, we say that the system has failed if all processors have failed in a manner such that they cannot be repaired. Then, if we define a reward structure such that

$$C(a) = 0, \forall \, a \in A$$

$$\mathcal{R}(v) = \begin{cases} 0 & \text{if } v = \{(A,0),(B,0)\} \\ 1 & \text{otherwise,} \end{cases}$$

$E[V_t]$ is the reliability (using the above definition of system failure) at time t.

Measures of availability [7, 8] can be represented just as easily using this reward structure type and an associated variable. If we consider the system to be available whenever there is at least one processor functioning, the reward structure

$$C(a) = 0, \forall\, a \in A$$

$$\mathcal{R}(v) = \begin{cases} 0 & \text{if } v = \{(B,0)\} \\ 1 & \text{otherwise,} \end{cases}$$

can be used to specify availability. For example, using this reward structure, the steady-state availability of the example multiprocessor system is $E[V_{t \to \infty}]$. The interval availability (i.e., the fraction of time the system is available during some interval of length l starting at time t) is $E[W_{[t,l]}]$ using the same reward structure. The distribution of availability, $F(t,l,x)$, is the probability that $W_{[t,l]} \le x$.

Performance-oriented measures can be specified using a stochastic activity network model that represents task arrivals and completions. A stochastic activity network model of a multiprocessor with N processors and M buffers is given in Figure 3. In this figure, each completion of activity *arrival* represents the arrival of a task to the buffer. The buffer is represented by place E. The marking of place E represents the number of tasks queued for service. Place F represents the status of each of the processors, where the number of tokens in F is the number of processors that are busy. The finiteness of the buffer is represented by gate $G1$. Gate $G1$ specifies (via its predicate) that activity *arrival* is enabled only when the number of tasks in the system is less than system capacity (i.e. the sum of the markings of E and F is less than the sum of the number of working processors and buffers).

The service of tasks is represented by activity *service*. Use of a marking dependent activity completion rate for *service* allows us to represent all processors via a single activity, due to the memoryless property of the exponential distribution. The rate for activity *service* is therefore the number of busy processors multiplied by the rate of a single processor. The number of busy processors is represented by place F.

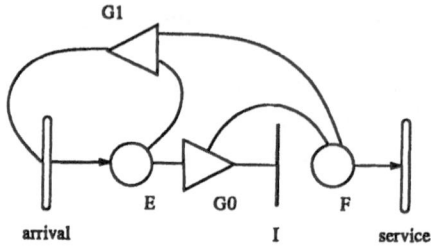

Gate	Enabling Predicate	Function
G0	MARK(F) < N and MARK(E) > 0	MARK(E) = MARK(E) - 1; MARK(F) = MARK(F) + 1;
G1	MARK(E)+MARK(F) < M+N	identity

Activity	Rate
arrival	α
service	$\beta * MARK(F)$

Figure 3. Multiprocessor performance model

If we define the throughput of the system during some interval $[t,t+l]$ as the number of tasks that are processed during the interval divided by the length of the interval, the throughput of the example system can be represented using a reward structure consisting only of impulse rewards. Specifically, consider the reward structure

$$C(a) = \begin{cases} 1 & \text{if } a = service \\ 0 & \text{otherwise} \end{cases}$$

$$\mathcal{R}(v) = 0, \ \forall \ v \in \mathcal{P}(P, \mathbb{N}).$$

Using this reward structure, the throughput is represented by the variable $W_{[t,t+l]}$. Steady-state throughput is given by the limit of this variable when $t=0$ and $l \rightarrow \infty$, i.e. $W_{[0,l]}$, $l \rightarrow \infty$.

An alternate representation of expected steady-state throughput can be formulated based on the arrival rate to the system and the probability that an incoming task is processed. Since tasks arrive as a Poisson process, the probability that an incoming task is processed is one minus the probability the buffer is full. This probability can be captured by the reward structure,

$$C(a) = 0, \forall\, a \in A$$

$$R(v) = \begin{cases} 0 & \text{if } v = \{(E,M)\} \\ 0 & \text{otherwise,} \end{cases}$$

and variable $E[V_{t\to\infty}]$. The expected steady-state throughput can then be written as

$$(1 - E[V_{t\to\infty}]) \times \alpha$$

where α is the rate of arrival of tasks to the system.

A representation of expected steady-state response time can be obtained using Little's result [15] and the expected number of tasks in the system in steady-state. The expected number of tasks in the system can be represented using the reward structure

$$C(a) = 0, \forall\, a \in A$$

$$R(v) = \begin{cases} i+j & \text{if } v = \{(E,i),(F,j)\} \\ 0 & \text{otherwise,} \end{cases}$$

if the variable is taken to be $E[V_{t\to\infty}]$. The expected steady-state response time is then $E[V_{t\to\infty}]$ divided by the rate at which tasks enter the system, i.e.

$$\frac{E[V_{t\to\infty}]}{\alpha}.$$

Processor utilizations can be obtained in a similar manner. Specifically, if the average processor utilization (in steady-state) is defined to be the fraction of the total number of processors that are busy, the reward structure

$$C(a) = 0, \forall\, a \in A$$

$$R(v) = \begin{cases} i & \text{if } v = \{(F,i)\} \\ 0 & \text{otherwise,} \end{cases}$$

can be used. Processor utilization is then $\dfrac{V_{t\to\infty}}{N}$, where N is the number of processors in the system. As can be seen by the previous examples, traditional performance related and dependability related variables can be easily represented in the reward framework presented. Other performance and dependability related variables can be constructed in a similar manner. In addition, performability measures can be formulated as variables within this type if performance and fault type activities are represented in a single SAN model.

To illustrate the applicability of this method to the specification of performability variables, we consider the performability evaluation of the multiprocessor just used to illustrate traditional performance and dependability variables. The reward structure type and variables developed, together with the decomposition technique of [21, 25], allows us to consider "bottom-line" performability measures that summarize aspects of system performance caused by both fault and workload environments. To see this, we characterize the "total benefit" derived from operating the system for some interval $[t, t + l]$. We assume that "benefit" is derived from the completion of tasks and that costs are associated with the repair of processors. To make the discussion more concrete, we attach a benefit of x dollars to each task completion and a cost of y dollars to each processor repair.

Regarding solution, we construct a performability model which differentiates between "performance" and "structure" related submodels [21, 25]. These submodels are just the two SANs considered in the previous examples linked by two common places. A stochastic activity network representing the multiprocessor is given in Figure 4. Places B and C are the common places. Since task completions are represented in the performance submodel, the rate of task completions (i.e. throughput) in each structure state serves as the basis for the determination of the rate component of the reward structure. Specifically, the rate of benefit derived for a structure state is the throughput in that state multiplied by the dollar benefit associated with each task completion. Clearly, the throughput is just the arrival rate of tasks to the system multiplied by the probability that a task which arrives will be processed. In terms of the SAN model of the system, an incoming task will be rejected if the sum of the number of tokens in places E and F is equal to the sum of the number of tokens in places B and C (i.e., the system is full). Since tasks arrive as a Poisson process, the probability that an incoming task is processed is one minus the probability that the system is full. This fact allows us to define a reward structure for the performance submodel that permits the determination of system throughput for each structural configuration of the system. In this case, different structural configurations are distinguished by the number of functioning buffers and processors. Specifically, when the number of functioning buffers is m, the expected throughput can be obtained using a reward structure where

$$C(a) = 0, \forall\, a \in A$$

$$\mathcal{R}(v) = \begin{cases} 1 & \text{if } v = \{(E,m)\} \\ 0 & \text{otherwise,} \end{cases}$$

and taking the variable to be

$$Thru(m,n) = \alpha \times (1 - E_{(m,n)}[V_{t \to \infty}]),$$

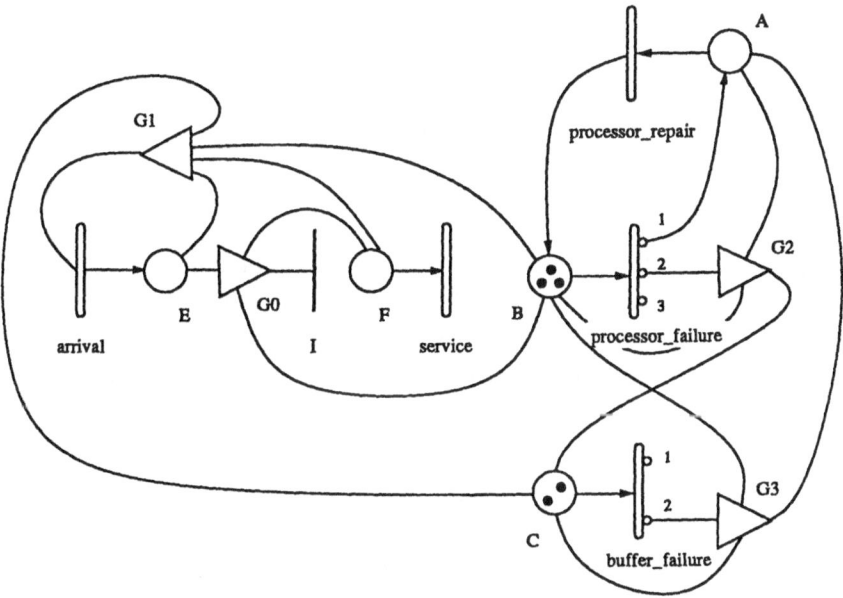

Gate	Enabling Predicate	Function
G0	MARK(F) < MARK(B) and MARK(E) > 0	MARK(E)=MARK(E)-1; MARK(F)=MARK(F)+1;
G1	MARK(E)+MARK(F) < MARK(B)+MARK(C)	identity
G2	-	MARK(A)=MARK(B)=0; MARK(C)=0;
G3	-	MARK(A)=MARK(B); MARK(C)=0;

Activity	Rate	Probability		
		case 1	case 2	case 3
processor_failure	$\gamma * MARK(B)$	cp1	cp2	cp3
buffer_failure	$\lambda * MARK(C)$	cb1	cb2	-
processor_repair	ζ	-	-	-
arrival	α	-	-	-
service	$\beta * MARK(F)$	-	-	-

Figure 4. Degradable multiprocessor model

where α is the rate of arrival of tasks to the system and $E_{(m,n)}$ is the expected value of the given variable when in there are m functioning buffers and n functioning processors. Costs associated with processor repairs are represented in the reward structure by associating a reward of $-y$ with each completion of activity *processor_repair*. Under these assumptions, the expected total benefit associated with operating the system for some utilization period $[0,t]$ can be found using the reward structure

$$C(a) = \begin{cases} -y & \text{if } a = processor\_repair \\ 0 & \text{otherwise,} \end{cases}$$

$$R(v) = \begin{cases} x \cdot Thru(m,n) & \text{if } v = \{(B,n),(C,m)\} \\ 0 & \text{otherwise,} \end{cases}$$

and variable $E[Y_{[0,t]}]$.

Explicit values for this reward structure can now be obtained by solving the stochastic process associated with the performance submodel to obtain the throughput for each structural configuration. The results are not given here, since the intent was to illustrate the measure definition process, but can found in [25].

5. Conclusions

As stated in the introduction, the objective of this work was to develop a unified method for specifying measures of performance, dependability, and performability. The framework proposed accomplishes this, we believe, and is particularly useful for critical systems where the measures used must capture what the user perceives as crucial to successful operation. In particular, the flexibility offered by first classifying measures based on their reward structure type and then based on categories within a particular type allows a user to specify precisely what is important in the evaluation process. The class of variables generated by the activity-marking oriented reward structure defined in Section 3 is rich in this regard, and permits specification of both traditional and non-traditional measures, as illustrated by the example variable instantiations given in Section 4.

The framework is also flexible in the sense that it allows room for definition of other reward structure types. In particular, while a large class of variables can be generated from instantiations of the activity-marking oriented reward structure type presented in this paper, these variables do not subsume all the variables that may be of interest to a user. Current work is directed toward defining additional reward structure types that quantify additional aspects of system behavior that may be of interest to a user.

Solution for the defined variables is also important issue, but beyond the scope of this paper. Depending on both the nature of the variable and model, the solution may be achieved by either simulation or analysis. We have investigated both of these solution approaches [26, 27], but have not yet implemented the methods in a software tool that allows one to directly specify variables using this framework. An effort to do this is currently underway, however, and should result in such a tool in the near future. This will allow us test these measure specification methods on larger and more realistic systems.

References

[1]   B.E. Aupperle, J.F. Meyer and L. Wei, "Evaluation of Fault-Tolerant Systems with Nonhomogeneous Workloads", in *Proc. 19th International Symp. on Fault-Tolerant Computing*, Chicago, IL, 1989.

[2]   A. Avizienis and J. C. Laprie, "Dependable computing:From concepts to design diversity", *Proc. of the IEEE*, vol. 74, no. 5, pp. 629-638, May 1986.

[3]   B. Ciciani and V. Grassi, "Performability evaluation of fault-tolerant satellite systems", *IEEE Trans. on Communications*, vol. COM-35, no. 4, pp. 403-409, April 1987.

[4]   L. Donatiello and B.R. Iyer, "Analysis of a composite performance reliability measure for fault-tolerant systems", *JACM*, vol. 34, no. 1, pp. 179--199, January 1987.

[5]   D. Ferrari, *Computer Systems Performance Evaluation*, Prentice-Hall, Englewood Cliffs, NJ, 1978.

[6]   D.G. Furchtgott and J.F. Meyer, "A performability solution method for degradable, nonrepairable systems", *IEEE Trans. on Computers*, vol. C-33, June 1984.

[7]   A. Goyal and S.S. Lavenberg, "Modeling and analysis of computer system availability", *IBM Journal of Research and Development*, vol. 31, no. 6, pp. 651-664, November 1987.

[8]   A. Goyal, S.S. Lavenberg, and K.S. Trivedi, "Probabilistic modeling of computer system availability", *Annals of Operations Research*, vol. 8, pp. 285-306, 1987.

[9]   A. Goyal and A.N. Tantawi, "Evaluation of performability for degradable computer systems", *IEEE Trans. on Computers*, vol. C-36, no. 6, pp. 738-744, June 1987.

[10] R.A. Howard, *Dynamic Probabilistic Systems, Vol II: Semi-Markov and Decision Processes*, New York: Wiley, 1971.

[11] B.R. Iyer, L. Donatiello, and P. Heidelberger, "Analysis of performability for stochastic models of fault-tolerant systems", *IEEE Trans. on Computers*, vol. C-35, no. 10, pp. 902-907, October 1986.

[12] H. Kobayashi, *Modeling and Analysis: An Introduction to System Performance Evaluation Methodology*, Addison-Wesley, Reading, MA, 1978.

[13] J.Laprie, "Dependable computing and fault tolerance: Concepts and terminology", in *Proc. 15th International Symp. on Fault-Tolerant Computing*, pp. 2-11, Ann Arbor, MI, June 1985.

[14] S.S. Lavenberg, *Computer Performance Modeling Handbook*, Academic Press, New York, NY, 1983.

[15] J. D. C. Little, "A Proof of the Queueing Formula $L = \lambda W$", *Operations Research*, vol. 9, pp. 383-387, 1961.

[16] R. Martinez, W.H. Sanders, Y. Alsafadi, J. Nam, T. Ozeki and K. Komatsu, "Performance evaluation of a picture archiving and communication system using stochastic activity networks", *in Proc. SPIE Medical Imaging IV*, Newport Beach, February 1990.

[17] J.F. Meyer, "On evaluating the performability of degradable computing systems", in *Proc. 1978 Int. Symp. on Fault-Tolerant Computing*, Toulouse, France, June 1978, pp. 44-49.

[18] J.F. Meyer, "On evaluating the performability of degradable computing systems", IEEE Trans. Comput., vol. C-22, pp. 720-731, Aug. 1980.

[19] J.F. Meyer, "Closed-form solutions of performability", *IEEE Trans. on Computers*, vol. C-31, pp. 648--657, July 1982.

[20] J.F. Meyer, "Performability modeling of distributed real-time systems", in *Mathematical Computer Performance and Reliability*, Amsterdam: North-Holland, 1984.

[21] J.F. Meyer, A. Movaghar, and W.H. Sanders, "Stochastic activity networks: structure, behavior, and application", in *Proc. International Workshop on Timed Petri Nets*, Torino, Italy, July 1985, pp. 106-115.

[22] J.F Meyer, K.H. Muraldihar and W.H. Sanders, "Performability of a token bus network under transient fault conditions", in *Proc. 19th Int. Symp. on Fault-tolerant computing*, Chicago, June 1989.

[23] J.F. Meyer and L. Wei, "Influence of workload on error recovery in random access memeories", in *IEEE Transactions on Computers*, Vol. 37, No. 4, April 1988.

[24] A. Movaghar and J.F. Meyer, "Performability modeling with stochastic activity networks", in *Proc. 1984 Real-Time Systems Symp.*, Austin, TX, Dec. 1984.

[25] W.H. Sanders, "Construction and solution of performability models based on stochastic activity networks", Computing Research Laboratory Technical Report CRL-TR-9-88, The University of Michigan, Ann Arbor, MI, August 1988.

[26] W.H. Sanders and J.F. Meyer, "METASAN: A performability evaluation tool based on stochastic activity networks", in *Proc. ACM-IEEE Comp. Soc. 1986 Fall Joint Comp. Conf.*, Dallas, TX, Nov. 1986.

[27]   W.H. Sanders and J.F. Meyer, "Reduced Base Model Construction Methods for Stochastic Activity Networks", in *Proc. Third International Workshop on Petri Nets and Performance Models*, Kyoto, Japan, Dec. 11-13, 1989.

[28]   C.H. Sauer and K.M. Chandy, *Computer Systems Performance Modeling*, Prentice-Hall, Englewood Cliffs, NJ, 1981.

[29]   R. M. Smith, K. S. Trivedi, and A. V. Ramesh, "Performability analysis: Measures, an algorithm, and a case study", *IEEE Trans. on Computers*, vol. C-37, no. 4, pp. 406-417, April 1988.

[30]   E. de Souza e Silva and H. R. Gail "Calculating availability and performability measures of repairable computer systems using randomization", *JACM*, vol. 36, no. 1, pp. 171-193, January 1989.

[31]   K. Trivedi, A. Reibman, and R. Smith, "Transient analysis of Markov and Markov reward models", in *Computer Performance and Reliability*, ed. G. Iazeolla, P.J. Courtois, and O.J. Boxma, North Holland, 1988.

[32]   L.T. Wu, "Operational models for the evaluation of degradable computing systems", in *Proc. ACM/SIGMETRICS Conf. on Measurement and Modeling of Computer Systems*, pp. 179-185, Seattle, WA, August 1982.

SUCCESSIVE OPERATIONAL PERIODS AS MEASURES OF DEPENDABILITY

Gerardo RUBINO - *Bruno* SERICOLA
I R I S A
Campus de Beaulieu, 35042 Rennes Cedex - France

Abstract

We consider fault-tolerant computing systems, that is, systems which are able to recover an operational state after a fault. We propose new measures of dependability to quantify the behaviour of such a system all along its lifetime. With respect to classical measures (point availability, reliability) we consider the successive periods during which the system is in operation. Under markovian assumptions, we give closed-form expressions of the distribution and the moments of these operational periods. These measures give more insight on the evolution of the system than classical ones. Their utilization is illustrated by means of a numerical example.

1. Introduction

A fault-tolerant computing system is characterized by its capability to restore automatically an operational state after a fault in its hardware or its software. Its evolution in time can be viewed as an alternance of operational states (the system delivers a service according to the specifications) and nonoperational states (reached after a failure) in which it is unable to meet the specifications. Let us say that the system is *up* in the first case and *down* in the second one. In a *down* state, it tries to put itself back in operation, eventually with a loss in performance. Real devices have a finite lifetime so, after a sequence of *up* and *down* states, a last (and fatal) failure occurs and the system will be no more able to restore an operational state.

The design or the analysis of such a system involves the utilization of a set of techniques corresponding to different aspects of the *dependability* concept [1], [2], [3] In this paper we are interested in the quantification of the properties associated with the fault tolerance capability. The most widely used mathematical tools to perform quantitative evaluations of these systems are Markov (or semi-Markov) models. The asymptotic analysis usually done, for instance in the performance evaluation area, is not adapted to the information which is needed here. The modeler will have to carry out a transient analysis rather than a stationary one.

Let us consider a fault-tolerant computing system modeled by a finite state Markov process X. Suppose that the user just wants to distinguish between three types of states. The first class (denoted here by B) contains the operational states in which the system delivers the specified service in a satisfactory way even if, eventually, its performance has been degradated from the beginning of its lifetime. In the second class (denoted here by B') the system is *down* as a consequence of some failure, that is, it performs no useful work (from the user point of view), but it tries to come back to the operational class B. The third class is composed by the states in which the system is *completly down* or *dead*. In such a state, it can neither perform nor come back to the classes B or B'. Since we are interested in the behaviour of the process on the first two classes B and B', we can consider that the third one is reduced to a single (absorbing) state denoted by a. Every other state of X is then transient.

Classically, the two basic measures to quantify the fact that a system performs correctly are the *reliability at time t*, defined by

$$R(t) \stackrel{\text{def}}{=} P(X_s \in B, \text{ for all } s \in (0,t))$$

and the *point availability at time t*,

$$PAV(t) \stackrel{\text{def}}{=} P(X_t \in B).$$

In the case of a fault-tolerant computing system, these metrics are not very efficient. The reliability $R(t)$ concerns just the "beginning" of the lifetime and the point availability is a local metric taking account of the system's behaviour at only a fixed instant.

A second set of measures which quantify the behaviour of the system over a fixed interval $(0,t)$ has been recently studied [4]. Let us denote by $U(t)$ the function

$$U(t) \stackrel{\text{def}}{=} \begin{cases} 1 \text{ if } X_t \in B \\ 0 \text{ if } X_t \notin B \end{cases}$$

The *cumulative operational time up to t* is defined by

$$O(t) \stackrel{\text{def}}{=} \int_0^t U(s) \, ds \, .$$

In particular, it is of interest the *total opeational time*

$$O(\infty) \stackrel{\text{def}}{=} \int_0^{+\infty} U(s) \, ds$$

which is finite (with probability 1) since every state is transient excepting state a. From this, the *availability up to time t* is defined by

$$AV(t) \stackrel{\text{def}}{=} \frac{O(t)}{t} \, .$$

The computation of this global measure ($AV(t)$ or $O(t)$) which takes into account the behaviour of X over the whole interval from 0 to t is rather expensive. In particular, there are no closed-form expressions available [4]. The user can limit himself to compute expectations. The expectation of $AV(t)$ is called *average availability up to time t*. We can write it in the following way.

$$E(AV(t)) \quad = \frac{1}{t} \int_0^t E(U(s)) \, ds$$

$$= \frac{1}{t} \int_0^t PAV(s) \, ds \, ,$$

that is, $E(AV(t))$ is the mean of the function PAV over the interval $(0,t)$.

The aim of this work is to propose another way to analyze the behaviour of such a system from the beginning of its lifetime. The main difference from the previous point of view is that in this approach, we will fix the number of *operational periods* instead of a fixed interval $(0,t)$. An *operational period* is defined as the time spent by the system in operation, that is, the delay between a recovery from a failure and the next failure (fatal or not). When the initial state is an operational one, the first *operational period* is equal to the time up to the first failure. Let us denote by $S_{B,n}$ the duration of the nth *operational period*, i.e. the random variable "nth sojourn time of X in B" (see the next section for some details about this definition). The distributions of these measures can be used by a designer to tune up some input parameters in order to obtain, for instance, the N first operational periods "large enough", that is, to get the probability $P(S_{B,n} > \tau)$ greater than $1 - \delta$ for $n = 1,2,...,N$, given the duration τ and the tolerance parameter δ.

Observe that $P(S_{B,1} > t \,/\, X_0 \in B) = R(t)$. We have also that

$$O\,(\infty) = \sum_{n \geq 1} S_{B,n}\,.$$

The consideration of *sojourn times* on subsets of the state space leads to many other interesting metrics. For instance, assume that failures causing transitions from B to B' are differentiated into two types, software and hardware failures. The states in B' can then be decomposed into two subsets, B'_s and B'_h, where B'_s is the set of *down* states reached after a software failure and ..B'h.. is the set of *down* states reached after a hardware one. Assume that when the system is down in the class B'_s, a hardware failure can put it in the class B'_h. Suppose that the user is interested in the duration of repairs in this last case, that is, the time necessary to repair itself from a hardware failure occuring while the system was trying to recover from a software one. This measure corresponds to a sojourn in B'_h with the supplementary condition that the process enters this set from a state of B'_s.

In this work, we limit the discussion to the successive operational times, that is, the sequence$(S_{B,n})$. We give closed-form expressions for the distribution of each $S_{B,n}$ and its moments. A related random variable is also studied, the number of visits of the process to the selected set B which is denoted by N_B.

In [5, Chapter 2], the authors consider a two-state system with one operational state and the other nonoperational (without absorbing state) and they give the distribution of the random variable $O(t)$, using the successive operational periods. They also show that the random variable $O(t)$ is, in this case, asymptotically normal.

Section 2 contains the derivation of the closed-form expressions and Section 3 is devoted to the application of the given results by means of an illustrative example. In the last section we propose some conclusions.

2. Model description — Results

We assume the system is modeled as a homogeneous Markov process $X = \{ X_t, t \geq 0 \}$. The finite state space is $E = \{1, ..., N, a\}$ where $1, ...,N$ are transient and a is absorbing. The partition $(B,B',\{a\})$ of E is given and, for simplicity, we assume that $B = \{1,2,...,L\}, 1 \leq L < N$. The process X is described by its transition rate matrix A (infinitesimal generator) and by its initial probability distribution α. For $i \in E$, we define $\lambda(i) \stackrel{\text{def}}{=} -A(i,i) \stackrel{\text{def}}{=} \sum_{j\neq i} A(i,j)$. The non negative real number $\lambda(i)$ is the output rate from state i. We denote by Λ the diagonal matrix whose ith element is $\lambda(i)$ and by P the transition probability matrix of the embedded Markov chain at the instants of state change of the process X, that is, $P = I + \Lambda^{-1} A$, where I denotes the identity matrix.

The partition $(B,B',\{a\})$ of E induces over the matrices A, Λ, P and over the initial probability vector α the following decomposition into submatrices and subvectors.

$$A = \begin{pmatrix} A_B & A_{BB'} & A_{Ba} \\ A_{B'B} & A_{B'} & A_{B'a} \\ 0 & 0 & 0 \end{pmatrix}$$

$$\Lambda = \begin{pmatrix} \Lambda_B & 0 & 0 \\ 0 & \Lambda_{B'} & 0 \\ 0 & 0 & 0 \end{pmatrix}$$

$$P = \begin{pmatrix} P_B & P_{BB'} & P_{Ba} \\ P_{B'B} & P_{B'} & P_{B'a} \\ 0 & 0 & 1 \end{pmatrix}$$

$$\alpha = (\alpha_B , \alpha_{B'} , \alpha_a).$$

Let t_i denote the instant of the ith transition of X with $t_0 = 0$. A *sojourn* or *visit* of X to B is a sequence of the form $(X_{t_m}, X_{t_{m+1}}, ... , X_{t_{m+k}})$ where $(X_{t_m}, X_{t_{m+1}}, ... , X_{t_{m+k-1}} \in B, X_{t_{m+k}} \notin B$ and if $m > 0$ then $X_{t_{m-1}} \notin B$. This *sojourn* begins at time t_m, finishes at time t_{m+k} and lasts $t_{m+k} - t_m$ units of time.

Let us define the random variable $S_{B,n}$ with values in $\mathbb{R}^+$, for $n \geq 1$, in the following way.

$S_{Bn} \overset{\text{def}}{=}$ **if** X visits B at least n times **then** duration of the nth visit of X to B **else** 0.

In the same way, we define, for $n \geq 1$, the random variable "state in which the nth visit of X to B begins" taking values in the set $B \cup \{ \Psi \}$ where the symbol Ψ means that there is no visit to B, that is

$V_n \overset{\text{def}}{=}$ **if** X visits B at least n times **then** state in which the nth visit of X to B begins **else** Ψ.

Define the row vector v_n with L entries by $v_n(i) \overset{\text{def}}{=} P(V_n = i)$, $i \in B$, $n \geq 1$. See that the distribution of V_n is given by the vector, of length $L + 1$, $(v_n, 1 - v_n 1^T)$ where 1^T denotes the column vector with all its elements equal to 1 (we will always use row vectors and $(.)^T$ denotes the transpose operator). Notice that $1 - v_n 1^T = P(V_n = \Psi)$.

Due to the strong Markov property, (V_n) is a homogeneous Markov chain with L transient states $\{1,2,...,L\}$ and one absorbing state Ψ. For every i and j in B, we define

$$G(i,j) \overset{\text{def}}{=} P(V_2 = j / V_1 = i) .$$

Observe that G is a sub-stochastic matrix. The following theorem summarizes the distributions of the random variables V_n and $S_{B,n}$. These results are analogous to those of [6] where the case of an irreducible and homogeneous Markov process is considered.

Theorem 2.1. *For every* $n \geq 1$,

$$v_n = v_1 \, G^{n-1} ,$$

$$P(S_{B,n} \leq t) = 1 - v_n \, e^{ABt} 1\text{T} \quad \forall t \geq 0$$

where

$$v_1 = \alpha_B + \alpha_{B'} \, (I - P_{B'})^{-1} P_{B'B}$$

$$= \alpha_B - \alpha_{B'} A_{B'}^{-1} A_{B'B}$$

and

$$G = (I - P_B)^{-1} P_{BB'}(I - P_{B'})^{-1}P_{B'B}$$

$$= A_B^{-1} A_{BB'} A_{B'}^{-1} A_{B'B}$$

Proof. The proof follows the proofs contained in [6]. We will just outline the necessary steps. First, it is verified that the matrices $I - P_B$ and $I - P_{B'}$ are regular. This is immediately done, for instance for the first one, by constructing an auxiliary homogeneous Markov chain with states $B \cup \{0\}$ and $(L + 1) \times (L + 1)$ transition probability matrix P_B^* decomposed in the following way.

$$P_B^* = \begin{pmatrix} P_B & (I - P_B)\,\mathbf{1}^T \\ 0 & 1 \end{pmatrix}$$

Then, it is enough to observe that every state in B is transient and the conclusion follows from the fact that $\lim_{n \to \infty} P_B^n(i,j) = 0, \forall i, j \in B$.

To derive the expression of G, we have to write and solve the set of equations satisfied by the conditional probabilities of the form $P(V_1 = j / X_0 = i), j \in B$, $i \in B \cup B'$. We will just display these equations. The basic tool to derive them is the Markov renewal theory. Let us denote by $P_i(.)$ the conditional probability $P(./X_0 = i)$.

For $i \in B'$ and $j \in B$,

$$P_i(V_1 = j)$$

$$= P(i,j) + \sum_{k \in B'} P(i,k)P_i(V_1 = j / X_1 = k)$$

$$= P(i,j) + \sum_{k \in B'} P(i,k)P_k(V_1 = j)$$

and for any $j \in B$,

$$P(V_1 = j) = P(X_0 = j)$$

$$+ \sum_{i \in B'} P_i(V_1 = j)P(X_0 = i).$$

This gives the expression of v_1. Now, for $i \in B$ and $j \in B$, we have:

$$
\begin{aligned}
G(i,j) &= P(V_2 = j/V_1 = i) = P_i(V_2 = j) \\[2mm]
&= \sum_{k \in B \cup B'} P(i,k) P_i(V_2 = j/X_1 = k) \\[2mm]
&= \sum_{k \in B} P(i,k) P_k(V_2 = j) \\[2mm]
&+ \sum_{k \in B'} P(i,k) P_k(V_1 = j) \\[2mm]
&= \sum_{k \in B} P(i,k) G(k,j) \\[2mm]
&+ \sum_{k \in B'} P(i,k) P_k(V_1 = j)
\end{aligned}
$$

and this gives, after some algebra, the expression of G.

As $(V_n)_{n \geq 1}$ is a homogeneous Markov chain, $v_n = v_1 G^{n-1}$; since all the states of B are transient, $\lim_{n \to \infty} v_n = 0$.

For the distribution of $S_{B,n}$, $n \geq 1$, we have

$$
\forall i \in B, \, P_i(S_{B,1} \leq t) = 1 - \sum_{j \in b} e^{ABt}(i,j)
$$

and $\forall i \in B'$,

$$
P_i(S_{B,1} \geq t) = \sum_{j \in B \cup B'} P(i,j) P_j(S_{B,1} \leq t).
$$

These two relations give

$$
P(S_{B,1} \leq t) = 1 - v_1 \, e^{ABt} 1T.
$$

For any $i \in B$ and for any $n \geq 1$, we can write

$$
P(S_{B,n} \leq t / V_n = i) = P_i(S_{B,1} \leq t).
$$

From this and the distribution of V_n we obtain the distribution of $S_{B,n}$, $n \geq 1$. $\square$

For every $k \geq 1$ and $n \geq 1$, the k-order moment of $S_{B,n}$ is given by the following expression where E denotes the expectation:

$$E(S_{B,n}^k) = (-1)^k k! \, v_n A_B^{-k} \, 1^T$$

Let us consider now the random variable "total time spent by X in the subset B until absorption" which has been denoted by $O(\infty)$ in the previous section. The distribution of $O(\infty)$ can be found in [7] for Markov processes and in [8] for semi-Markov reward processes. It is given in the following theorem.

Theorem 2.2.

$$P(O(\infty) \leq t) = 1 - v_1 e^{Mt} 1^T \qquad \forall t \geq 0$$

where v_1 has been given in the previous theorem and

$$M = - \Lambda_B (I - P_B - P_{BB'}(I - P_{B'})^{-1} P_{B'B})$$

$$= A_B - A_{BB'} A_{B'}^{-1} A_{B'B} \, .$$

Proof. See [7] ❑

The following lemma gives another expression of this distribution using the previously defined matrix G.

Lemma 2.3.
$$P(O(\infty) \leq t) = 1 - v_1 e^{A_B(I-G)t} 1^T \qquad \forall t \geq 0$$

Proof.

$$M = - \Lambda_B (I - P_B - P_{BB'}(I - P_{B'})^{-1} P_{B'B})$$

$$= - \Lambda_B (I - P_B - (I - P_B)G)$$

$$= - \Lambda_B (I - P_B)(I - G)$$

$$= - A_B (I - G) \text{ (since } P = I + \Lambda^{-1} A).$$ ❑

For every $k \geq 1$, the k-order moment of $O(\infty)$ is:

$$E(O^k (\infty)) = (-1)^k \, k! \, v_1 (A_B(I - G))^{-k} \, 1^T.$$

Since $E(O(\infty))$ gives the expected cumulative operational time until absorption, it can be interesting to compute the proportion of operational time until absorption $r \overset{def}{=} E(O(\infty)) / E(LT)$ where LT denotes the total lifetime of the process. Recall that

$$P(LT) \leq t) = 1 - \overline{\alpha} \, e^{\overline{A} \, t} \mathbf{1}^T$$

where $\overline{\alpha} = (\alpha_B, \alpha_{B'})$ and $\overline{A}$ is the submatrix of A obtained by deleting the last row and the last column which correspond to the absorbing state, that is

$$\overline{A} = \begin{pmatrix} A_B & A_{BB'} \\ A_{B'B} & A_{B'} \end{pmatrix}$$

The expectation of the total lifetime is then given by

$$E(LT) = -\overline{\alpha} \, (\overline{A})^{-1} \mathbf{1}^T$$

Last, consider the random variable $N_B \overset{def}{=}$ "total number of visits to the subset B until absorption". From the definition of V_n we have $\{N_B \geq k\} = \{S_{B,k} > 0\}$ for $k \geq 1$. It follows from Theorem 2.1. that for $k \geq 1$, $P(N_B \geq k) = v_k \mathbf{1}^T$ and

$$P(N_B = 0) = 1 - v_1 \mathbf{1}^T,$$

$$P(N_B = k) = v_k \mathbf{1}^T - v_{k+1} \mathbf{1}^T$$

$$= v_k (I - G) \mathbf{1}^T.$$

From this formula, we have

$$E(N_B) = \sum_{k \geq 1} v_k \mathbf{1}^T = v_1 (I - G)^{-1} \mathbf{1}^T.$$

Observe that if we know that the process has visited the set B at least n times $(n \geq 1)$, that is, given that $S_{Bn} > 0$, the evaluation of the duration of the nth sojourn in B changes. The conditional distribution of S_{Bn} given that $S_{B,n} > 0$ can be derived as follows.

$$P(S_{B,n} \leq t \, / \, S_{B,n} > 0)$$

$$= \frac{P(S_{B,n} \leq t \text{ and } N_B \geq n)}{P(N_B \geq n)}$$

$$= \frac{P(S_{B,n} \leq t) - P(S_{B,n} \leq t \text{ and } N_B \geq n)}{P(N_B \geq n)}$$

$$= \frac{P(S_{B,n} \leq t) - P(N_B \geq n)}{P(N_B \geq n)}$$

$$= \frac{(1 - v_n e^{A_B t} \mathbf{1}^T) - (1 - v_n \mathbf{1}^T)}{v_n \mathbf{1}^T}$$

$$= 1 - \frac{v_n e^{A_B t} \mathbf{1}^T}{v_n \mathbf{1}^T} \, .$$

For instance, the conditional expectation of the duration of the nth sojourn is

$$E(S_{B,n} \, / \, S_{B,n} > 0) = - \frac{v_n A_B^{-1} \mathbf{1}^T}{v_n \mathbf{1}^T} \, .$$

3. An illustrative example

We will illustrate the previous results by means of a simple model taken from the distributed algorithms area. Consider a set of processes organized in an unidirectional ring and assume that there is a special message (the token) used to control some distributed application. For instance, one can think of a network of processors sharing a common resource (a peripherial device) in mutual exclusion. A processor may use the resource iff it possesses the token. Of course, some kind of mechanism is provided to avoid the monopoly on the resource by one or a few processors. We are interested in the behaviour of such a system from a fault-tolerant point of view. We assume that the token is vulnerable only when traveling from a node (here, node = process = site) to the next one on the ring. The modelling assumptions are the following. The different sojourn times of the token in each site are i.i.d. random variables exponentially distributed with common parameter λ_r (the mean *resource-token speed*). When passing from a process to the following one, the token may be lost (unsuccessful transmission, perturbations, ...) with probability p. If the transmission is successful, the time of the transfer is negligible (that is, zero in the model). The results of each transmission are independent events. The lifetime of the token is then exponentially distributed with mean $1/(\lambda_r p)$.

Concerning the problem of detecting the token loss and regenerating it, a first solution was proposed in [9] using local clocks. A different approach was considered in [10]. A second token is used to increase the lifetime of the system. The two tokens carry some additional information and leave a trace of their visits when leaving the sites. When one of them is lost, there always exists a node in the ring such that if it is reached by the remaining token, the detection and regeneration take place at that moment in that site. In [11] the previous solution was improved and generalized to support any number of tokens. It must be observed that these solutions need no local clock managing in the nodes. Here, we will consider this last algorithm when there are two tokens in the ring. The only functional aspect we need to know here is that the site in which a detection and regeneration can take place follows exactly the last one visited by the lost unit (for technical details concerning the algorithms, see [11] and [12]).

We assume that the stochastic behaviour of the second unit (the *system-token*) is as for the first one, with mean speed λ_s . Furthermore, while there are two units on the ring their respective evolutions are independent. For the illustrative purposes in this paper, let us consider only the case of three sites in the ring. The lifetime of such a system is the absorbing time of the homogeneous Markov process exhibited in Figure 1.

The bidirectional arrows mean that transitions are possible in the two directions. In this case, the rate of each transition is labelled near the arriving state. At any instant, the distance from token x to token y (when they are both "alive") is defined as the number of lines that token x has to pass through in order to join y (remember that the ring is unidirectional). When one of the tokens is lost, the distance between them is the number of lines that the remaining unit has to pass through to regenerate the lost one.

States D_i, $i = 0,1,2$, correspond to the case of two tokens alive, where i equals the distance from the resource-token to the system-token. States R_i (respectively S_i) correspond to the resource-token alive (resp. the system-token alive) and the other unit lost, i representing the distance between them, $i = 1,2,3$. The arcs without endpoints represent the transitions to the absorbing state (system completely down). We define the operational states as the states in which the resource-token is alive. This gives $B = \{D_0, D_1, D_2, R_1, R_2, R_3\}$. The set of nonoperational states is $B' = \{S_1, S_2, S_3\}$.

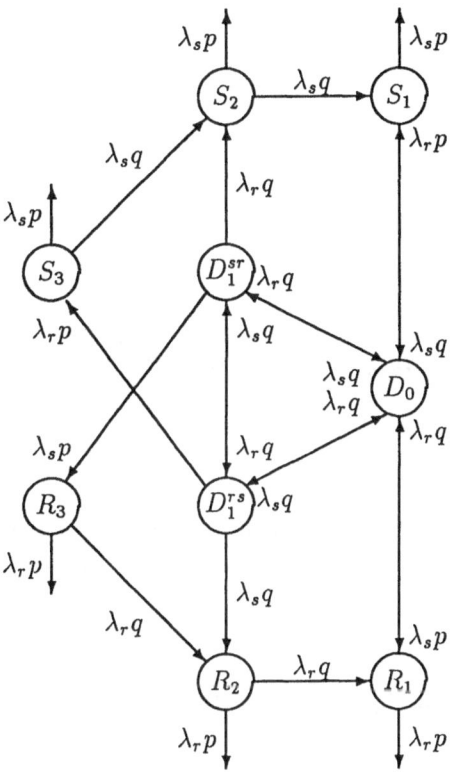

Figure 1. Markov model for three nodes.

Assume that the two parameters p and λ_r are input data and that the user wants to tune up the value of the speed of the system-token in order to satisfy the following "informal" constraints: the lifetime of the system must be "high" and the system must spend "as much time as possible" in the operational states.

Thanks to the particular structure of matrix A_B, we get the following simple expressions.

$$P(S_{B,n} \leq t) = 1 - g^{n-1}\, e^{-\lambda_r pt} \qquad n \geq 1,$$

$$P(S_{B,n} \leq t \,/\, S_{B,n} > 0) = 1 - e^{-\lambda_r pt} \qquad n \geq 1,$$

$$P(N_B = k) = g^{k-1}\,(1 - g) \qquad k \geq 1,$$

where $g = G(1,1)$. The mean cumulative operational time until absorption is

$$E(O(\infty)) = \frac{E(N_B)}{\lambda_r p} = \frac{1}{\lambda_r p (1 - g)} \ .$$

Let us assign numerical values to the input parameters. We take $\lambda_r = 1.0$ (say 1.0/secs) and $p = 10^{-3}$. It can be easily verified that $\lambda_s \to E(LT)$ and $\lambda_s \to P(S_{B',n} > 0)$ (for any fixed value of t) are decreasing functions.

Also, $P(S_{B',n} > t \,/\, S_{B',n} > 0)$ does not depend on n in this particular case:

$$P(S_{B',n} > t \,/\, S_{B',n} > 0) = \frac{v'_1 e^{A_{B'} t} 1^T}{v'_1 1^T}$$

where $v'_1 = - (1, 0, 0) A_B^{-1} A_{BB'}$.

To formalize the optimization problem, we look for values of λ_s such that the mean lifetime of the system is greater than a given value T_{min} and the conditional probability $P(S_{B',n} > t \,/\, S_{B',n} > 0)$ is less than a given level ε. For instance, let us set $T_{min} = 259200$ secs (three days). Since $E(LT)$ decreases with λ_s, we get

$$E(LT) > 259200 \text{ secs} \Leftrightarrow \lambda_s \in]\, 0, 0.9387 [.$$

Assume that the other specification is $t = 5.0$ with a 20% level ($\varepsilon = 0.2$). This leads to

$$P(S_{B',n} > 5 \,/\, S_{B',n} > 0) < 0.2 \Leftrightarrow \lambda_s > 0.7940.$$

The solution interval is

$$\lambda_s \in]\, 0.7940, 0.9387 [$$

and the mean lifetime belongs to the interval $]259200, 280130[$.

4. Conclusions

The main contribution of this work is a closed-form expression of the distribution of the successive operational periods in computing systems modeled by Markov processes. These quantities are dependability measures. With respect to classical metrics (availability, reliability) they allow a detailed analysis of the transient behaviour. A designer is therefore able to follow the evolution of a system alternatively in the operational and in the nonoperational states all along its lifetime.

Other measures to analyze in further work concern, for instance, more detailed classifications of the model states (that is, partitions of the state space containing more than three classes). For example, this can be the case when several types of failures are to be considered.

Acknowledgements

We thank the post-conference referee for its detailed comments, and in particular, for the observations on terminology.

References

[1] J.C.Laprie, "Dependable Computing and Fault Tolerance: Concepts and Terminology", In *15th IEEE Int. Symp. on Fault Tolerant Computing (FTCS-15)*, Ann Arbor (Michigan), 1985.

[2] A.Avizienis, J.C.Laprie, "Dependable Computing: from concepts to design diversity", *Proc. of the IEEE*, 74(5): 629-638, August 1986.

[3] B.Parhami, "From Defects to Failures: a View of Dependable Computing", *Computer Architecture News*, 16(4), September 1988.

[4] E. de Souza e Silva, H. R. Gail, "Calculating cumulative operational time distributions of repairable computer systems", *IEEE Transactions on Computers*, C.35: 322-332, April 1986.

[5] B. V. Gnedenko, Yu. K. Belyayev, A. D. Solovyev, *Mathematical Methods of Reliability Theory*, Academic Press, 1969.

[6] G. Rubino, B. Sericola, "Sojourn times in finite Markov processes", *J. Appl. Prob.*, 27: 744-756, December 1989.

[7] R. Marie, B. Sericola, "Distribution du temps total de séjour dans un sous-ensemble d'états transitoires d'un processus Markovien homogène à espace d'état fini", Technical Report 585, I.N.R.I.A., Campus de Beaulieu, 35042 Rennes Cedex, France, November 1986.

[8] G. Ciardo, R. Marie, B. Sericola, K. Trivedi, "Performability analysis using semi-Markov process", Technical Report CS-1988-9, Duke Computer Science, 1988, to appear in IEEE Trans. on Comp.

[9] G. Le Lann, "Algorithms for Distributed Data-Sharing Systems which use Tickets", In *Proc. 3rd Berkeley Workshop on Distributed Data Base and Computer Networks*, Berkeley, USA, 1978.

[10] J. Misra, "Detecting Termination of Distributed Computations Using Markers", In *Proc. 2nd annual ACM Symposium on Principles of Distributed Computing*, Montreal, Canada, 1983.

[11] M. Raynal,G. Rubino, "An Algorithm to Detect Token Loss on a Logical Ring and to Regenerate Lost Tokens", In *Proc. of the International Conference on Parallel Processing and Applications,* L'Aquila, Italie, North-Holland, 1987.

[12] M. Raynal, G. Rubino, Détecter la perte de jetons et les régénérer sur une structure en anneau, Technical Report 428, I.N.R.I.A., Campus de Beaulieu, 35042 Rennes Cedex, France, Juillet 1985.

Architecture

Chair: T.B. Smith (IBM T.J. Watson Research Center, Yorktown Heights, New York, USA)

A DATA-DRIVEN DEPENDABILITY ASSURANCE SCHEME WITH APPLICATIONS TO DATA AND DESIGN DIVERSITY

Behrooz PARHAMI
Dept. of Electrical & Computer Engineering - University of California
Santa Barbara, CA 93106 - USA

Abstract

A new scheme for dependability assurance in digital systems that allows highly selective use of redundancy with low additional overhead is introduced. This scheme, which can be described as data-driven, contrasts with and complements the conventional structure-driven and function-driven approaches to dependability procurement for digital systems. The proposed approach is based on attaching a dependability tag (d-tag) to each data object and updating pertinent d-tag values as a computation unfolds. Normal operations on data objects tend to lower d-tag values while comparisons and voting on redundant versions of a result work in the opposite direction. Judicious intermixing of dependability-lowering and dependability-raising operations, driven dynamically by the dependability requirements for various data objects, can lead to a desired overall dependability for computation results. Following an exposition of basic concepts of the proposed method with the assumption of perfect d-tags and operations, various issues in the formulation of a strategy for dealing with erroneous d-tags and imperfect operations are outlined. Applications of the proposed data-driven approach in connection with data and design diversity are presented through simple examples. Some benefits of the proposed scheme for fault tolerance in a class of critical systems are discussed.

1. Introduction

Numerous techniques have been proposed for enhancing the dependability of computations through the avoidance and/or tolerance of undesired conditions such as defects, faults, errors, malfunctions, and degradations that can potentially lead to result-level failures [4]. Proposed dependability procurement methods fall into two categories: Structure-driven and function-driven. Structure-driven methods employ static and dynamic hardware redundancy (among other schemes) to make the system highly dependable, so that any process run on it can be assumed to produce dependable results, given that the process itself is trustworthy. The resulting increase in dependability is universal and applies to every process or algorithm, regardless of whether or not this level of dependability is needed in all cases. Function-driven methods, on the other hand, are applied to specific processes or algorithms and take advantage of particular properties of the problem at hand to provide an explicit cost-dependability tradeoff. They are typically more efficient and/or selective in their use of redundancy but imply increased design effort for setting up computations that deal with the specially modified data structures or for providing multiple versions of the required processes.

In many situations, one can benefit from an even more selective approach to the application of redundancy, not only from one process to another but also within a single application process. For example, the degree of replication of a data object manipulated by a process should ideally depend on its:

1. *Value or criticality:* Critical data must be protected through a higher degree of replication.

2. *Regeneratability:* Error detection with good coverage may be sufficient for easily regenerated data.

3. *Size:* High degree of replication is less desirable for very large data objects.

4. *Resilience:* The higher the resilience, the lower the needed degree of replication.

5. *Amenability to consistency check:* Checked duplication may be an acceptable substitute for greater replication.

6. *Extent and type of use:* Rarely referenced or restricted- access (e.g., read-only) data may need less protection.

Unfortunately, incorporating various degrees of replication for data objects in each application process is hopelessly complex, even if all of the above aspects could be reasonably quantified. Furthermore, certain of these aspects (e.g., size of data object and its usage) cannot be accurately predicted as they may depend on run-time conditions. It follows that a general framework for handling varying (non-uniform) degrees of replication along with a capability for automatically deciding on the required replication factors for programs and data is needed.

2. The data-driven approach

2.1. Dependability tags for data objects

Suppose that a dependability tag (d-tag) is attached to each data object as an indicator of the data object's correctness probability. Thus, a data object D and its d-tag d will comprise a composite tagged object $\langle D, d \rangle$. The d-tag d assumes values from a finite set of dependability designations

$$d \in \{0, 1, \dots, \delta-1\}, \tag{1}$$

where δ is an application-dependent constant. Associated with each d-tag value d, are constants π_d and π'_d such that the d-tag d in the composite tagged data object $\langle D, d \rangle$ signifies:

$$\pi_d \le \text{prob}[D \text{ is correct}] \le \pi'_d \tag{2}$$

We will assume that $\pi_j < \pi_{j+1}$ and $\pi'_j \le \pi'_{j+1}$, so that a larger d-tag value implies higher confidence in the correctness of the associated data object. Unless otherwise stated, the upper bound π'_j will be assumed to be 1 in the remainder of this paper. We also assume $\pi_0 = 0$, $\pi_{\delta-1} = 1$; i.e., the d-tag values 0 and $\delta-1$ are reserved for hopeless and perfect values, respectively. Finally, we will assume that errors in various data objects are random and statistically independent, although extension of the method to correlated errors is possible, given the availability of data on the probability of such non-independent errors.

Note that as defined, d-tags essentially represent a general and very flexible discretization scheme for correctness probabilities. In other words, the π_j values need not be selected to conform to any particular rule or pattern. However, in practice, it is desirable to have a capability for greater discrimination at the high end of dependability values. This is because correctness probabilities 0.99 and

0.999 are significantly different while the values 0.4 and 0.5 need not be distinguishable as they both represent practically useless values. The following is an example of 8-valued d-tags ($\delta = 8$):

j:	0	1	2	3	4	5	6	7	(3)
π_j:	0	0.75	0.883	0.9612	0.9894	0.9973	0.9999	1	

The above d-tag values will be used in all numerical and application examples that follow in subsequent sections of the paper. A justification for selecting these particular values for the constants π_j will be provided in Subsection 2.4.

Although theoretically it is possible to attach d-tags to data objects at any level, practical considerations such as data storage redundancy and computational overhead will probably restrict meaningful applications to high-level data objects with complex structures and associated operations. Of course, regardless of the level at which d-tags are applied, the problem of determining d-tag values is non-trivial and must be dealt with in depth. However, given correctly assigned d-tags and ignoring for now the possibility that the d-tags themselves may be corrupted in the course of our computations, we can discuss the manipulation of tagged data in terms of dependability-lowering and dependability-raising operations.

2.2. Dependability-lowering operations

Normal operations on data objects tend to lower the d-tag values. Assuming that operations are themselves perfectly dependable, the dependability of each result is only a function of the operands' dependabilities. A unary operator u tansforms the data object D into $u(D)$. In our scheme, we define for each u, a corresponding unary operator u^* such that :

$$u^*(\langle D , d\rangle) = \langle u(D) , d\rangle \tag{4}$$

This simply means that the dependability of the result $u(D)$ is the same as the dependability of the operand D. In the case of a binary operator b in$D = b$ (D', D''),we define the corresponding binary operator b^* which operates on composite data objects giving $\langle D, d\rangle = b^*(\langle D', d'\rangle, \langle D'', d''\rangle)$. The key to this extension is a procedure for determining d from d' and d''. Thus we write:

$$b^*(\langle D' , d'\rangle , \langle D'' , d''\rangle) = \langle b(D', D'') , g(d', d'')\rangle, \tag{5}$$

where the function g is the dependability evaluation function associated with binary operators. More generally, one can consider a dependability evaluation function $g_b(d', d'')$ for each binary operator b or even $g_b(d', d'', D', D'')$. However, let's keep things simple for now.

Assuming that $b(X, Y)$ depends on both X and Y (i.e., b is not actually a unary function) we can define $g(d', d'')$ as follows:

$$g(d', d'') = d \quad \text{such that} \quad \pi_d \le \pi_{d'}\pi_{d''} < \pi_{d+1} \tag{6}$$

In other words, d is the largest possible value for the index j such that $\pi_j \le \pi_{d'}\pi_{d''}$.

Because of the way d-tags are defined, the value of d thus obtained satisfies $d \le \min(d', d'')$. The equality occurs when one operand has the "perfect" d-tag of $\delta - 1$. To make sure that the d-tags are always lowered by the smallest possible amount (thus gauranteeing the highest possible values for result d-tags), we may impose the requirement that

$$d \ge \min(d', d'') - 1 \tag{7}$$

The worst-case reduction occurs when $d = d'$. Thus, to ensure d-tag reduction of no more than 1 in every case, we must have $\pi_{j-1} \le \pi_j{}^2$ for all values of j. The π_j values given in (3) satisfy this requirement as evident from the first three lines of (8) below:

j:	0	1	2	3	4	5	6	7	
π_j:	0	0.75	0.883	0.9612	0.9894	0.9973	0.9999	1	
$\pi_j{}^2$:	0	0.56	0.780	0.9239	0.9789	0.9946	0.9998	1	(8)
$\pi_j{}^3$:	0	0.42	0.688	0.8881	0.9685	0.9919	0.9997	1	
$\pi_j{}^4$:	0	0.32	0.608	0.8536	0.9583	0.9892	0.9996	1	

Had we opted for the minimal reduction in correctness probabilities as well as in d-tag values, we would have selected one of the two sets of π_j values in (9), depending on which of the values π_1 (0.75) or π_6 (0.9999) we wanted to keep as before:

j:	0	1	2	3	4	5	6	7	
π_j:	0	0.75	0.866	0.9306	0.9647	0.9822	0.9911	1	(9)
π_j:	0	0.9968	0.9984	0.9992	0.9996	0.9998	0.9999	1	

In these two examples, each π_j is exactly the square root of π_{j-1} for $2 \le j \le 6$.

The above can easily be generalized to a k-variable function. The dependability of the result is never more than the smallest d-tag value involved (this is why we call all such operations "dependability-lowering" operations) and the condition for minimal d-tag reduction becomes:

$$\pi_{j-1} \le \pi_j{}^k \tag{10}$$

The numerical values in (8) indicate that our example d-tag values of (3) would satisfy this condition for $k = 3$ if π_1 were slightly lowered (say, to 0.68), but for $k = 4$, they violate the condition for most values of j. Obviously, Condition (10) is violated for all larger values of k as well.

In the proposed scheme of computation, if the final data objects (computation results) end up with acceptable d-tags after all such lowerings, then nothing more needs to be done. Otherwise, we need to structure the computation in such a way that it also includes dependability-raising operations at some points. Such operations are discussed in the next subsection. In general, the dependability-raising mechanisms will be built into the computation but they will be invoked only if needed; the "need" is determined dynamically by the d-tags generated at the end or in the course of computation, depending on implementation details.

2.3. Dependability-raising operations

Suppose that we obtain a result in two different ways using some form of hardware, software, or time redundancy. Let the two results with their corresponding d-tags be $\langle D_0, d_0 \rangle$ and $\langle D_1, d_1 \rangle$. We wish to draw the conclusion that the result is $\langle D, d \rangle$ where d is the highest possible d-tag that can be attached to a value D inferred from the inputs. Obviously, if $D_0 \ne D_1$, then $\langle D, d \rangle = \langle D_i, d_i \rangle$, $i \in \{0, 1\}$, such that $d_i \ge d_{1-i}$. On the other hand, if $D_0 = D_1$, then $D = D_0 = D_1$ and d is computed as follows. Let p_0 and p_1 be the actual correctness probabilities for D_0 and D_1, respectively. By definition, $p_i \ge \pi_{d_i}$ for $i = 0, 1$. Then the correctness probability for D is:

$$p = p_0 p_1 / [p_0 p_1 + (1 - p_0)(1 - p_1)] \tag{11}$$

The value of p given above is a nondecreasing function of both p_0 and p_1. Thus:

$$p \ge \pi_{d_0} \pi_{d_1} / [\pi_{d_0} \pi_{d_1} + (1 - \pi_{d_0})(1 - \pi_{d_1})] \tag{12}$$

To maximize the d-tag of the result, d must be selected such that:

$$\pi_d \leq \pi_{d_0}\pi_{d_1} / [\pi_{d_0}\pi_{d_1} + (1 - \pi_{d_0})(1 - \pi_{d_1})] < \pi_{d+1} \tag{13}$$

It is easily shown that in both cases of $D_0 \neq D_1$ and $D_0 = D_1$ considered above, $d \geq \max(d_0, d_1)$ provided that $\pi_j \geq 0.5$ for all $j > 0$. This is true by definition in the case of $D_0 \neq D_1$. For the case of $D_0 = D_1$, we rewrite (13) as:

$$\pi_d \leq \pi_{d_0} / [1 - (1 - \pi_{d_0})(2 - 1/\pi_{d_1})] \tag{14}$$

$$= \pi_{d_1} / [1 - (1 - \pi_{d_1})(2 - 1/\pi_{d_0})] < \pi_{d+1}$$

With the assumption $\pi_j \geq 0.5$, the term $2 - 1/\pi_j$ is positive and the middle part of (13) is no less than π_{d_0} (and by a similar argument, no less than π_{d_1}).

However, in general, Inequality (13) does not guarantee that d will be strictly higher than $\max(d_0, d_1)$. To analyze such dependability-raising operations, we assume $d_1 = d_0 + \varepsilon$ for some $\varepsilon \geq 0$ and seek conditions under which $d > \max(d_0, d_1) = d_1$; i.e., there is some increase in dependability as a result of the match in our comparison. For a minimal increase of one unit with respect to the larger d-tag value d_1, Inequality (13) yields the following necessary condition:

$$\pi_{d_1+1} \leq \pi_{d_1-\varepsilon}\pi_{d_1} / [\pi_{d_1-\varepsilon}\pi_{d_1} + (1-\pi_{d_1-\varepsilon})(1-\pi_{d_1})] \tag{15}$$

Let us denote the right-hand side of (15) by $r(d_1, \varepsilon)$; i.e., define:

$$r(j, \varepsilon) = \pi_{j-\varepsilon}\pi_j / [\pi_{j-\varepsilon}\pi_j + (1 - \pi_{j-\varepsilon})(1 - \pi_j)] \tag{16}$$

Then the condition for obtaining an increase of at least one unit in dependability as a result of matched comparison becomes:

$$\pi_{j+1} \leq r(j, \varepsilon) \quad \text{for } \varepsilon < j < \delta - 2 \tag{17}$$

The values $j \leq \varepsilon$ are excluded from (17) because $j = \varepsilon$ leads to the appearance of $\pi_0 = 0$ in (15) and thus no increase in dependability and $j < \varepsilon$ leads to undefined values. Similarly, $j = \delta - 2$ is excluded because $r(j, \varepsilon)$ is always less than 1 and thus $\pi_{\delta-1} = 1 \leq r(\delta - 2, \varepsilon)$ cannot be satisfied. In the special case of $\varepsilon = 0$, the satisfaction of Inequality (17) for $1 \leq j < \delta - 2$ implies:

$$\pi_{j+1} \leq r(j, 0) = \pi_j^2 / [2\pi_j^2 - 2\pi_j + 1] \tag{18}$$

As an example, let us compute $r(j, \varepsilon)$ for different values of ε in the case of d-tags defined by (3):

j :	0	1	2	3	4	5	6	7	
π_j :	0	0.75	0.883	0.9612	0.9894	0.9973	0.9999	1	
$r(j, 0)$:	0	0.90	0.983	0.9984	0.9999	1.0000^-	1.0000^-	1	(19)
$r(j, 1)$:	--	0	0.958	0.9947	0.9996	1.0000^-	1.0000^-	1	
$r(j, 2)$:	--	--	0	0.9867	0.9986	0.9999^-	1.0000^-	1	

We see that (17) is satisfied by $r(j, 0)$ but not by $r(j, 1)$ or $r(j, 2)$. However, even in the latter cases, (17) is satisfied for sufficiently large values of j. One can conclude that in general, dependability improvement is likely as a result of matched comparison as long as the d-tag values associated with the comparands are sufficiently high.

The above is easily generalized to n versions of a result. Let the n versions be $D_0, D_1, \ldots, D_{n-1}$, with the corresponding d-tags $d_0, d_1, \ldots, d_{n-1}$. If all of these data objects are different, then the output is taken to be $\langle D_i, d_i \rangle$ such that $d_i \geq d_j$ for $0 \leq j < n$. At the other extreme, if all of these n data objects are identical, then the output is set to one of them and assigned the d-tag value d such that

$$\pi_d \leq (\Pi_i \, \pi_{d_i}) / [\Pi_i \, \pi_{d_i} + \Pi_i \, (1 - \pi_{d_i})] < \pi_{d+1}, \tag{20}$$

where in all cases the index i of the product terms varies from 0 to $n - 1$. In general, the n data objects can be partitioned into classes of identical objects. For each class, a d-tag value is computed and the class with the largest associated d-tag determines the output.

2.4. Tradeoffs in the discretization scheme

It should be evident from the previous discussions that various discretization schemes (the constants π_j) can be selected to satisfy various requirements of, or to provide desirable features for, dependability-lowering and dependability-raising operations. For example, if the satisfaction of (17) with $\varepsilon = 0$ is the only requirement, then the following discretization scheme is optimal in that it results in maximal dependability increase with each matched comparison (of two results having identical d-tags):

j :	0	1	2	3	4	5	6	7	(21)
π_j :	0	0.75	0.90	0.987	0.9997	$1 - 10^{-7}$	$1 - 10^{-14}$	1	

In addition, selection of the value of δ must be based on detailed cost-benefit analysis with respect to the level of redundancy implied and the improvement in dependability that can be achieved with a larger δ. The optimal value of the constant δ for each application must initially be selected based on estimates for the various overheads involved and the requirements of dependability-lowering and dependability-raising operations. Intuitively, a larger value of δ, with the correspondingly finer subdivisions defined by π_j, causes smaller reductions in d-tag values while a smaller δ may yield potentially larger increases in comparison-type operations. In general, a range of values for δ may satisfy the requirements of dependability-lowering and dependability-raising operations. Smaller values in the acceptable range obviously imply lower direct storage and processing overheads. However, selecting a larger value may eventually translate into lower replication requirements for computations and data in view of the higher likelihood of smaller decreases in d-tag values and thus a potentially lower overall cost.

We will not discuss this last topic any further in this paper, but rather focus on the selection of π_j values for a given value of δ. To achieve the desirable properties of minimal dependability reduction (as discussed in Subsection 2.2) and guaranteed dependability increase (Subsection 2.3), Inequalities (10) and (17) must be satisfied simultaneously for reasonable values of k and ε. To show that this is possible, we consider the special case of $k = 2$ with $\varepsilon = 0$ as an example. In this case we must have

$$\pi_j < \pi_{j+1}{}^2 \leq \pi_{j+1} \leq \pi_j{}^2 / (2\pi_j{}^2 - 2\pi_j + 1) \tag{22}$$

which is equivalent to:

$$\sqrt{\pi_j} < \pi_{j+1} \leq \pi_j{}^2 / (2\pi_j{}^2 - 2\pi_j + 1) \tag{23}$$

A chain of probability values satisfying (23) can be found if for each π_j in the chain, we have:

$$\pi_j < [\pi_j{}^2 / (2\pi_j{}^2 - 2\pi_j + 1)]^2 \tag{24}$$

Straightforward manipulation converts Inequality (24) to:

$$\pi_j{}^3 > (2\pi_j{}^2 - 2\pi_j + 1)^2 = 4\pi_j{}^4 - 8\pi_j{}^3 + 8\pi_j{}^2 - 4\pi_j + 1 \tag{25}$$

Thus, the following Inequality must be satisfied for each π_j:

$$(1 - \pi_j)(4\pi_j^3 - 5\pi_j^2 + 3\pi_j - 1) > 0 \tag{26}$$

The first term on the left-hand-side of (26) is positive and thus the second term must also be positive. This condition is satisfied for values of π_j greater than the single root of the corresponding cubic equation at $\pi_j = 0.688^+$. In practice, values of interest for π_j are greater than 0.688 and thus (26) is always satisfied.

As an example, starting with $\pi_1 = 0.75$, Inequality (23) yields the constraints $0.866 < \pi_2 \leq 0.900$ for π_2. Any value in the above range can be selected for π_2. Proceeding in this manner, we can construct many valid sequences of probability values. Asuming that the selection of an average value in the range of possibilities is a reasonable compromise, we obtain $\pi_2 = 0.883$ and the subsequent values as given in (3). In fact, this is how the example values in (3) were obtained in the first place.

3. Refinements and extensions

3.1. Dealing with imperfect d-tags

Obviously, d-tags are themselves subject to errors and cannot be trusted completely. An erroneous d-tag may be lower or higher than the correct value. In the first case, the error is "safe" in the sense that at the end, a *negative d-tag error* can lead to one of three things:

1. A correct result with erroneously low, but acceptable, d-tag value that is trusted and used.

2. A correct result with unacceptably low d-tag value that is either discarded or used cautiously.

3. An incorrect result with correct d-tag that is either discarded or used with appropriate care.

The third possibility arises because erroneously low d-tag values for correct results may cause incorrect values to prevail in comparison-type operations. However, such incorrect values will carry with them appropriate d-tags that indicate their levels of dependability, provided that no error of the second type occurs.

It is, therefore, sufficient to guard against errors that improperly raise d-tag values. Such *positive d-tag errors* have three causes:

1. Incorrect storage and/or transmission of d-tag values.

2. Errors during dependability-lowering operations.

3. Errors during dependability-raising operations.

Storage and transmission errors are the easiest ones to deal with. One can of course use error codes in a straightforward manner for protection. However, several properties of this particular application can be exploited to devise more effective methods. The fact that we are only interested in protecting against the more damaging positive d-tag errors suggests that any asymmetry of errors (e.g., higher likelihood of 1-to-0 compared to 0-to-1 errors) can be exploited by proper encoding. Also, since the arithmetic value of a d-tag is significant, arithmetic error codes are potentially useful. Finally, one can use a gray-type encoding scheme to limit the damage caused by an uncaught error.

Determination of d-tag values during dependability-lowering and dependability-raising operations essentially involves simple function evaluations. Assuming that these functions are evaluated by table lookup, many different methods can be used for detection and/or correction of potential errors. For example, with 4-bit d-tags, a 256-word by 4-bit table will be needed in a binary operation. For protection against errors, one can store the two 4-bit tags with each 4-bit table entry and then encode the resulting 12-bit entries in some error code. Alternatively, the use of self-checking circuits for the manipulation of d-tags may be contemplated.

3.2. Dealing with imperfect operations

Even when the data objects D' and D'' are perfectly dependable, the data object $D = b(D', D'')$ may have potential errors due to imperfect hardware or software implementing the binary operator b. To deal with this problem, appropriate dependabilities must be assigned to various operations and used in determining the d-tag values for operation results. The key issue here is to use dependability estimates that are pessimistic (so that any resulting d-tag error is on the safe side) but not too pessimistic so as to require excessive redundancy to overcome the resultant decrease in d-tags associated with the data objects involved.

Comparison and voting operations may also be imperfect. For example, in the comparison of two versions of a result leading to $\langle D, d \rangle = c(\langle D', d' \rangle, \langle D'', d'' \rangle)$, an error may occur in judging the equality of D' and D''. If $D' = D''$ but it is erroneously determined that they are unequal, the error is on the safe side in that dependability is not raised. If, on the other hand, $D' \neq D''$, dependability may be improperly raised by a comparison error. Therefore comparison mechanisms (hardware or software) must be designed to have asymmetric error modes if possible. Similar observations apply to voting errors when more than two copies or versions of a data object are involved. Obviously, the criticality of the method for dealing with imperfect comparison and voting operations increases directly with any increase in the dependability of original data sources.

Imperfect operations on composite tagged data objects can be modelled in several distinct ways. One possibility is to view the operation as simply another element of the computation with its associated d-tag. Then, for example, a binary operation with the associated d-tag d_b performed on data objects having d-tags d_1 and d_2 will be like a perfect ternary operation on data objects having the d-tags d_b, d_1, and d_2. This conceptually simple scheme does not solve all of our problems as the assignment of d-tags to various operations is nontrivial. Another possibility is to attach to each data object a second "operation count" tag. Then, assuming that all operations have roughly equal complexities, the operation count tag provides a second indication of how dependable the data object is by showing how many transformations it has undergone. Operation count tags can be periodically reset to zero and their effects incorporated into d-tags by suitable adjustment algorithms.

The above discussion must be viewed as a preliminary attempt to address the issue of imperfect operations on tagged data objects. Details of the suggested schemes remain to be worked out. Also, the possibility of data-dependent reliabilities for operations could be taken into account.

3.4. Modeling, evaluation and extensions

Although the development of mathematical models for evaluating the cost-effectiveness of the proposed data-driven approach to dependability assurance should proceed in parallel with the resolution of the problems discussed in Subsections 3.2 and 3.3, it is still necessary to integrate the various aspects of the modeling process into a "clean" mathematical model and to evaluate and fine-tune the proposed techniques. The development of such an integrated mathematical model constitutes part of the author's future research plans in this area. An aspect of the evaluation phase of this research program is the identification of application areas where the proposed approach is likely to lead to improved cost-effectiveness compared to current design paradigms and to formulate a research plan for dealing with these application areas in more detail. Some preliminary results along this line are presented in Sections 4 and 5 of the paper.

Other applications where data-driven dependability assurance may prove useful are in dataflow computation [25] and distributed systems [22]. In the first area, data-driven dependability assurance is a natural method that nicely matches the main computational scheme. In the second area, the independence of failures in various sites and links can be exploited to satisfy our scheme's requirement for statistical independence of errors in different data objects. Once the promising application areas such as the above have been identified, design issues must be examined in greater detail for selected areas in order to develop verifiable quantitative results on the gains and benefits versus the accrued cost. It may also be desirable to examine the areas where data-driven dependability procurement appears to be impractical or uneconomical to discover the underlying reasons and to look for extensions and/or variations of the technique (or combinations with other methods) that may overcome the perceived problems.

There is also a potential for using the data-driven dependability procurement paradigm to build a unified framework for the automatic synthesis of ultrareliable systems. This is because several previously proposed redundancy techniques can be viewed as degenerate cases of our data-driven scheme. Consider, for example, the two extremes of binary (two-valued) and continuous (real-valued) d-tags. With binary d-tags, voting on multiple versions of a result is achieved by ignoring all versions that have d-tags of 0. This can be viewed as being equivalent to the self-purging redundancy scheme [13]; a generalization of hybrid redundancy [14] which itself is a combination of voting and standby sparing techniques. With binary results (bits as data objects) and real-valued d-tags denoting exact correctness probabilities (rather than a lower bound on them; i.e., $\pi_j = \pi_j'$) our scheme for voting on multiple versions is equivalent to the use of optimal adaptive vote-takers [20]. In fact, almost any adaptive redundancy scheme can be modelled by specifying suitable d-tag manupulation algorithms.

4. Data and design diversity

In this section, we present simple pedagogical examples to show how d-tags can be used in conjunction with data and design diversity as an aid for the management of resources in a selective approach to redundancy. These "toy" problems adequately serve the purpose of conveying the essence and advantages of the proposed approach. Issues relating to the use of d-tags in some realistic applications are dealt with in Section 5.

4.1. Use of d-tags with diverse data

Amman and Knight [1] have formulated the concept of data diversity for software fault tolerance by taking advantage of the fact that a program failing on one set of data may produce correct results when the input data is reformulated before running the program. Depending on the application and implementation details, program results with the reformulated data may be identical to, approximations of, or equivalent to (representing in a different form the same information as) the original results that were saught. For example a program computing the area s of a rectangle from the length z of its two diagonals and the angle α between them may fail (obtain incorrect result) if run to compute

$$s = b(z, \alpha) = 0.5z^2 \sin \alpha \tag{27}$$

but may succeed in computing $s = b(z, \pi - \alpha)$ which yields the exact value for the desired result or $s' = b(z/100, \alpha)$ which yields the result in square meters rather than in square centimeters. Failure or success is determined by running an acceptance test on the results. This acceptance test is similar in nature to the acceptance tests required in implementing the recovery block scheme [21] for software fault tolerance.

Using the discretization scheme defined by (3), suppose that the input data z and α have d-tags of 7 (i.e., that they are perfectly dependable) and that the operation b and the reformulations to obtain $\pi - \alpha$ and $z/100$ can be performed with dependability levels 5, 6, and 6, respectively. Then, we can compute the d-tags associated with various results as:

$$b^*(\langle z, 7 \rangle, \langle \alpha, 7 \rangle) = \langle b(z, \alpha), g(7,7,5) \rangle = \langle s_1, 5 \rangle \tag{28}$$

$$b^*(\langle z, 7 \rangle, \langle \pi - \alpha, 6 \rangle) = \langle b(z, \pi - \alpha), g(7,6,5) \rangle = \langle s_2, 4 \rangle \tag{29}$$

$$b^*(\langle z/100, 6 \rangle, \langle \alpha, 7 \rangle) = \langle b(z/100, \alpha), g(6,7,5) \rangle = \langle s_3, 4 \rangle \tag{30}$$

In the proposed data-driven dependability assurance scheme, if the d-tag value 5 obtained from the primary computation (28) is unacceptably low, the secondary computation (29) will be invoked. In the case of a matching result s from the two computations, the (perfectly dependable) comparison c yields:

$$c(\langle s, 5\rangle, \langle s, 4\rangle) = \langle s, 6\rangle \tag{31}$$

The d-tag obtained in (31) has the highest possible value for this particular example. However, if it had turned out to be inadequate or if there had been a mismatch between $b(z, \alpha)$ and $b(z, \pi - \alpha)$, the tertiary computation (30) could have been invoked and the result determined through voting. Here are some possible outcomes for the voting process:

$$\text{if } s_1 = s_3 \neq s_2 \text{ then } v(\langle s_1, 5\rangle, \langle s_2, 4\rangle, \langle s_3, 4\rangle) = \langle s_1, 6\rangle \tag{32}$$

$$\text{if } s_1 \neq s_2 = s_3 \text{ then } v(\langle s_1, 5\rangle, \langle s_2, 4\rangle, \langle s_3, 4\rangle) = \langle s_2, 5\rangle \tag{33}$$

$$\text{if } s_1 = s_2 = s_3 \text{ then } v(\langle s_1, 5\rangle, \langle s_2, 4\rangle, \langle s_3, 4\rangle) = \langle s_1, 6\rangle \tag{34}$$

$$\text{if } s_1 \neq s_2, s_2 \neq s_3, s_3 \neq s_1 \text{ then } v(\langle s_1, 5\rangle, \langle s_2, 4\rangle, \langle s_3, 4\rangle) = \langle s_1, 5\rangle \tag{35}$$

The difference between the above suggested method and the data diversity scheme as defined by Amman and Knight [1] is that in the latter, alternate or "reformulated" data enter the picture when an acceptance test is not satisfied. Because acceptance tests seldom provide perfect coverage, it may be advisable to perform the computation with several sets of data before making the final decision. Dependability tags provide a convenient mechanism for keeping track of potential errors and inaccuracies and for algorithmically combining a multiplicity of potentially incorrect values into dependable final results.

4.2. Use of d-tags with diverse designs

We now reconsider the example introduced in Subsection 4.1 with diverse designs b_1, b_2, and b_3 for the binary operator b. The three designs or versions [3] may be assigned various levels of dependability derived from data collected on their past performance or based on information about the designer or design team that produced the version. Let us assume dependability levels 5, 4 and 4 for the three versions b_1, b_2 and b_3, respectively. Then:

$$b_1{}^*(\langle z, 7\rangle, \langle \alpha, 7\rangle) = \langle b_1(z, \alpha), g(7,7,5)\rangle = \langle s_1, 5\rangle \tag{36}$$

$$b_2{}^*(\langle z, 7\rangle, \langle \alpha, 7\rangle) = \langle b_2(z, \alpha), g(7,7,4)\rangle = \langle s_2, 4\rangle \tag{37}$$

$$b_3*(\langle z,7 \rangle, \langle \alpha,7 \rangle) = \langle b_3(z, \alpha), g(7,7,4) \rangle = \langle s_3,4 \rangle \tag{38}$$

Here, we assume statistical independence of errors in the outputs produced by the three versions. We will not try to justify this assumption which is the subject of an ongoing debate in the field of design diversity. However, we note that it is not difficult to extend the proposed technique to deal with correlated errors if data on their probabilities is available. With this assumption, the results of (36), (37), and (38) can be combined through the use of comparison and/or voting algorithms in much the same way as was done for (28), (30), and (31) in the case of data diversity.

In all the discussions and examples thus far, a result d-tag was computed as a function of the operation d-tag and input data d-tags. We now present an example where it is beneficial to make the result d-tag also dependent on the nature of the operation and the input data *values* (not just their d-tags). Consider the "consistent comparison problem" defined by Brilliant, Knight, and Leveson [7]. The problem, simply stated, is as follows. If multiple versions of a system with diverse designs are used for performing an inexact computation, slight computation errors do not necessarily produce correspondingly small inaccuracies in the final results of the versions. One reason is that when a version reaches a decision point where one of two algorithm paths must be selected based on a comparison of inexact values that happen to be nearly equal, either path may be selected with non-negligible probability and the result becomes unpredictable.

We can solve the consistent comparison problem through the use of d-tags if the d-tag for the result of an inexact comparison is made a function of the relative difference of the two values being compared; i.e., if the comparison result is assigned a relatively high d-tag value when one operand is much larger than the other and a very low d-tag value when the two are nearly equal. The way in which the magnitude of the difference affects the d-tag of the comparison result will of course be application-dependent. When such a comparison result becomes the basis for a path selection, its dependability will affect the d-tags of all data that are dealt with subsequently. Again, routine comparison of and/or voting on multiple results and their d-tags yield the final result along with an indication of its dependability.

4.3. Combined data and design diversity

We again return to the example introduced in Subsection 4.1 and consider the availability of the lengths x and y of the two sides of the rectangle, and the radii r and R or its inscribed and circumscribed circles, in the composite tagged forms $\langle x,7 \rangle$, $\langle y,7 \rangle$, $\langle r,7 \rangle$, and $\langle R,7 \rangle$; i.e., all with perfect dependabilities. The values x, y, r, and R may have been measured independently or they may have been obtained from known z and α through the reformulations $x = z \sin(\alpha/2)$, $y = z \cos(\alpha/2)$, $r = 0.5 \min(x, y)$, and $R = 0.5z$. In either case, the pairs of values (z, α), (x, y), and (r, R) are three diverse representations of the same rectangle. The area of the rectangle can be computed by any of the three diverse computations:

$$b_1(z, \alpha) = 0.5z^2 \sin \alpha \tag{39}$$

$$b_2(x, y) = xy \tag{40}$$

$$b_3(r, R) = 4r \sqrt{(R^2 - r^2)} \tag{41}$$

These diverse computations also use diverse data. Let us assume dependability levels 5, 6, and 4 for the three versions b_1, b_2, and b_3, respectively. Then:

$$b_1^*(\langle z,7 \rangle, \langle \alpha,7 \rangle) = \langle b_1(z, \alpha), g(7,7,5) \rangle = \langle s_1,5 \rangle \tag{42}$$

$$b_2^*(\langle x,7 \rangle, \langle y,7 \rangle) = \langle b_2(x, y), g(7,7,6) \rangle = \langle s_2,6 \rangle \tag{43}$$

$$b_3^*(\langle r,7 \rangle, \langle R,7 \rangle) = \langle b_3(r, R), g(7,7,4) \rangle = \langle s_3,4 \rangle \tag{44}$$

Results obtained in this way can again be combined to raise the dependability level if required.

One complication that must be dealt with is the comparison of and voting on multiple inexact results. This problem is also present when design diversity is used without data diversity. However, the use of data diversity may intensify the problem, particularly if diverse representations of the results are obtained. In the latter case, there may be a need for sophisticated "voting" schemes that deal with non-atomic data objects. Simple voting is based on retaining the matching majority and discarding the non-matching minority. However, for complex data objects, this approach is overly pessimistic because it leads to the invalidation of a non-matching data object in its entirety as a result of a minor error. Component-by-component voting may be inapplicable because the objects are incomplete or non-identical in structure. Again d-tags provide a mechanism for keeping track of error probabilities and for algorithmically combining several potentially incorrect composite data objects into dependable final results.

From the above discussion, it is evident that the proposed data-driven dependability assurance scheme, when combined with data and design diversity, provides a powerful mechanism for selective use of redundancy that is both simple and efficient. The claimed efficiency stems from the fact that redundant computations are invoked only when needed to raise the dependability levels. A final example will show why this is conceptually an important step forward. Consider a real-time system that has to complete a particular critical computation by a given deadline.There may be several algorithms that can accomplish the task and several processors that can run the algorithms. If the worst-case running time of an algorithm is known, it can be initiated at the proper instant to meet the deadline. At that instant, the required (diverse) data for the algorithm may have low dependabilities, so the algorithm produces low-dependability results. However, such results are far from worthless, because several low-dependability values can reinforce one another and thus yield a highly dependable final result. The alternative to the above scheme would have been to wait until the required data had acquired higher dependabilities through dependability-raising operations. But if this occurred too late, the specified deadline might have become impossible to meet even by the fastest algorithm available.

5. A class of critical systems

Certain critical applications are characterized by a data flow path starting at a large number of sensors and terminating at one or more actuators. Data collected by sensors are forwarded (periodically) to data logging equipment. A number of evaluators manipulate the data in the course of computing various conditions. The condition outcomes are combined to reach certain decisions. Finally, the decisions form the inputs to the actuators. The data flow path is depicted below:

$$\text{Sensors} \rightarrow \text{Loggers} \rightarrow \text{Evaluators} \rightarrow \text{Deciders} \rightarrow \text{Actuators} \qquad (45)$$

Redundancy is typically applied to all system components, from sensors to actuators, in an effort to increase the overall system dependability. Two examples that have been well-documented in the literature are safety shutdown systems for nuclear reactors [27], [6] and interceptor launch enabler for a radar tracking system [12], [9].

In what follows, we sketch the application of the proposed data-driven dependability assurance scheme to the design of such a critical system and point out some of the advantages as well as potential problem areas. Refinements of these techniques as well as application of similar methods to other classes of critical systems are currently under investigation.

5.1. Sensors and data loggers

Data from each sensor can be tagged to reflect its dependability. The relevant d-tags can come from tables compiled based on past experience with each type of sensor. Such tables can be stored in ROMs within data loggers in order to avoid accidental modification or erasure. Although dynamic updating of such tables is conceptually feasible, it may cause problems and is not recommended in this context. There are certain advantages to this approach. First, it provides the flexibility of dealing with devices having varying degrees of dependability with a uniform strategy. In other words, the system design framework does not change from one application to another; only the required replication factors may be affected. Second, the design of the rest of the system becomes independent of the sensor subsystem. For example more reliable sensors may be provided later on to improve the overall system reliability or to increase its performance (because with more dependable sensor data, fewer dependability procurement features may need to be invoked in subsequent stages), all with no modification in the structure of the system. Such modularity of design is clearly desirable.

Data loggers do not perform complicated functions and thus the main issue in their operation is that of hardware reliability. Replication with identical or diverse designs can be used for dependability assurance. Associated with each logger, is a d-tag that is combined with d-tags for sensor data before passing the data on to the condition evaluators. In fact, for added protection, the d-tag combination function may be performed distributively. Here is one possible scheme that is applicable throughout the system. Each unit performing a computation $f(D)$, passes two d-tags to the next unit: d-tag of the input data D and the combined computed d-tag for the result $f(D)$. The receiving unit combines the d-tag for D with a stored d-tag for the function f as performed by the sending module, compares the result to the d-tag suggested for $f(D)$, and chooses the smaller value as the d-tag to be used for the value $f(D)$. This scheme provides protection against improperly raised d-tags in the sending or receiving unit.

The problem of assigning suitable d-tags to various hardware and software entities is inherently difficult. However, this problem exists in all multi-version design paradigms and is not unique to our approach (i.e., how does one decide that 3 or 5 versions should be used without somehow quantifying the dependability of each version?). From a practical viewpoint, the initial assignment of d-tags to data objects is quite similar to the quantification problem encountered in certain expert systems such as those applied in the area of medical diagnosis [8] where a "strength" factor should be assigned, based on the opinion of an expert, to the belief that a given evidence (symptom) supports a particular hypothesis (diagnosis).

5.2. Condition evaluators

Condition evaluators detect the existence of certain conditions based on their input data. This is normally the most compute-intensive part of the application and where software diversity is most needed. The output may be a set of binary true/false indicators or numerical values signifying the degree to which various conditions are satisfied. It may be argued that these conditions are not significant in themselves and that it suffices to compare and combine the decider outputs; i.e., that design diversity should be applied to the combination of evaluator and decider functions. However, there is always a danger in comparing unreliable values from a small space of possibilities. In fact several researchers that have experimented with diverse designs for such systems have been forced to compare intermediate values as well as final outputs to guard against faults in the decision logic (e.g., [27], p. 38, and [12], p. 99). By making the conditions explicit, separating the evaluation and decision functions, and applying design diversity to each part, dependable results can be obtained with a potentially lower level of redundancy.

We illustrate this last point by means of a simple example. Suppose that three binary conditions x, y, and z need to be evaluated and that there are four versions of the corresponding software. Initially, three versions are selected and run, each producing a set of satisfied conditions as its output. If, for example, all three versions produce the set $\{x, y\}$, then the deciders will attach high d-tag values to x and y and will proceed to compute their decisions. This is the expected situation in most cases, and thus we usually avoid running the fourth version. However, suppose that the three output sets are:

$$\{x, y\}, \quad \{x, y, z\}, \quad \{y\} \tag{46}$$

That is, all three versions agree that y is satisfied, two versions agree that x is satisfied, and one version "thinks" that z is satisfied. Even in this case, the deciders may be able to function reliably despit the low d-tag value for z. If the decision function is such that the value of z is critical (this will show up as a low d-tag value for the decision result), then the fourth version that was initially ignored for efficiency reasons may be invoked and its output set used to raise the operands' d-tags.

In this context, it may even be possible to use unverified versions of some evaluators in the system by assigning very low d-tag values to them and using idle system resources (extra processors in a parallel processor or available capacity between real-time iterations) to compile data on their performance. This scheme is similar to an established method for evaluating the suitability of multiple-choice questions in standardized tests by hiding among the "real" questions, special "trial" questions which are ignored in scoring the test. As confidence builds up in the dependability of a particular version, its associated d-tag can be modified by the system administrator and the newly configured system will continue to run with no need for modifying its structure.

5.3. Deciders and actuators

Deciders are relatively simple procedures that deal with a limited amount of data (the conditions). In the case of binary conditions, a decider may be characterized by a Boolean expression such as $xy + yz$. It is easy to see that sometimes dependable decisions may be reached with conflicting data. For example, if with the conventions discussed in Subsection 5.2 the condition sets for three evaluator versions are $\{x,y\}$, $\{x,y,z\}$, and $\{y,z\}$, then the resulting "true" decision will have a high d-tag value despite the relatively low d-tag values associated with x and z on which all three versions do not agree.

A potential complication here arises when a decision expression involving negated conditions is to be evaluated. For example, if the decision is represented by $xy + y'z$ and the condition sets of the three versions are $\{x,y\}$, $\{x,y,z\}$, and $\{y,z\}$ as before, then it is not clear what d-tag value can be attached to the truth of y'. The difficulty lies in the fact that a computed d-tag for the truth of y provides a lower bound on the probability that y is true. This can be used to compute an *upper bound* on the probability that y' is true but not a lower bound on the same. One solution is to have each version compute for each condition c the truth of both c and c' with corresponding d-tags.

The design of dependable actuators to take the data d-tag values into account is relatively straightforward based on simple extension of standard design practices for such devices. Typically, an actuator has a "safe" state. If the safe state corresponds to no action, then the actuator may be designed to operate only if all versions or a large majority agree that it should do so (e.g., through the coincidence of several forces, the sum of which exceeds the device's operating threshold). On the other hand, if the safe state corresponds to some action such as nuclear reactor shutdown, then a single or a relatively small number of versions should be able to trigger the actuator. In rare situations where there is no preference with respect to the safety of the two states, simple majority might be the rule. The inclusion of d-tags adds some complexity to the process. A reasonable approach would be to reject values whose d-tags fall below a certain threshold and then consider the remaining values as "equally good" in computing the final decision in order to simplify the electromechanical decision component of the actuator.

6. Conclusion

Much of the literature on dependable computing is built on the binary distinction between correct and incorrect data. Thus, any data object, such as a word or a record, is either completely dependable or totally incorrect. Various levels of data accuracy are sometimes taken into account, but accuracy is quite different from dependability. Whenever "gray-level" dependability is considered, it is with respect to functional units that generate or transform and channels that carry the data, rather than the data objects themselves. A notable exception is the notion of "signal reliability" [16] which was introduced for the analysis of digital logic circuits to determine the probability of a correct output being generated. However, the application of such analyses is limited to bit-level results and even then it is practically restricted to relatively small logic circuits. Clearly, different data objects, or even diverse representations of the same object, can have different "error characteristics" when manipulated by identical functional units or transmitted over identical channels. Thus, the desirability of a data-driven approach to dependability assurance.

In a way, the data-driven dependability assurance method is nothing but replication plus comparison and/or voting. The major difference with existing methods of dependability procurement based on the same concepts is that the replication factors and the invocation of voting and consistency check procedures are dynamically controlled by the d-tags associated with various data objects. The replication factor may be not only a function of the particular data object (structure and contents) but also dependent on the access method and frequency for a given critical computation. Although custom design based on a-priori estimates of data object size and access patterns may be effective in some cases, it is both expensive and ineffective when there are wide dynamic variations in the relevant parameters. Dependability tags, as defined here, provide a convenient mechanism for keeping track of potential errors and for algorithmically combining a multiplicity of potentially incorrect data into dependable results.

In addition to data and design diversity discussed in Section 4 of the paper, at least two other application areas can greatly benefit from the proposed method; namely dataflow computations and distributed systems. In the area of dataflow architectures, where dependability studies [15], [11], [24], [10] have been few, sketchy, and inconclusive, data-driven dependability procurement is a natural method and is expected to lead to a satisfactory approach for ultrareliable dataflow computation. One can envisage data tokens carrying one or more dependability-related tags that are manipulated according to specific rules as the computation unfolds. At the very end, results will have dependability indicators attached to them. If in the course of a computation, dependabilities fall below a certain threshold, program segments holding backup or alternate computations are activated in order to check or verify the obtained results. Such verifications will raise the dependability values and will allow the computation to proceed.

For distributed computer systems, important theoretical contributions have established the difficulty of the dependability procurement problem but few have led to practical design strategies. The use of d-tags (along with suitable distributed manipulation algorithms) is likely to lead to a solution to this problem as well. Consider for example a distributed database consisting of various data objects (e.g., relations in the relational database model), each with multiple copies stored at different sites. Each processor may have copies of several data objects with their associated d-tags. Exchanging information about the data objects between processors can modify the d-tag values dynamically, restoring the gradually deteriorating dependabilities to acceptable levels and correcting erroneous information which may have aquired high d-tag values due to malicious faults. Developing the detailed protocols and decision algorithms for this class of applications constitutes a fruitful area for further investigation. Note in particular that d-tags provide a convenient mechanism for implementing probabilistic agreement protocols. Each processor can be viewed as working on the construction of an "agreement" data object that it initializes with its own view of the world and a low associated d-tag value. As information is exchanged between processors, "agreement" data objects gradually acquire higher associated d-tag values. The process of information exchange, which itself is driven by the d-tags, continues until desired dependability levels are reached.

In Subsection 3.4, we discussed the potential for using the data-driven dependability procurement paradigm to build a unified framework for the automatic synthesis of ultrareliable systems. Given that such a framework can be established, reduction of the design time and cost will appear as a new factor in comparing the cost-effectiveness of various approaches to dependability assurance and may well tip the balance in favor of the data-driven approach in additional application areas.

Currently, research is in progress on methods for dealing with erroneous d-tags and imperfect operations, as discussed in Subsections 3.1 and 3.2. To deal with erroneous d-tags, coding schemes that take advantage of the asymmetric nature of d-tag errors (i.e., safety of negative errors and criticality of positive errors) must be examined. As for imperfect operations, data manipulation as well as comparison and voting functions must be dealt with. Work is also in progress on the design of voting schemes for composite data objects that take advantage of correct or partially correct components of an incorrect composite object to yield a more dependable result [19].

In this paper, we have considered comparison and voting as the only dependability-raising operations. Clearly, consistency checks can be viewed as performing similar functions. In order to quantify the effect of a consistency check, dependability parameters must be attached to both the check itself (the function) and the procedure implementing the check. A simple acceptance test may be functionally unreliable but easy to implement whereas a more "complete" check may be highly dependable but correspondingly more difficult to implemet. The design of fail-safe comparison and voting schemes constitutes another line of research that can bring the proposed data-driven dependability assurance scheme closer to practical use.

References

[1] P.E. Ammann and J.C. Knight, "Data Diversity: An Approach to Software Fault Tolerance", *IEEE Transactions on Computers*, Vol. 37, No. 4, pp. 418-425, Apr. 1988.

[2] T. Anderson, P.A. Barrett, D.N. Halliwell and M.R. Moulding, "Tolerating Software Design Faults in a Command and Control System", in [26], pp. 109-128.

[3] A. Avizienis and J.P.J. Kelly, "Fault Tolerance by Design Diversity: Concepts and Experiments", *Computer*, Vol. 17, Aug. 1984, pp. 67-80.

[4] A. Avizienis, H. Kopetz and J.-C. Laprie (Editors), *The Evolution of Fault-Tolerant Systems* (Dependable Computing and Fault-Tolerant Systems, Vol. 1), Springer-Verlag, Wien, 1987.

[5] A. Avizienis, M.R.T. Lyu, W. Schutz, K.-S. Tso and U. Voges, "DEDIX 87 -- A Supervisory System for Design Diversity Experiments at UCLA", in [26], pp. 129-168.

[6] P.G. Bishop, "The PODS Diversity Experiment", in [26], pp. 51-84.

[7] S.S. Brilliant, J.C. Knight and N.G.Leveson, "The Consistent Comparison Problem in N-Version Software", *Software Engineering Notes*, ACM SIGSOFT, Vol. 12, No. 1, pp. 29-34, Jan. 1987.

[8] B.G. Buchanan and E.H. Shortliffe, *Rule-Based Expert Systems: The MYCIN Experiments of the Stanford Heuristic Programming Project*, Addison-Wesley, 1984 (in particular Chapters 10-12, pp. 209-271).

[9] J.R. Dunham, "Experiments in Software Reliability: Life-Critical Applications", IEEE Transactions on Software Engineering, Vol. SE-12, No. 1, pp. 110-123, Jan. 1986.

[10] J.L. Gaudiot and C.S. Raghavendra, "Fault Tolerance and Data-Flow Systems", *Proc. of the International Conf. on Distributed Computing Systems*, San Francisco, May 1985, pp. 16-23.

[11] J.L.A. Hughes, "Error Detection and Correction Techniques for Dataflow Systems", *Proc. of the International Symp. on Fault-Tolerant Computing*, Milano, June 1983, pp. 318-321.

[12] J.C. Knight and N.G. Leveson, "An Experimental Evaluation of the Assumption of Independence in Multiversion Programming", *IEEE Transactions on Software Engineering*, Vol. SE-12, No. 1, pp. 96-109, Jan. 1986.

[13] J. Losq, "A Highly Efficient Redundancy Scheme: Self-Purging Redundancy", *IEEE Transactions on Computers*, Vol. C-25, No. 6, pp. 569-578, June 1976.

[14] F.P. Mathur and A. Avizienis, "Reliability Analysis and Architecture of a Hybrid-Redundant Digital System: Generalized Triple Modular Redundancy with Self-Repair", *AFIPS Conf. Proc.*, Vol. 36 (Spring Joint Computer Conf.), AFIPS Press, Montvale, NJ, 1970, pp. 375-383.

[15] D.P. Misunas, "Error Detection and Recovery in a Data-Flow Computer", *Proc. of the International Conf. on Parallel Processing*, Aug. 1976, pp. 117-122.

[16] R.C. Ogus, "The Probability of a Correct Output from a Combinational Circuit", *IEEE Transactions on Computers*, Vol. C-24, No. 5, pp. 534-544, May 1975.

[17] B. Parhami, "From Defects to Failures: A View of Dependable Computing", *Computer Architecture News*, ACM SIGARCH, Vol. 16, No. 4, pp. 157-168, Sep. 1988.

[18] B. Parhami, "A New Paradigm for the Design of Dependable Systems", *Proc. of the International Symp. on Circuits and Systems*, Portland, OR, May 1989, pp. 561-564.

[19] B. Parhami, "Voting Schemes for Composite Data Objects and Their Reliability Analyses", in preparation.

[20] W.H. Pierce, "Adaptive Decision Elements to Improve the Reliability of Redundant Systems", *IRE International Convention Record*, Mar. 1962, pp. 124-131.

[21] B. Randell, "System Structure for Software Fault Tolerance", *IEEE Transactions on Software Engineering*, Vol. SE-1, No. 2, pp. 220-232, June 1975.

[22] M. Raynal, *Networks and Distributed Computation: Concepts, Tools, and Algorithms*, MIT Press, 1988.

[23] E.H. Shortliffe and B.G. Buchanan, "A Model of Inexact Reasoning in Medicine", in [23], pp. 233-262.

[24] V.P. Srini, "A Fault-Tolerant Dataflow System", *Computer*, Vol. 18, pp. 54-68, Mar. 1985.

[25] A.H. Veen, "Dataflow Machine Architecture", *Computing Surveys*, Vol. 18, No. 4, pp. 365-396, Dec. 1986.

[26] U. Voges, (Editor), *Software Diversity in Computerized Control Systems* (Dependable Computing and Fault-Tolerant Systems, Vol. 2), Springer-Verlag, Wien, 1988.

[27] U. Voges, "Use of Diversity in Experimental Reactor Safety Systems", in [26], pp. 29-49.

IMPLEMENTING DISTRIBUTED CAPABILITIES WITHOUT A TRUSTED KERNEL

Maurice P. HERLIHY [1] *, J. D. TYGAR*
Computer Science Dept., Carnegie Mellon University
Pittsburgh, PA 15213 - USA

Abstract

Capabilities are well-known to be a simple and efficient technique for implementing protection in centralized systems. In decentralized distributed systems, however, implementing capabilities can be considerably more difficult. Two problems stand out: (1) how to communicate information about capabilities across an insecure communication network, and (2) how to revoke capabilities in the presence of failures such as message delays, crashes, and network partitions. This paper describes a new scheme for managing capabilities in a distributed system that incorporates novel solutions to both problems. The communication problem is addressed by a new and efficient protocol that exploits recent developments in "zero-knowledge" authentication protocols. The revocation problem is solved by new protocols that rely on approximately synchronized real-time clocks to create the illusion that revocation occurs instantaneously, even in the presence of failures.
This research was sponsored by the Defense Advanced Research Projects Agency (DOD), ARPA Order Numbers 4976, monitored by the Air Force Avionics Laboratory under Contracts F33615-84-K-1520. J.D. Tygar received additional support from National Science Foundation Presidential Young Investigator Grant CCR-8858087.

---

[1] Author's current address: Digital Equipment Corporation, Cambridge Research Laboratory, One Kendall Square, Cambridge, MA 02139

1. Introduction

Capabilities are well-known to be a simple and efficient technique for implementing protection in centralized systems. In decentralized distributed systems, however, implementing capabilities can be considerably more difficult. Two problems stand out: (1) how to communicate information about capabilities across an insecure communication network, and (2) how to revoke capabilities in the presence of failures such as message delays, crashes, and network partitions. This paper describes a new scheme for managing capabilities in a distributed system that incorporates novel solutions to both problems. The communication problem is addressed by a new and efficient protocol that exploits recent developments in "zero-knowledge" authentication protocols. The revocation problem is solved by new protocols that rely on approximately synchronized real-time clocks to create the illusion that revocation occurs instantaneously, even in the presence of failures.

2. Model

A *distributed system* consists of a collection of computers, called sites, that are geographically distributed and connected by a communications network. A distributed program is organized as a collection of modules, each of which resides at a single site. Modules communicate through messages, acting as clients and as servers. A *client* makes use of services provided by other modules, while a *server* provides services to others by encapsulating a resource, providing synchronization, protection, and crash recovery. The client/server model is hierarchical: a particular module may be both a client and a server.

Not every client is allowed access to every server. Each client has a set of rights, where a right is permission to make use of services provided by certain servers. A client may delegate those rights to other clients, and rights may be revoked. A well-known way to implement such protection is to associate each right with a *capability*, a token which is produced by a client having that right. Certain priviledged servers are allowed to create capabilities and to distribute them to deserving clients. The server that created the capabilities is called their *owner*. In addition, clients may delegate rights by exchanging capabilities among themselves. A capability's owner may revoke clients' rights by rendering that capability invalid. If client *A* has transferred a capability to client *B*, then if *A*'s right to use that capability is revoked, so is *B*'s. It has long been recognized that capability systems are as powerful as full access control matrices for protection

[17]. For this reason, numerous operating systems have used capability mechanisms as the basic means of implementing protection [36, 6, 8, 18, 35].

In a distributed environment, capability systems have typically been implemented by storing capability based information in a distributed kernel, or by having trusted kernels exchange information about capabilities [37, 8] . By contrast, in this paper, we propose a capability scheme where each client stores its own authentication information. When a client wishes to begin a transaction with a server, the client must first present the security information to the server. We call this approach *self-securing*, since each client and server is responsible for monitoring security concerns, and not the underlying operating system kernel. In this paper, we do not presume that the data network is secure; neither do we presume that the base operating system provides security other than address space protection. (We do not, however, attempt to address denial of service problems here. For some theoretical contributions to this problem, see [13].) This paper addresses three problems: (1) What form do capabilities take and how are they distributed? (2) How can we guarantee that individual clients will not forge or steal capabilities? (3) How can capabilities be revoked?

To illustrate the issues involved, consider the following unsatisfactory technique for implementing capabilities in a distributed environment. Each capability is associated with a password. Each site keeps a public list of capabilities, where each capability is associated with a password encrypted by a one-way function (as in [24]). A client who wishes access to a server transmits the password corresponding to that capability. Access is permitted only if the encrypted version of the capability password matches the version in the published list. Clients exchange capabilities simply by transferring passwords. This simplistic protocol has two problems:

- Since messages are not presumed to be secret, an eavesdropper could steal the password, and hence the capability.

- There is no provision for revocation.

Nonetheless, this example suggests that an authentication method, if it could be made secure and revocable, could act as the foundation for constructing self-securing programs in a capability based system. This paper presents a new class of efficient authentication methods which do not suffer from drawbacks existing authentication schemes have, as well as an examination of several revocation schemes.

2.1. Zero knowledge authentication

Authentication is at the heart of the security system for any loosely-coupled distributed operating system. How can client A and server B prove their identities to one another? The problem is difficult because A and B may reside on different sites, thus they must communicate by exchanging messages across a potentially vulnerable communications network. Since messages transmitted over the network may be intercepted by a third party, C, A and B must be able to prove their identities without revealing information which would allow C to successfully feign an identity as A or as B.

How well do existing authentication methods accomplish this goal? In practice, not very well. For example, Rivest, Shamir, and Adleman proposed an authentication method based on the RSA public-key signature methods [32]. In their protocol, values are encrypted according to a public-key encryption e function $E(m)=m^e \bmod n$, where m is a message, e is an encryption key, and n is the product of two large primes p and q. Decryption is accomplished through the function $D(c)=c^d \bmod n$, where C is the ciphertext, d is chosen so that $ed=1 \bmod (p-1)(q-1)$. It is true there is no published method for quickly decrypting messages given only e and n, and not the factorization of n. Nevertheless, this method leaks information. For example, Lipton points out that the well known Legendre function L satisfies the relation $L(m,n)=L(E(m),n)$ [19]. Indeed, the problem is much worse. Alexi, Chor, Goldreich, and Schnorr recently proved that if an adversary can find the low order bit of m $50\%+\varepsilon$ of the time given $E(m)$, she can invert arbitrary RSA encryptions [1, 5]. A corollary to this result is that the usual query-response methods for encryption, such as the family of protocols described in [23], an adversary can 2 emulate another client or server after engaging in $O((\log n)^2)$ authentications.

Needham and Schroeder have suggested an authentication method which uses private-key cryptographic methods [26]. Needham and Schroeder's work presupposes a secure key distribution method and a private-key cryptosystem. Recent work by Luby and Rackoff suggests that authentication methods depending on DES [25] are vulnerable to a "low-bit" attack similar to the one mentioned above for the RSA cryptosystem [20]. For example, the second author has found a method to subvert the authentication scheme used by the Andrew File System VICE [34, 33], which uses a strategy similar to Needham and Schroeder's.

To give users confidence in a system, we would like to be able to prove that an authentication method does not leak information. Several researchers have independently proposed protocols, termed *zero-knowledge protocols*, which satisfy this constraint, given a complexity assumption that P ≠ NP [9, 3, 10, 7]. To convey a flavor of arguments used, we summarize a zero-knowledge protocol below by which A can prove to B that some graph G with n vertices known to both A and B contains a k-clique (that is, a set Q of k vertices such that between every two vertices in Q there exists an edge). (This version of the proof is due to M. Blum.) Let G be a graph with n vertices known to both A and B. Suppose that A knows a k-clique in G. Since the problem of finding a k-clique in an arbitrary graph is NP-complete, B can not in general find the k-clique. This protocol will allow A to prove to B that G has a k-clique without revealing any information about the vertices in Q.

1. A secretly labels each vertex of G with random unique integer from 1 to n.

2. A prepares $n(n-1)/2$ envelopes labeled uniquely with a pair of integers $\langle i,j \rangle$, $i<j$. A puts "Yes" in the envelope labeled $\langle i,j \rangle$ if an edge exists between the vertices labeled i and j, and "No" otherwise.

3. A seals the envelopes and presents them to B. B flips a coin and reports its value to A. If the value is heads, A must open all the envelopes and show the numbering of the vertices of G. B then verifies that the descriptions are correct. On the other hand, if the value is tails, A must then open only the envelopes which are labeled $\langle i,j \rangle$ where i and j belong to Q. B then verifies that all envelopes contain "Yes".

4. The above protocol is repeated t times (with an independent random numbering assigned each time in step 1). If A successfully responds B's queries, the probability that A does know a proof is 2^{-t}.

Clearly, if A knows a k-clique and correctly follows the above protocol, A will succeed. On the other hand, suppose C is trying to masquerade as A. Since C does not know a k-clique, C has two choices: it can correctly perform step 2 (in which case it is caught whenever B gets tails) or it can put false values in the envelopes (in which case it is caught whenever B gets heads). Hence in each of the t iterations of the protocol the probability that C's ignorance revealed is 1/2. After t iterations, C will be caught with probability $1-2^{-t}$. Finally, notice that B acquires no information about the location of the clique. If B could find information about the clique from the above protocol, it could generate the same information by flipping a coin and generating a random numbering of the graph when it gets heads and a random numbering of a complete graph on k vertices when it gets tails.

Notice that since any problem lying in NP can be reduced to the NP-complete problem k-clique [14], A could use this protocol to prove to B that it had a proof or disproof of, for example, Fermat's Last Theorem. At the end of this protocol, B would be convinced that A did, in fact, have a proof or disproof without having any idea which way the problem was resolved, much less any idea of the technique used to solve the problem. In principle, this protocol could be used to generate authentication proofs: A would publish the graph G in a public white pages directory. To prove its identity, A would give a zero-knowledge proof of the existence of a k-clique in the graph.

In practice, however, this protocol would not work well. First, A would have to find a graph G in which it was computationally intractable to find a k-clique. While it is true that our complexity hypothesis guarantees that such graphs must exist, most random graphs with k-cliques, those cliques can be found through efficient heuristics. Second, A and B would have to develop a good cryptographic scheme for implementing "envelope exchange". For even a modest security level, the size of data involved here is on the order of 10^{200} bytes. Using the highest bandwidth transmission techniques available today, execution of this protocol would exceed the time remaining before the heat death of the universe.

2.2. An efficient protocol

In research described in [38], we have developed a family of zero-knowledge protocols which are efficient for real use in applications. That paper also gives timing figures for those algorithms. Below we give a simplified (and slightly less efficient) version of the protocol.

The protocol we use can depend on one of two complexity assumptions: that factoring large integers can not be done in polynomial time, or that it is hard to invert messages encrypted by random keys under DES. Other similar complexity assumptions may be used instead. The protocol described below depends on the complexity of factoring integers. We recall the following lemma by Rabin [28]:

> If there exists a polynomial time algorithm for finding square-roots modulo $n=pq$, where p and q are large primes, then we can factor pq in polynomial time.

Rabin observed that we can take a random integer r between 1 and pq-1; check that GCD$(r,n) \neq 1$ (if this value is p or q, then we have factored n). Calculate $x = r^2$ mod n and find a square-root s, so that $s^2 = x = r^2$ mod n. A simple number-theoretic argument demonstrates that x has four square-roots modulo n, including r and $-r$. Since r is chosen at random, there is a 50% chance that $s \neq \pm r$ mod n. If $s = \pm r$ mod n, then we can pick a new r and repeat the algorithm. If $s \neq \pm r$ mod n, then it is the case that GCD$(r+s,n) = p$ or q. Hence finding square-roots is equivalent to factoring.

In this protocol, we assume that the system manager publishes a product of two large primes $n = pq$, keeping the factorization secret. This n can be used for all authentication protocols, and no one need ever know its factorization. To 2 initialize its puzzle, A picks a random r and publishes $x = r^2$ in the white pages. I will prove it knows a square-root of x without revealing any information about the value.

Here is the protocol:

1. A computes t temporary random values, $v_1, v_2, ..., v_t$, where each v_i satisfies $1 \leq v_i \leq pq$-1. A sends to B the vector $\langle v_1^2$ mod n,v_2^2 mod n,...,v_t^2 mod n$\rangle$.

2. B flips t independent coins and sends back a vector of t random bits $\langle b_1,...,b_t \rangle$ to A.

3. For $1 \leq i \leq t$ A computes:

$$z_i = \begin{cases} v_i & \text{if } b_i = 0 \\ rv_i \text{ mo } n & \text{otherwise} \end{cases}$$

   A transmits the vector $\langle z_1, ..., z_t \rangle$.

4. B verifies that for $1 \leq i \leq t$, that:

$$z_i^2 = \begin{cases} v_i^2 \text{ mod } n & \text{if } b_i = 0 \\ xv_i^2 \text{ mo n)} & \text{otherwise} \end{cases}$$

   If the equalities hold, A has authenticated its identity to B with probability $1 - 2^t$.

Once again note that if A knows a value r such that $r^2 = x$ mod n, then it can easily follow the above protocol. Suppose C is trying to masquerade as A. Since C

doesn't know such a value r, it can not know both v_i and $rv_i \bmod n$, since $r = rv_i (v_i)^{-1} \bmod n$. Finally, all that B sees is a series of random values of the form $\langle z_i, z_i^2 \bmod n \rangle$. If B could find any information about r from the above protocol, it could do so by generating a set of t random values and squaring them modulo n, and thus factoring n.

The above protocol uses only an expected $3t$ multiplications to generate security of $1\text{-}2^{-t}$. Our improved protocol uses only expected $1.5t$ multiplications to achieve the same level of security.

2.3. Self-securing programs

As mentioned above, our algorithm presumes communication networks which are potentially vulnerable. A and B need to use methods to protect the privacy and integrity of their data. If we extend an authentication algorithm to support key exchange, A and B can transmit their messages though highly secure private-key encryption methods. We have adapted our algorithms to also perform this operation -- A can send a temporary key e_A and B can send a temporary key e_B. Both parties can then use a trusted private-key encryption method, such as DES with the key $e = e_A \oplus e_B$, where $\oplus$ is bit-wise exclusive-or. Hence, if A and B have protected address spaces, we can make all messages transmitted in the system public, since no observer can find the encryption key used. Also, sending messages encrypted by the key removes the need to re-authenticate until either party decides to establish a new temporary key. This approach yields a *self-securing program* which requires only a minimal amount of security in our base operating system.

Indeed, our algorithm obviates the need to ever use public-key cryptography. If A wishes to transmit a message to B without having a shared private key, A can simply authenticate itself to B exchanging a temporary encryption key. All further communications are protected by encryption. The signature functions of public-key cryptography can be performed by the fingerprinting algorithm described in the next Section.

3. Fingerprinting

Karp and Rabin introduced an algorithm which computes a *cryptographic checksum* [30, 15]. Their algorithm takes a bit string s of arbitrary length and secret key k of d bits (where d-1 is prime) and returns a *fingerprint* sequence of

d-1 bits $\Phi_k(s)$. Each key k defines a *fingerprint* function, and if the keys are k chosen with uniform distribution, the family of fingerprint functions Φ_k can be viewed as a provably good random hash function in the style of [4, 31]. Without the secret key, computing a fingerprint given a string of bits is intractable. On the other hand, if the secret key is known, it is easy to compute the fingerprint.

Given the fingerprint algorithm, the problem of protecting the integrity of data from alteration becomes much simpler. For example, to protect a file F, we could store $\langle F, \Phi_k(F) \rangle$. If an adversary attempts to alter the file by k replacing it with F', he will need to calculate $\Phi_k(F')$. But since the k adversary does not know k, he can not compute the fingerprint of F'. Even if the adversary attempts to find a F' with the same fingerprint $\Phi_k(F)=\Phi_k(F')$, he will be thwarted, since the problem of finding an k k input which generates a given fingerprint is intractable without the key value k.

The fingerprinting algorithm views a sequence of bits s as a polynomial $f_s(x)$ over the integers modulo 2. For example, the bit sequence $s=$ "100101001" is taken to be the polynomial $f(x)=x^8 +x^5 +x^3 +1$. The secret key for this algorithm corresponds to a random irreducible polynomial $g(x)$ of degree d-1 over the integers modulo 2. It is extremely easy to generate these polynomials (several approaches are outlined in [29, 30]) and we have implemented two different efficient routines for doing so. Compute $r(x) = f_s(x) \mod g(x)$. $r(x)$ is a polynomial over the integers modulo 2 of degree at most d-1. Both the polynomial $r(x)$ and the key k can be represented as a string of d bits. The bits produced by the algorithm define the Φ function.

4. Revocation

At any time, a capability can be revoked by its owner, rendering it invalid for all clients to which it has been transferred. Ideally, revocation should take effect instantaneously, ensuring that any subsequent attempt to use that capability will be refused. In a decentralized distributed system, however, it is difficult to make such a guarantee in the presence of variable message delays, site crashes and communication failures. Instead, we guarantee that revocation appears to be instantaneous to each individual client or server in the system, in the sense that no one should be able to observe that a capability has been used after it has been revoked.

To illustrate the meaning of this guarantee, consider the following simple example. A professor issues a homework assignment to each student in her class, together with a capability for turning in the completed assignment. The professor will accept late homeworks up until the time she gets around to publishing the sample solution. To keep students from resubmittting the published solution, the professor simply revokes the capability before publishing the sample. If revocation appears to be instantaneous, a student who has read the published solution no longer has the right to turn in that assignment.

As discussed in Section 5, our revocation scheme is an adaptation of an algorithm originally developed for a very different purpose: managing orphans in distributed transaction systems [22].

4.1. The basic mechanism

Each server has a clock. Our revocation method relies on the assumption that clocks can be kept approximately synchronized: the maximum skew between the clocks at two servers protected by the same capability is always less than some constant Δ. It is important that no adversary can trick two servers' clocks into diverging by more than Δ, thus if clock synchronization is maintained by periodic exchange of messages (e.g., [21]), then all such messages must be authenticated.

Each server protected by a capability maintains a *quiesce time* and a later *refusal time*. Subject to constraints described below, different servers may have different quiesce and refusal times for the same capability. If an invocation arrives at a server before the capability's local quiesce time has elapsed, the capability is considered valid, and the operation is allowed to proceed. If it arrives after the refusal time has elapsed, however, it is considered invalid, and the operation is refused. If the invocation arrives between the quiesce and refusal time, then the capability's status is indeterminate. The server postpones a decision to accept or reject the operation until either (1) the capability's lifetime is extended, in which case the operation is permitted, or (2) the refusal time elapses, in which case the operation is refused.

Let *First(Refusal(c))* denote the earliest refusal time for c at any server protected by *c*, and let *Last(Quiesce(c))* denote its latest quiesce time. The capability's quiesce and refusal times are subject to the following *revocation invariant*:

$$\text{Last(Quiesce(c))} - \text{First(Refusal(c))} > \Delta$$

The difference between a capability's latest quiesce time and its earliest refusal time must exceed Δ, the maximum clock skew. This invariant ensures that there is always some moment of real time, call it t_0, at which the 0 capability's quiesce time has elapsed at every server but its refusal time has not yet arrived. This t_0 is the moment at which revocation appears to take 0 effect.

We assume that the owner knows a lower bound for First(Refusal(c)) and an upper bound for Last(Quiesce(c)). In the next two sections, we describe a *revocation protocol*, a fault-tolerant protocol for revoking capabilities, and a *refresh protocol*, a fault-tolerant protocol for extending capabilities' lifetimes.

4.2. The revocation protocol

The simplest way for an owner to revoke a capability is just to wait for the earliest refusal time to elapse, at which time the capability will no longer be considered valid anywhere in the system. Nevertheless, if the refusal time is far in the future, it will be desirable to hasten revocation by active means. This section describes a *revocation protocol* that can be used to adjust the refusal time without violating the revocation invariant.

The revocation protocol is a two-phase protocol similar to the two-phase commit protocol [11]. For brevity, we use "all servers" as shorthand for "all servers protected by the capability in question". In the first phase, the owner sends a message to all servers, directing each one to reset its local quiesce time to the present. Each server returns a confirmation. The first phase is successful if the owner receives confirmations from all servers. If the first phase is successful, the owner sends a second message, directing all servers to reset their refusal times to the present. The owner can then resume execution. Although no confirmation messages are needed for phase two, they might reduce the likelihood of unnecessary delay at servers that missed the phase two message. If the first phase fails, revocation is blocked until the earliest refusal time elapses.

This protocol is *fail-safe*. By adjusting the quiesce and refusal times in two phases, we ensure that the revocation invariant is preserved at all times, even in the presence of site failures, lost messages, and network partitions. Such failures would adversely affect performance, but they cannot affect the correctness of revocation. In the presence of failures such as network partitions, it is difficult to imagine how one could make a stronger guarantee.

4.3. The refresh protocol

A disadvantage of the scheme as described so far is that capabilities have fixed lifetimes. If a server's quiesce time arrives before a legitimate client has finished, the client must request a new capability from the owner. To avoid this difficulty, an owner may periodically undertake a *refresh* protocol to advance each server's quiesce and refusal times. The interval between a site's current time and the quiesce time for any server is the *quiesce interval*, and the interval between the quiesce and refusal times is the *refusal interval*. The interval between refresh protocols is the *refresh interval*. These terms are illustrated in Figure 1. Unnecessary revocation will be unlikely if clocks are closely synchronized and if the refresh interval is significantly less than half the quiesce interval.

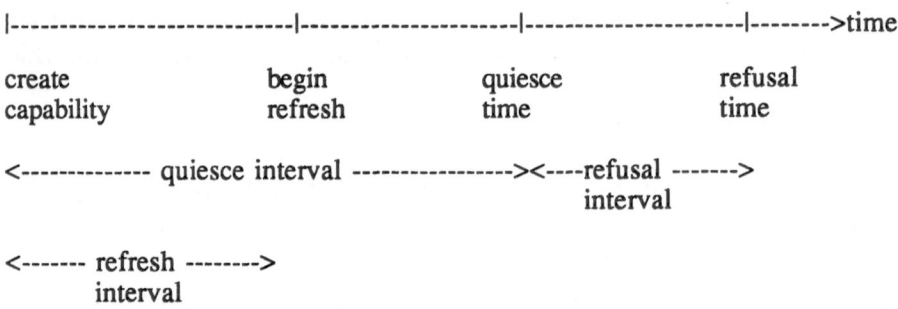

Figure 1. Quiesce, Refusal, and Refresh Intervals

The refresh protocol is a two-phase protocol similar to the revocation protocol In the first phase, the owner attempts to advance the capability's refusal time at all servers. If the first phase is successful, i.e., if all server confirm they have done so, the owner attempts to advance the capability's quiesce time at all servers. Here, too, the two phases preserve the revocation invariant, and hence the protocol is fail-safe in the presence of arbitrary site failures and network partitions. As before, confirmation messages are not needed for the second phase, but they might enhance performance by reducing the likelihood of spurious delays. In practice, the refresh and quiesce intervals may have to be tuned to incorporate such factors as lost refresh messages and the retransmission rate.

4.4. Crash recovery

A simple way to preserve the revocation invariant across site crashes is to keep quiesce and refusal times in non-volatile storage, perhaps in a small "stable cache". If this technique is impractical, an alternative technique is to set a system-wide maximum value for the *quiesce interval*, the duration between a server's current clock value and the quiesce time for any capability (see Figure 1.). When a server recovers, it reinitializes its clock, and refuses all operation invocations until the maximum quiesce interval (plus Δ) has elapsed at every server, ensuring that all capabilities valid at the time of the crash have quiesced.

4.5. Lazy revocation

An interesting "lazy" variant of this revocation method arises if, instead of using clocks to drive revocation, we use revocation to drive clocks. Real-time clocks are replaced by logical clocks [16] satisfying the following properties:

1. Each server's clock generates successively increasing timestamps.

2. When a message is sent from one server to another, the time at which it is received (by the receiver's clock) is later than the time at which it was sent (by the sender's clock).

The second property is ensured if, whenever one server sends a message to another, the sender includes its current logical time, and the recipient advances its own logical clock beyond the observed value.

It must be emphasized, however, that properties 1 and 2 can be guaranteed only if all sites communicate via trusted network interface modules that can be relied upon to affix valid logical timestamps to outgoing messages. The lazy scheme is not secure in systems where an adversary can transmit messages with forged logical timestamps.

As before, each capability has a quiesce and refusal time at each server, satisfying a slightly different revocation invariant.

$$Last(Quiesce(c)) < First(Refusal(c))$$

This invariant ensures that there exists some logical *time* before which the capability has quiesced at all servers, and after which it will be refused. As before, the capability is considered valid before its quiesce time has elapsed; it is considered invalid after its refusal time has elapsed, and otherwise its status is indeterminate. The difference here is that all times are logical times, not physical times.

The major advantage of lazy revocation is that it is "demand-driven" rather than "time-driven". To revoke a capability, an owner simply advances its local logical clock to its refusal time, and resumes execution. It is never necessary to wait for a capability's refusal time to elapse, nor is it necessary to undertake a revocation protocol, simply because a server's logical clock can be advanced instantaneously. Similarly, it is not necessary to undertake refresh protocols at fixed intervals. Instead, refresh protocols are triggered by revocations. When an owner revokes a capability, it advances its local clock, which eventually advances other clocks as messages propagate through the system. If a server's quiesce time suddenly elapses, it may send a message to the owner asking it to initiate a refresh protocol.

Whether the eager scheme's combination of periodic refresh protocols with delays is more cost-effective than the lazy scheme's demand-driven refresh protocols without delays depends on the expected frequency of revocations and the relative costs of delay and of message traffic. Of course, one disadvantage of the lazy scheme is that it provides no real-time guarantees about revocation. A related disadvantage is that this scheme is vulnerable to information flowing through "covert channels". In the homework example, if the professor's machine is partitioned from the machine where the homework is to be handed in, a student could read the posted solution in one partition, walk over to a machine in the other partition, and submit the stolen solution.

5. Discussion

How should quiesce and refusal times be chosen in practice? Naturally, this choice depends on the expected rates of failures and the expected frequency of revocation. The message traffic associated with refresh protocols can be kept to a minimum by choosing quiesce and refusal times relatively far in the future. Although it would then take a long time for the earliest refusal time to elapse, most revocation would be accomplished by the two-phase revocation protocol. An owner would wait for the earliest refusal time to elapse only as a fall-back measure when network failures cause the revocation protocol to fail, presumably a rare occurrence.

Our revocation scheme is easily extended to allow the set of servers protected by a capability to change dynamically. To create a new server protected by an existing capability, it is enough to leave a "forwarding address" at an existing server. The revocation and refresh protocols are extended by having the older server forward the phase one message to the newer server, and making its own phase one confirmation conditional on the newer server's confirmation.

The revocation method proposed here is an adaptation of an algorithm originally developed to detect and eliminate *orphans* [22] in distributed transaction systems. An orphan is an activity executing on behalf of an aborted transaction. Orphans are undesirable because they waste system resources, and because they may observe inconsistent data. At first, it may seem curious that revocation and orphan elimination are so closely related, but the connection becomes clearer if we think of an orphan as a transaction whose "right to exist" is revoked when it is aborted.

It is natural to ask whether other orphan elimination algorithms can also be adapted for revocation. Orphan elimination schemes for non-transactional systems [27] do not seem appropriate, since they do not provide any consistency guarantees. The algorithm used by Argus [12] requires that each process include all the orphan information it "knows" in each message, relying on various optimizations to keep message sizes under control. Unfortunately, the Argus scheme, like the "lazy" scheme described above, requires the cooperation of all processes in the system, and hence it is vulnerable to an adversary who neglects to pass on the required information. Our scheme circumvents this difficulty by relying on approximately synchronized real-time clocks; in effect, using the passage of time as a secure message channel.

6. Conclusions

We have seen how we can build a capability system in a distributed environment without placing extensive support code in the kernel. While this paper primarily addresses systems that implement the client-server model, we believe that our algorithms may also benefit other approaches to distributed systems. We have presented a new zero-knowledge authentication protocol and a new revocation protocol, each of which can be provided through library routines in individual programs running on a loosely-coupled system. Although we have not attempted to solve the denial of service problem, recent work has shown that distributed client/server systems which use replication to achieve fault tolerance can be efficiently integrated with secure systems. We intend to pursue research in this direction, to see if self-securing programs can protect themselves against denial of service attacks through the use of replication.

Our self-securing capability system depends on some assumptions. We assume that address spaces are protected, and that certain problems, such as factoring large integers or inverting text enciphered by DES, are intractible. Our self-securing approach has several important implications. It implies that we can build systems with only minimal security support, knowing that we can retrofit the systems later. It means that we can verify and validate software which can meet arbitrarily high standards of security without using the traditional nested security approach, in which a distributed secure application relies on a distributed secure library of programs which in turn rely on a distributed secure file system, which in turn relies on a distributed secure kernel, etc. [2]. Instead we can move security out of the kernel and into user space. This approach may help tame the computer security problem for distributed systems and reach high levels of security with relatively small amounts of development effort.

References

[1]   W. Alexi, B. Chor, O. Goldreich, and C. P. Schnorr, RSA and Rabin Functions "Certain Parts are as Hard as the Whole", in *Proceedings of the 25th IEEE Symposium on the Foundations of Computer Science*, November 1984. To appear in SIAM J. on Computing.

[2]   J. P. Anderson, "Computer Security Technology Planning Study", Technical Report ESD-TR-73-51, USAF Electronic Systems Division, October 1972.

[3]   Laszlo Babai, "Trading Group Theory for Randomness", in *Proceedings of the 17th ACM Symp. on Theory of Computing*, pp. 421-429, May 1985.

[4]   J. Carter and M. Wegman, "Universal Classes of Hash Functions", in *Proceedings of the 17th IEEE Foundations of Computer Science*, pp. 106-112, May 1976.

[5]   Ben-Zion Chor, *ACM Distinguished Dissertations: Two Issues in Public Key Cryptography: RSA Bit Security and a New Knapsack Type System*, MIT Press, 1986.

[6]   E. Cohen and D. Jefferson, "Protection in the Hydra Operating System", in *Proceedings 5th Symp. on Operating Systems Principles*, pp. 141-160, November 1975.

[7]   Uriel Feige, Amos Fiat, and Adi Shamir, "Zero Knowledge Proofs of Identity", in *Proceedings of the 19th ACM Symp. on Theory of Computing*, pp. 210-217, May 1987.

[8]   E. F. Gehringer and R. J. Chansler, Jr., "Star OS User and System Structure Manual", Technical Report, Carnegie Mellon University, 1981.

[9] Shafi Goldwasser, Silvio Micali, and Charles Rackoff, "The Knowledge Complexity of Interactive Proof Systems", in *Proceedings of the Seventeenth Annual ACM Symposium on Theory of Computing*, May 1985.

[10] S. Goldwasser and M. Sipser, "Arther Merlin Games versus Zero Interactive Proof Systems", in *Proceedings of the 17th ACM Symp. on Theory of Computing*, pp. 59-68, May 1985.

[11] J.N. Gray, *Notes on Database Operating Systems,* Lecture Notes in Computer Science 60, Springer-Verlag, Berlin, 1978, pp. 393-481.

[12] M.P. Herlihy, N.A. Lynch, M. Merritt, and W.E. Weihl, "On the correctness of orphan elimination algorithms", in *17th Symposium on Fault-Tolerant Computer Systems (FTCS)*, July 1987. Abbreviated version of MIT/LCS/TM-329.

[13] Maurice P. Herlihy and J. D. Tygar, "How to Make Replicated Data Secure", in *Advances in Cryptology, CRYPTO-87*, Springer-Verlag, August 1987.

[14] R. M. Karp, "Reducibility among Combinatorial Problems", *Complexity of Computer Computations*, Plenum Press, New York, 1972, pp. 85-103.

[15] Richard M. Karp, 1985 Turing Award Lecture, "Combinatorics, Complexity, and Randomness", *Communications of the ACM* 29(2): 98-109, February 1986.

[16] L. Lamport, "Time, clocks and the ordering of events in a distributed system", *Communications of the ACM* 21(7): 558-565, July 1978.

[17] B. W. Lampson, "Protection", *ACM Operating Systems Review* 8(1): 18-24, January 1974.

[18] Henry M. Levy, *Capability-Based Computer Systems*, Digital Press, 1984.

[19] R. Lipton, Personal communication.

[20] Michael Luby and Charles Rackoff, "Pseudo-random Permutation Generators and Cryptographic Composition", in *Proceedings of the 18th ACM Symp. on Theory of Computing*, pp. 356-363. May 1986.

[21] K. Marzullo and S. Owicki, "Maintaining time in a distributed system", in *Proceedings of the second ACM Symposium on Principles of Distributed Computing*, pp. 295-305, August 1983.

[22] M.S. McKendry and M.P. Herlihy, "Time-driven orphan elimination," in *Fifth Symposium on Reliability in Distributed Software and Database Systems*, January 1986. Also CMU-CS-85-138.

[23] C. Meyer and S. Matyas, *Cryptography*, Wiley, 1982.

[24] R. Morris and K. Thompson, "Unix Password Security", *Communications of the ACM* 22(12): 594-596, December 1979.

[25] NBS Standard, "Data Encryption Standard", Technical Report FIPS Publication 46, National Bureau of Standards, January 1977.

[26] Roger M. Needham, Michael D. Schroeder, "Using Encryption for Authentication in Large Networks of Computers", *Communications of the ACM* 21(12): 993-999, December 1978. Also Xerox Research Report, CSL-78-4, Xerox Research Center, Palo Alto, CA.

[27] B. Nelson, "Remote Procedure Call", Technical Report CSL-79-3, Xerox Palo Alto Research Center, 1981.

[28] Michael Rabin, "Digitalized Signatures and Public-Key Functions as Intractable as Factorization", Laboratory for Computer Science Technical Report MIT/LCS/TR-212, Massachusetts Institute of Technology, January 1979.

[29] Michael O. Rabin, "Probabilistic Algorithms in Finite Fields", *SIAM Journal on Computing* 9:273-280, 1980.

[30] Michael Rabin, "Fingerprinting by Random Polynomials", Center for Research in Computing Technology, Aiken Laboratory TR-81-15, Harvard, May 1981.

[31] J. Reif and J. D. Tygar, "Efficient Parallel Pseudo-Random Number Generation", in *Advances in Cryptology: CRYPTO-85*, pp. 433-446, Springer-Verlag, August 1985. To appear in SIAM J. on Computing.

[32] R. Rivest, A. Shamir, and L. Adleman, "A Method for Obtaining Digital Signatures and Public-Key Cryptosystems", *Communications of the ACM* 21(2): 120-126, February 1978.

[33] M. Satyanarayanan, "Integrating Security in a Large Distributed Environment", Technical Report CMU-CS-87-179, Carnegie-Mellon University, November 1987.

[34] M. Satyanarayanan, John H. Howard, David A. Nichols, Robert N. Sidebotham, Alfred Z. Spector, Michael J. West, "The ITC Distributed File System: Principles and Design", in *Proceedings of the Tenth Symposium on Operating System Principles*, pp. 35-50, ACM, December 1985. Also available as Carnegie-Mellon Report CMU-ITC-039, April 1985.

[35] B. J. Walker, R. A. Kemmerer, and G. J. Popek, "Specification and Verification of the UCLA Unix Security Kernel", *Communications of the ACM* 23(2): 118-131, February 1980.

[36] M. V. Wilkes and R. M. Needham, *The Cambridge CAP Computer and its Operating System*, North-Holland, 1987.

[37] W. A. Wulf, R. Levin, and S. P. Harbison, *HYDRA/C.mmp: An Experimental Computer System*, McGraw-Hill, 1981.

[38] Bennet S. Yee, J. D. Tygar, and Alfred Z. Spector, "A Self-Securing Protection System for Distributed Programs", Technical Report CMU-CS-87-184, Carnegie-Mellon University, December 1987.

RESYNCHRONIZATION INTERFACES:

SOURCES OF METASTABILITY

ERRORS IN COMPUTING SYSTEMS

D. DEL CORSO*, F. MADDALENO*, M. MINICHINO **, E. PASERO ***
* Politecnico di Torino, Dipartimento di Elettronica -
Corso Duca degli Abruzzi 24 - I-10129 Torino - Italy
** ENEA - Casaccia, Roma - Italy
*** Dip di Ing. Elettronica - II Università, Roma - Italy

Abstract

In computing systems, bistable flip-flops are used to store binary variables and to resynchronize asynchronous signals in the various interfaces. These bistable devices can be driven by specific input conditions into a metastable state, where they may remain for a time predictable only in a statistical way. An output in metastability may introduce random errors in the connected circuits. Being related with basic time-decision ambiguity, this phenomenon is unavoidable. Viable ways to decrease the metastable hazard are the technology improvement, a correct design methodology, and/or an increase of the decision delay. The metastable hazard increases with the evolution of the technology, as the relative indetermination of propagation delays becomes more significant. This work presents the fundamental concept of time-decision ambiguity and shows that significant error bursts may occur in presence of "almost equal" clock frequencies; the interface specifications of some multiprocessor buses (VME, Multibus 2, Nubus, Futurebus) and serial communications (Ethernet) are then analyzed from this point of view.

1. Introduction

Interfaces among asynchronous parts, such as processors and external resources, or different processors are one of the most critical points in today complex systems. Before entering clocked sequential logic circuits, external signals must be synchronized with each local clock to assure correct timing. Also internal signals exchanged among parts synchronized by different oscillators are affected by the same problem. Resynchronization is achieved by means of a bank of D flip-flops, and the process introduces a significant hazard of metastable errors.

A synchronizer may enter a metastable state when input signals and the clock have transitions at the "same" time, or better within a narrow time window. The time required by the metastable output to resolve into an unambiguous logic state is unbounded. If these outputs are used as inputs for other logic circuits, the allowed resolving time is determined by the clock period, and this introduces a finite error probability. In this case incorrect states or results can be generated inside the system. It is not easy to trace these errors owing to their randomness, and to recover from them could be even more difficult.

Metastability is unavoidable because it is related to the time-decision energy indetermination. Standard error correction techniques, such as redundancy, are not effective against metastability errors, but a correct design methodology and the use of a suitable technology can decrease the metastable hazards.

In the continuous process of performance improvement, handling metastability is a task switched between device and system designers. The basics of this phenomenon are known since a long time [5]. The device designer tries to make the resolving time of the flip-flops as short as possible. If this delay is much smaller than the clock period, the FF output settles to a correct state by the next clock edge and metastability problems appear very seldom. As the system designers learn to use the same devices at higher clock speed, the FF output cannot resolve in one clock period and metastability-related errors pop up.

The technology evolution will then make available faster FFs, which make it possible for the system designer to push the metastability error rate down to an acceptable level; these same improvements will also allow them to use faster clocks, and the problem appears again.

The increasing speed of logic devices and the constant propagation speed of electromagnetic waves make however propagation delays and associated skews more and more significant in the overall delay balance of a logic system. This in turn increases the probability of timing violations. It is therefore predictable that the hazard of metastability will increase, and handling the problem will be concern of the device designers and of the system engineers.

This work presents a study of critical synchronization sections in processing systems, resuming the criteria for quantitative evaluations of error rates from system and device specifications. Section 2 describes the basic concepts of metastability and section 3 specifies how and where such problems can arise in complex systems. In section 4 some busses are analyzed: VME, Nubus, Multibus II, Futurebus and Ethernet. Section 5 presents some basic design rules to keep as low as possible the probability of metastable errors.

2. Fundamentals of metastability

Any physical system having at least two stable states has also a third state of unstable equilibrium, named *metastable state*. A qualitative proof for the existence of this metastable state is obtained considering that a stable state is associated to a local minimum of the system energy, represented versus a system state variable. In a physical system the energy is finite, and this implies that, if the function is continuous, a maximum exists between two minima (Figure 1).

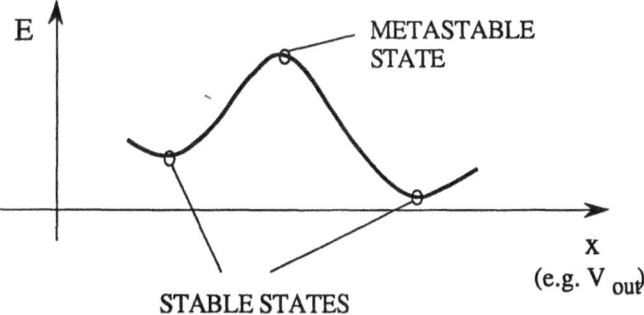

Figure 1. Energy diagram with two minima (stable states)
and one maximum (metastable state).

The energy maximum represents the metastable state. If the system is driven exactly into the metastable state and not disturbed, it will stay in that state for an infinite time. This happens with a probability equal to zero, because the domain measure of the metastable state is zero, being a single point on a real curve.

If the system is driven near the maximum, it will take an unbounded long time to evolve towards a stable state. This time depends on the initial point and on the function shape near the maximum. The nearer is the system to the metastable point, the longer is the time it will take to evolve into a stable state. Detailed discussions of metastable flip-flop behaviour are in [6] and [13].

A digital flip flop is an example of a system having two stable states, and hence it has also a metastable state. Such a flip flop can be trapped for a non-zero time interval near the metastable state, and in this situation its output signal is halfway between the two correct levels, which correspond respectively to the logical one and zero. In this situation the flip flop is said to be in metastability, even if, strictly speaking, it is only near the metastable state (Figure 2). After some time the system evolves towards a stable state. The system is considered out of metastability when its outputs fall within the correct logical ranges, even if they are still moving towards the final steady level.

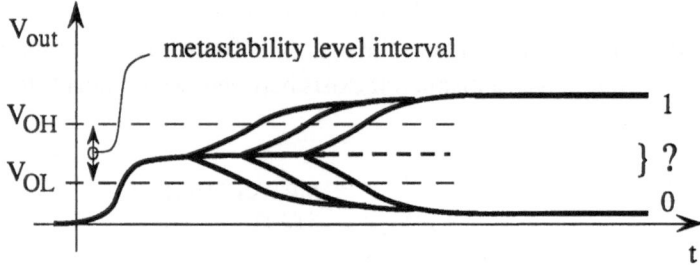

Figure 2. Evolutions from the metastable state.

This situation is critical for the circuits receiving information from the flip flop in metastability, because an intermediate level (neither one, neither zero) can be differently interpreted by the various parts of the circuit (Figure 3).

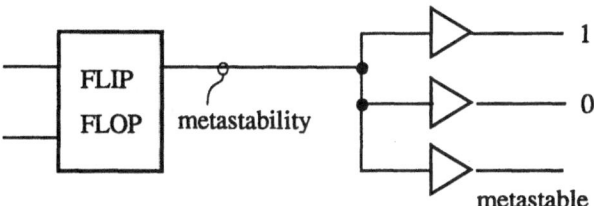

Figure 3. Misinterpretation of a metastable level.

A gate receiving at its input a metastable level can consider it as a zero or a one, depending on its electrical characteristics, which vary slightly from a gate to another, even within the same integrated circuit. Moreover, this gate could go in its linear region, in which case it propagates to the output the incorrect level. It is obvious that the same signal interpreted in opposite ways may cause serious malfunctions. For instance, if an arbiter goes into metastable state, two requesters can interpret the answer as a grant and try to use the resource at the same time.

A bistable circuit can be driven in metastability by violating the level or timing requirements at its inputs. If for instance a D flip flop samples an electrical level halfway between the correct ranges representing the logical zero and one, it can enter in metastability. This incorrect level can be generated by a faulty interface, but also by a preceding flip flop in metastability: this means that also a bistable can propagate a metastable state.

The second way to drive a bistable in metastability is the violation of input timings (set-up and hold times, tsu and th respectively). Examples of these violations are given by a signal changing while the flip flop is sampling the signal, or within a very short time (hazard window, tz) from clock edge; the hazard window can be placed anywhere in the setup-hold window, as shown in Figure 4.a. An SR flip-flop may enter a metastable state when the deactivation of the two S and R pulses are too close each other, that is within the hazard window, as in Figure 4.b.

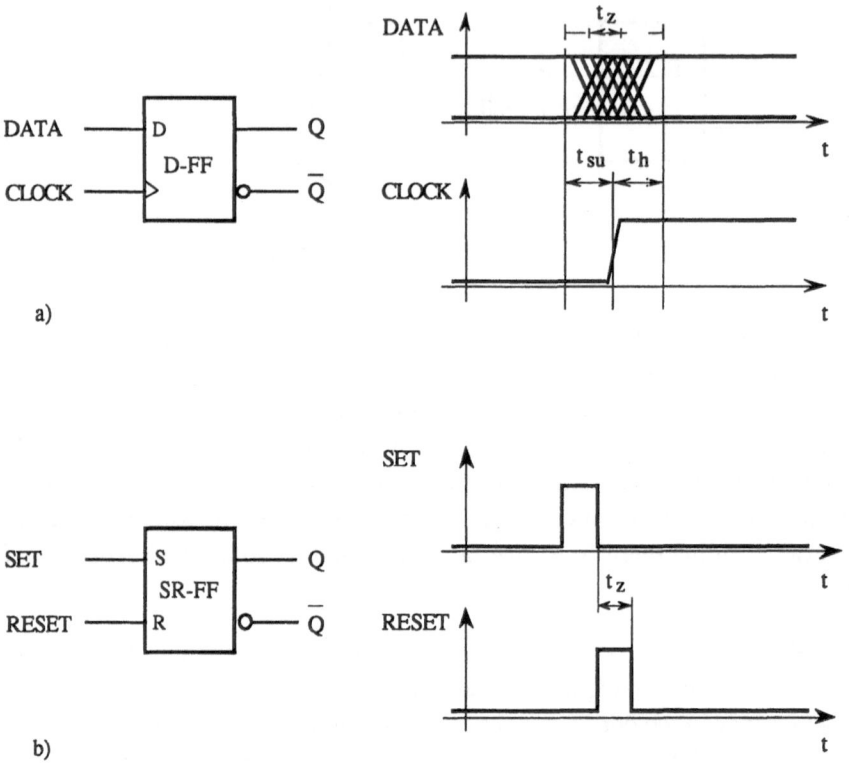

Figure 4. Input changes in the hazard windows for D (a) and RS (b) flip flops.
(Data change during the hazard window for the D FF, and the
SET/RESET pulses are too close for the RS FF)

Such timing violations are unavoidable when the digital system must communicate with an external world not synchronized with its internal clock. This happens for instance for any computer receiving external input signals.

The length of a metastable state can be characterized only in a statistical form. A simple but accurate model for the metastable duration is discussed in [14] and [9]; it states that if the bistable is in metastability at t=0, the probability to have a metastable state longer than a predefined time interval T is the following:

$$P_{ms} = e^{-T/\tau} \tag{1}$$

The parameter τ is typical of each device family, and can be measured or (seldom) obtained from the device data sheet. If a D flip flop with a clock frequency f_{ck} samples an uncorrelated signal with transition frequency f_d, the frequency of metastable states Fms which take more than T to evolve into a stable and recognizable state is given by

$$F_{ms} = f_{ck} * f_d * T_0 * e^{-T/\tau} \tag{2}$$

where the parameters T_0 and τ are typical of the flip flop technology.

Another way to describe this behaviour is to consider that a metastable state longer than T is generated only if a data transition occurs within a temporal window of width T_w

$$T_w = T_0 * e^{-T/\tau} \tag{3}$$

placed across the clock sampling edge. This hazard window is inside a wider temporal interval defined by the circuit manufacturer and indicated with the set-up and hold time specifications. The exact position of the hazard window is unknown inside that interval. The window width depends on the bistable technology (parameters T_0 and τ), and on the time T given to the flip flop to resolve its metastability: the longer the resolution time, the narrower the hazard window.

The T_0 parameter indicates the (extrapolated) width of the hazard window which generates metastable states, regardless of their length. The τ parameter is the decay time constant of the number of metastable states: its effect is the same as to have a narrower hazard window. The values of T_0 and τ for D flip-flops of different families are given in Table 1.

Family	T_0	τ
HC	2 ms	2.7 ns
LS	300 s	1.3 ns
F	100 μs	0.8 ns
AS	1.5 s	0.34 ns

Table 1. Typical values of T_0 and τ parameters of '74 flip flop
(Texas Instruments and Fairchild data)

A straightforward way to decrease the probability that a metastable state may cause errors is to increment the time T, i.e. to wait a longer time for the metastability resolution. This correspond to reduce the system operating speed (e.g. by reducing the clock frequency).

Another way is to reduce the width of the hazard window, which in turn lowers the probability to enter the metastable state. This correspond to reducing the T_0 and τ parameters, and can be achieved by choosing a logic family with smaller values for T_0 and τ. The most sensitive parameter is τ, because even a small variation on it is expanded by the exponential function. In fact a change in τ causes a variation on F_{sk} greater by a factor T/τ with respect to a similar relative change on T_0: this explains also the spread of T_0 values, ranging from microseconds to seconds.

The behaviour above described is referred to as analog metastability; the same class of phenomena may also cause oscillations: a rectangular wave appears at the bistable outputs, which oscillate in phase a number of times between zero and one before settling in a stable state (Figure 5). This kind of oscillatory metastability is triggered by the same violations previously described (input timing or level), and the metastable duration has the same statistical distribution of the analog metastability.

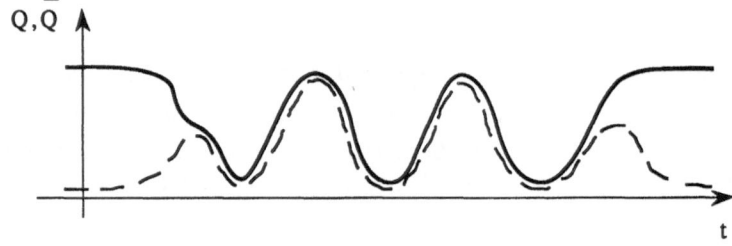

Figure 5.   Oscillatory metastability. The continuous and the dashed line corresond respectively to Q and Q* outputs.

The analog or oscillatory behaviour depends on the logic family, and more precisely on the ratio between the gate propagation delay and the signal rise and fall times: a family with intrinsic long propagation delays with respect to the switching times presents an oscillatory metastability [12], [8].

3. Metastability in complex systems

In a complex sytem the metastability hazard occurs in those sections where signals generated with different time references interact. This happens at the interface section between parts clocked by different oscillators or at the interface with the external world (usually not synchronized with the internal clocks). In a complex system, like a multiprocessor, typical critical interfaces are the global bus arbiter, the arbiter of dual port memories, the input interface and the interface between the CPU and coprocessors, if these last are clocked by independent oscillators. Moreover, if the bus has a clocked protocol, the conditions for metastability may occur also at the module/bus interfaces. Figure 6 shows these critical locations.

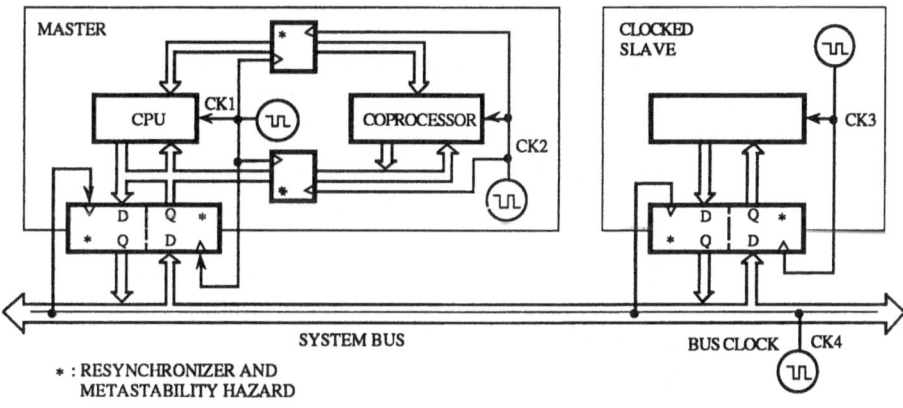

Figure 6. Critical resynchronization sections.

It is impossible to introduce hardware redundancy to mask metastability errors: it is however possible to design a system in such a way that it can recover from the effects of a metastable error. Metastability can occur on the control circuits or, more seldom, on the data path. Metastability errors in a data path have less impact on the system reliability because they can be treated as computation errors and recovered with standard techniques, like duplication. Such an error generates wrong results, but does not stop the computer activity: from the hardware point of view, it is a transient error.

On the other hand, a metastability in the control part of a computing system can, beside producing data errors, also bring a state machine in an inconsistent state, from which the system cannot recover: in this case this can be considered as a permanet fault until the next reset command.

For example, a metastability in an arbiter can result in a grant sent to two requesters, or a short grant, which however starts the bus access operation but is removed by the arbitration circuitry before that the access cycle could be completed. In the former case (double grant) two processors are allowed to access at the same time the same memory. Both processors receive simultaneous grants driving a tri-stated or open collector bus at once, resulting in a Wired-Or configuration. This generally screws up both processors. When address and data multiplexers are used only one processor is connected to the memory bus. One of the processors will access the wrong location, and both read or write wrong data.

A system can be designed in such a way as to survive to these errors, either using software methods, either detecting the double access by hardware and starting a recovery procedure. In any case, the design of a system able to recover from these metastable errors requires the knowledge of the error rate and the error statistical distribution, to evaluate the probability of consecutive errors.

The case of two mutually asynchronous interacting subsystems clocked by oscillators at the same nominal frequency (or integer multiple frequency) is extremely critical from this point of view, due to the frequency tolerance and drift of quartz oscillators. These frequency changes generate a continuous phase shift between the two clocks, similar to a slow beat, at a frequency fb equal to the frequency difference between the two clocks at that instant. The maximum frequency differences will of course be related to the maximum tolerance numbers, provided that each crystal is within tolerance. This is related to the frequency tolerance but typical instantaneous frequency differences are often much better than this. The number of metastable errors in a burst can thus be very long. The problem is particularly sever in the case where the systems are not co-located and the temperature is changing at one end of the link. with resultant changes in the clock frequency of that clock from one side to the other with respect to the other clock. There is thus a point in which the frequencies are very close to one another for a prolonged period of time. The resulting beat frequency can be of very low frequency, well under one hertz, with hundreds of consecutive metastable errors possible.

At every cycle of the beat frequency, the relative phase of the two clocks can bring the signals of the two subsystems in a temporal relation where the data changes within the hazard window of the clock, with a high probability of generating a metastable state. The error probability depends on

- the frequency of communications,

- the width of the hazard window during which the data must be stable,

- the time required by an oscillator to change its phase through the window.

As an example, two subsystems clocked at 10 MHz, with a relative tolerance of 2 ppm (e.g. with medium-quality cristal oscillators) are considered. The phase of the two systems changes continuously, and returns at the same relative position with a frequency of 20 Hz. This means that 20 times per second the subsystems pass through the critical temporal relationship. From a 10 MHz cycle to the following one, a clock shifts with respect to the other of 200 fs: if the hazard window given by (3) is larger than this value, there will be many potential consecutive cycles with metastable states. For instance, a hazard window of 2 ps gives 10 consecutive metastable errors.

This result shows that the metastability errors generated by two asynchronous systems driven at the same nominal frequency at a rate given by (2), can have a burst distribution. This fact must be considered during the system design, in order to provide mechanisms able to recover from such errors, or to avoid burst errors. Standard techniques to cope with metastability are designed for single errors, since with true asynchronous signals the probability of sequences of metastable errors is very low.

Burst errors can be avoided by slight changes in clock frequencies, in order to have a single metastable per beat period.

4. Case studies

This section points out which parts of some existing system buses may be affected by metastability errors. The buses analyzed (VME, NUBUS, MULTIBUS II, and FUTUREBUS) are used in complex multiprocessor systems, whit a number of subsystems timed by independent clock oscillators. All these buses are asynchronous; two of them have no global time reference for bus operations

(VMA, FUTUREBUS); the other two use a 10 MHz clock to time bus driving and sensing. A serial communication system (ETHERNET) is also analyzed to show the possible critical points of these structures.

4.1. VME

The data transfer protocol of VME is asynchronous, therefore, if the bus interfaces guarantee an upper bound to propagation skew, and if the slaves are properly designed, defined time margins among data or address and related strobes can be achieved. The slave should simply acknowledge the transfer cycle only after deskewing, set-up, and hold delays.

Modules with clocked logic have to face metastability both from external world and from the bus which acts as asynchronous inputs; the bus specification specify time margins between strobes and static signals (related with the upper bound on skew), but synchronizers are left to the board designer.

A protocol layer which is subject to this malfunction is bus arbitration. The arbitration cycle uses a reserved set of lines and can be concurrent with data transfer. Arbitration is based on a four-level daisy chain, as shown in Figure 7. The signals used are:

BR<0..3>	Bus Request
BGIN<0..3>	Bus Grant In; daisy chain in
BGOUT<0..3>	Bus Grant Out; daisy chain out
BBSY	Bus busy (wired-OR from all masters)

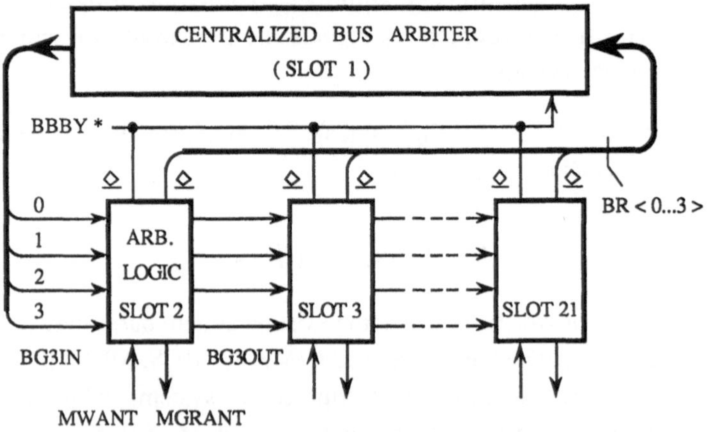

Figure 7. The VME bus arbitration structure.

The BRx/BGINx work as four REQ/GRANT pairs toward/from a centralized arbiter. These lines are wired in the backplane as four independent daisy chains, so that requesters sharing a common request line are prioritized by slot position.

The chain BRx - BGINx - (BGOUTx) - BBSY corresponds to an asynchronous handshake; however a request can be raised while a grant removal is propagating on the chain, as shown in Figure 8.

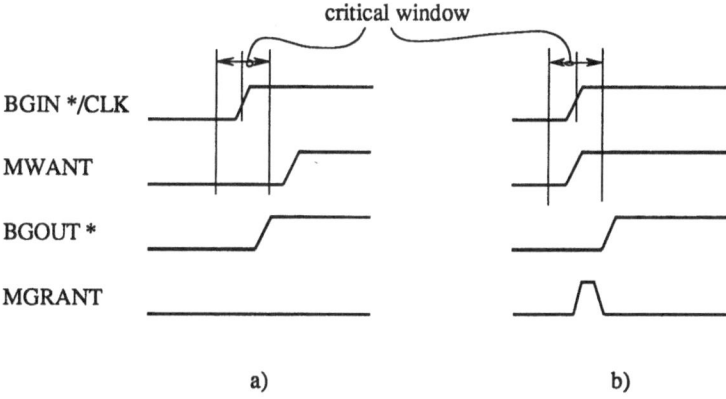

Figure 8. Correct timing (a) and metastable error (b) in a daisy chain arbiter.

Some solutions are proposed in the appendix D of the VMEbus specification. The following diagrams shows explicitly the input synchronizer, which freezes the access request so that the selection logic could operate on steady signals.

Since the arbitration system does not use any global clock or synchronization command, requests are synchronized either by a clock or by the grant signal BGIN which propagates in the chain. These two solutions are in Figure 9. In both cases a delay is inserted between the sampling of the request and the activation of local grant and chain out, in order to mask possible metastable states at the output of the request synchronization FF, which could cause transient incorrect states on these lines (explicitly forbidden by the bus specs).

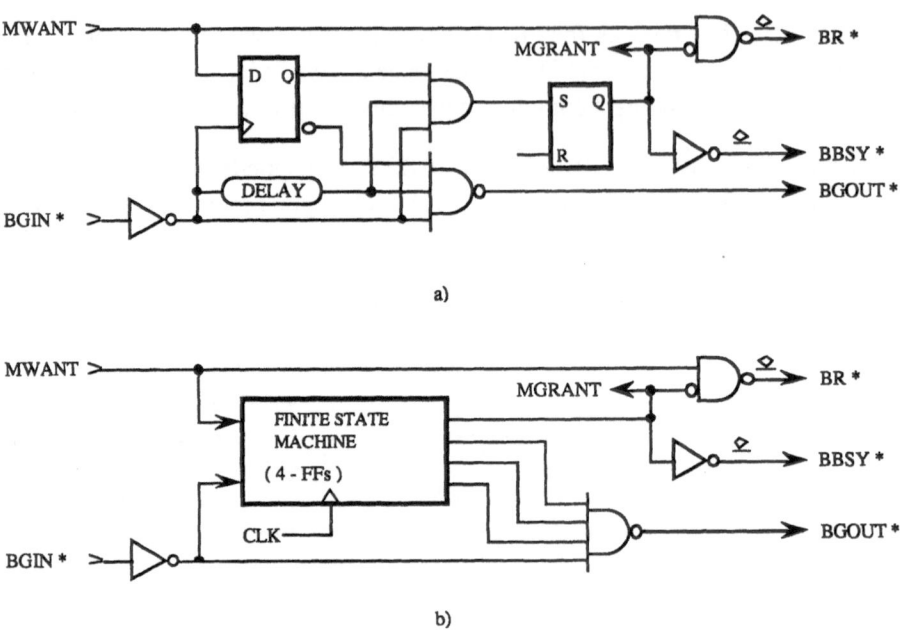

Figure 9. Chain synchronization in VME arbiters:
a) asynchronous circuit;
b) clocked circuit.

A similar problem may occur in the centralized bus arbiter, where incoming requests are sampled at any bus exchange (which is asynchronous with new incoming requests). Again, also the interrupt logic is subject to metastability, because it uses a distributed daisy chain selection circuit. The VME-IEC 821 bus specification [1] includes the appendix D on how to handle signal resynchronization in the arbitration circuits.

4.2. Nubus and Multibus II

These two buses are discussed together because their protocols have a similar approach to data transfer and arbitration [3], [2].

The information transfer protocols of Multibus II and of Nubus are clocked asynchronous: the active master sends a command (e.g. the address and appropriate operation qualifiers), then waits for an answer from the selected slave, as shown in Figure 10. Operations are timed by a system-wide clock.

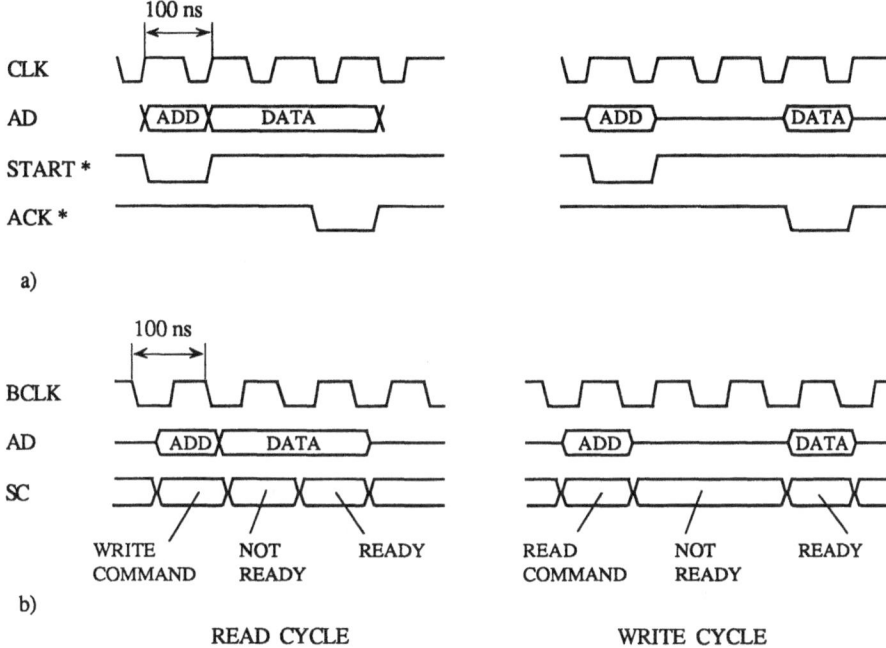

Figure 10. Nubus (a) and Multibus II (b) data transfer protocol.

Since on-board circuits use their own clock generators, any signal must be resynchronized at the bus interface. The frequency of the bus clock (10 MHz) is very low from the point of view of metastability hazard. As an example, a single-FF board-to-bus synchronizer (clocked at 10 MHz) in TTL-F logic has a metastability MTBF orders of magnitude greater than the universe lifetime, for a T=100 ns.

On the other hand, a bus-to-board synchronizer with 50 MHz board clock (which is a reasonable figure for a local device clock), still in TTL-F, has a MTBF of only about 2.5 ms; this last figure can be improved with a double-FF synchronizer (up to several tens years), at the expense of delayed effect of bus signals on internal board circuits.

It is therefore possible to design this interface layer in such a way as to keep error probability within any low bound, at the expense of speed. The problem of burst errors may occur if the frequencies of board clocks are m/n times 10 MHz, where m and n are small integers.

The complexity of Multibus II protocol makes mandatory the use of Intel interface ICs, which run on one side with signals timed by the bus clock, and with board signals on the other. Board resynchronization is therefore handled inside these devices.

For bus arbitration Multibus II and Nubus use a distributed self selection priority circuit. Operations are again timed by the central system clock CLK (10 MHz). The duration of priority contention depends on transmission time on the backplane (which should not change over the life of a bus), and on gate delays (which probably will change). The central clock cuts the probability of metastable errors at the bus level, but freezes the arbitration speed.

4.3. Futurebus

Futurebus (IEEE 896.1, [4]) has asynchronous protocol with the possibility of N-partner handshake (three-wire) for all operations. Non-standard line drivers reduces bus propagation delays and skew; time margins for setup and hold on ADD/DATA lines are responsibility of destination modules (slaves for write, masters for read).

The three-wire handshake technique is shown in Figure 11. In the addressing and data transfer cycles it is used to guarantee correct information exchange also in broadcast and broadcall operations. In the arbitration this protocol synchronizes the distributed control state machine taking in to account the speed requirements of all participating units.

A proper design can guarantee that no metastable condition could arise both in the data transfer and in the arbitration, and the asynchronous protocol keeps in both cases the delays to the minimum required by involved modules and by interconnections.

4.4. Ethernet

Ethernet is a serial protocol of communication between systems that are typically located in different physical environment. This is the case described in section 3 where two systems that are not co-located communicate at the same nominal frequency but with different internal clocks. The most common implementation has the receiver phase locked to the incoming bit stream and the transition to the local clock occurs when deserialized bytes are passed to the local CPU for processing. With Ethernet at a nominal bit rate of 10 MHz the resynchronization occurs at somewhat over 1 MHz, the byte rate of the transmission.

The occurrence of metastable errors at this 1 MHz rate can be kept fairly well bounded but this problem is getting worse. The FDDI standard operates at 100 MHz with a 10 MHz byte clock. The circuitry to process these bytes is CMOS: this choice is dictated by the complexity of the FDDI standards but is not a good solution. This technology, as explained in [10], shows its limits when used with high speed clocks. A typical critical structure is the master slave flip flop. The data transfer from master to slave may fail if the transition time of the internal clock exceeds some critical limit.

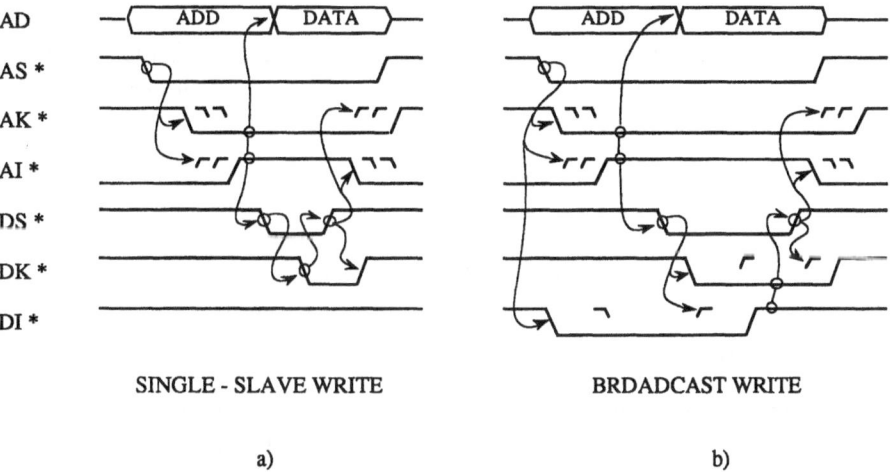

SINGLE - SLAVE WRITE   BRDADCAST WRITE

a)   b)

Figure 11.   Three-wire handshake in Futurebus:
          a) single-slave write
              (three-wire handshake only on addressing);
          b) broadcast write
              (three wire handshake in the address and in the data cycles).

5. Design guidelines

As general guidelines to the design of complex systems, two basic rules can be given to keep as low as possible the probability of metastable errors:

1) reduce the number of synchronization interfaces

2) increase the ratio T/τ in (1)

To apply the first rule the system architecture must be carefully analyzed in order to find the structure allowing for a minimum of asynchronous interfaces (which require resynchronization). Once these critical sections are reduced to a minimum, the only way to decrease the metastability error rate is the increase of the ratio T/τ in (1). This can be obtained in two ways:

- by giving more time to the synchronizer to solve its metastable states (e.g. lower the clock frequency, or use a multiple stage synchronizer, or mask the outputs with delays);

- by using a technology with a smaller τ.

Rule 1) impacts the system architecture, while rule 2) must be taken into account both at the system level (value of T, related with the clock period), and for selection of devices (value of τ). The time spent for metastability solution is in any case a delay introduced in the signal path.

If the asynchronous signal is connected to a state machine, the internal flip flops could be driven in metastability: a synchronization is then mandatory, either at the input or at the output of the state machine; the delay added for synchronization is the same in both cases. If the synchronization is done at the state machine outputs, the designer must carefully study the correspondence between the states and their representations in terms of zeros and ones, in order to obtain a state sequence numbered in a Gray code.

In this way if a metastable state occurs, it will evolve towards the previous or the following state, without reaching wrong or undefined states where the state machine could be trapped. Even in this case, to avoid the generation of anomalous short pulses coming from the state machine, the designer is forced to add some idle states, and the delay introduced by the additional idle states is equivalent to the delay of a synchronizer added on input signals.

In the cases of "nominally equal" clock frequencies, a definite phase relations among the oscillators must be guaranteed. Phase-lock-loops have been already proposed to synchronize bus accesses between a CPU and its coprocessor to avoid bus driving conflicts [11]. The same approach and circuit could be profitably used for module synchronization, resulting in extremely low metastable error probability with almost no time penalty.

A redundant synchronizer, for example composed of three flip flops and a majority voter, is not effective against metastability errors. Such a system, even if composed of several parts, has at least two stable states corresponding to the 0 and 1 output states, and between these stable states a metastable exists. Moreover, the increased number of component used to implement a redundant synchronizer decreases the system reliability. Infact the high number of gates is more susceptible to classical faults such as "stuck-at" faults and the reliability of the global digital system will be strongly decreased.

In case of timing violation, a flip flop can switch towards the zero state, a second one towards the 1 state, and the third flip flop can stay in metastability. The voter receives a 0, a 1 and it must wait for the third flip flop to settle before to give a stable output. The metastability characteristics of a 'redundant' synchronizer are hence the same of the single flip flops constituting the circuits. Moreover the voter circuit introduces a time delay, and this time must be subtracted to the flip flop resolving time.

References

[1]   ANSI/IEEE Std 1014-1987, *Standard for a Versatile Backplane Bus: VMEbus*, IEEE, 1987.

[2]   ANSI/IEEE Std 1296-1987, *High Performance Synchronous 32-bit Bus: MULTIBUS II*, IEEE, 1987.

[3]   ANSI/IEEE Std 896.1-1987, *Backplane Bus Specification For Multiprocessor Architectures: Futurebus*, IEEE, 1987.

[4]   ANSI/IEEE Std 1196-1987, *Simple 32-bit Bus: NuBus*, IEEE, 1987.

[5]   I. Catt, "Time Loss Through Gating of Asynchronous Logic Signal Pulses", *IEEE Transactions on Electronic Computers*, February 1966, pp. 108-111.

[6]   T.J. Chaney, C.E. Molnar, "Anomalous Behaviour of Synchronizer and Arbiter Circuits", *IEEE TC*, n° 4, April 1973, pp. 421-422.

[7]   D. Del Corso, H. Kirrmann, J.D. Nicoud, *Micromputer Buses and Links*, Academic
       Press, 1986.

[8]   D. Del Corso, L. Reyneri, B. Sacco, "Oscillatory Metastability in Homogeneous and
       Inhomogeneous Flip-flops", submitted to *IEEE JSSC*.

[9]   J.H. Hohl et al., Prediction of Error Probabilities for Integrated Digital Synchronyzers,
       IEEE JSSC, Vol. 19, n° 2, April 1984, pp. 236-244.

[10]   J.U. Horstmann et al.,"Metastability Behavior of CMOS ASIC Flip-Flops in theory and
       Test", *IEEE JSSC*, Vol. 24, n° 24, February 1989, pp. 146-157.

[11]   M.G. Johnson, E.L. Hodson, "A Variable Delay Line PLL for CPU-Coprocessor
       Synchronization", *IEEE JSSC*, Vol. 23, n° 5, October 1988, pp. 1218-1223.

[12]   T. Kacprzak, "Analysis of Oscillatory Metastable Operation of an RS Flip-flop", *IEEE
       JSSC*, Vol. 23, n° 1, February 1988, pp. 260-266.

[13]   L. Kleeman, A. Cantoni, "Metastable Behaviour in Digital Systems", *IEEE Design and
       Test of Computers*, December 1987, pp. 4-19.

[14]   L. Marino, "General Theory of Metastable Operation", *IEEE TC*, Vol. 30, n° 2, February
       1981, pp. 107-115.

Design for Security

Chair: R. Kemmerer (UCSB, Santa Barbara, California, USA)

FRAGMENTED DATA PROCESSING: AN APPROACH TO SECURE AND RELIABLE PROCESSING IN DISTRIBUTED COMPUTING SYSTEMS

Jean-Michel FRAY - Jean-Charles FABRE
Laboratoire d'Automatique et d'Analyse des Systèmes du C.N.R.S.
7, Avenue du Colonel Roche, 31077 Toulouse Cedex - France

Abstract

This paper presents a new approach for solving the problem of reliable processing of sensitive information in distributed computing systems [1] . Distributed systems offer redundancy that can be used to enhance reliability by means of adequate fault tolerance mechanisms. For security requirements, a given computer cannot be used to run any application . Consequently, only a given set of *trusted* processors can be used, thus reducing the flexibility of the network.

In this paper, we first discuss the problem of ensuring both security and reliability in distributed systems. Then we present the principles of a new approach (*Fragmented Data Processing*), which is currently being investigated at LAAS, and that seems to be a promising approach to solve, at least partially, the problem of Secure and Reliable Processing in Distributed Computing Systems.

---

1 This study is partially supported by DRET under convention n°88.34.051.00.470.75.01

1. Introduction

The development of Distributed Computing Systems based on large local area networks allows organizations to take advantage of a distributed architecture within their management and production structures. Facilities such as computing resource sharing, inter-process communication, file transfer, parallelism, fault tolerance, and incremental computing power are of interest for any engineer, designer or manager, from factories to military headquarters. Thanks to the evolution of distributed operating systems, the flexibility of using a large number of computers located in a given local geographical area allows cooperation, to improve their utilization. For instance, it is possible to dramatically increase the available computing power if each computer is dedicated to one part of a global task. On the other hand, the redundancy of computing resources is naturally available to run multiple copies of the same task in order to increase reliability.

Dependability with respect to reliability [9], can take advantage of the redundancy offered by a distributed architecture, on which fault-tolerant mechanisms can rely.

From another point of view, computer security is of interest for many applications and must be taken into account despite the distributed nature of the computing architecture. One of the basic techniques for achieving security is to implement applications on physically and logically isolated computers since it is difficult to totally protect sensitive information during processing, especially when intrusions are considered at the physical level. Consequently, it does not seem realistic to implement secure applications by using untrusted remote computers on the network. This precludes making these applications reliable, using fault-tolerance mechanisms based on the available (untrusted) computer redundancy (for instance, several copies of the application).

Dependability with respect to security [9], suffers from the fact that confidence cannot be placed on all the computers of the network.

According to a few assumptions on the distributed architecture presented here, we propose an original solution to the problem of **Secure and Reliable Processing in Distributed Computing Systems**. Our approach consists in executing code with fragmented data, using processors scattered in a distributed system, thus taking advantage of available computers, even untrusted, to support both reliable and secure applications.

Fragmentation and Scattering techniques [5, 6, 13] and *Information Dispersion for Security* [12] have already been proposed to realize a secure and reliable file system. However, this only addresses the problem of long term secure data storage (archives) and not sensitive information processing.

Fragmented Data Processing (FDP) is discussed here from a general point of view. We outline the interest of such a method and the topology and specific characteristics of the distributed environment that it requires. Various granularity levels of fragmentation are considered, offering a large field of possibilities for implementing FDP techniques.

This paper is composed of two main sections. Section 2 summarizes the problem statement and shows various solutions based on known techniques such as cryptography. Section 3 describes the key features of the FDP approach.

We conclude by discussing the advantages and drawbacks of FDP to achieve both secure and reliable processing in a distributed environment based on a local area network, and by giving the future prospects and guidelines of our work on FDP techniques.

2. The problem of secure and reliable processing

2.1. Distributed computing environment

The distributed system architecture we consider is composed of a large number of computers (or processing elements) interconnected through a local area network. From a geographical point of view, the computers are distributed in one or several buildings of a company or a large campus, thus being physically scattered in different rooms and/or departments. Nevertheless, we suppose that the network is restricted in scale, and offers efficient communication throughput.

Organizations where computer security is a major consideration are usually divided into different sectors, i.e. the buildings are partitioned into different areas, each of them dedicated to one specific activity. The notion of partitioning is not only applied to topics of interest but also to people who cannot move freely by from one sector to another. Every user can be authenticated by an object (a magnetic card) or personal authentication parameters (finger print) and can only enter a subset of the various areas. Separation is one of the standard methods to ensure security. In this context, it is assumed that a given area does not trust the others. This assumption should be true for most commercial activities and is usually realized in military environments. As far as security is concerned, any sensitive information relevant to one area cannot be accessed (read, modified or destroyed) by people from other areas, unless they are so authorized by the specified security policy. In such working environments, users must trust the computers located in their own security area as well as the inter-area data exchanges. For example, due to the presence of physical controls, only authorized users can get near a computer even when nobody is working on it. On the other hand, it has been shown in [16], that it is possible to manage data exchanges through the network (between different security areas) according to the defined security policy, so as to ensure security at the logical level.

Finally, we suppose that conspiracy between more than a given number of different security area users or administrators is impossible. This is a classical notion related to the notion of threshold schemes such as in [17]. In our case the threshold depends on the application. We assume that it is not possible for more than the threshold of intruders belonging to different security areas, to cooperate to obtain information from a single security area, either by trapping and exchanging information between security areas or by implementing cooperating Trojan Horse programs running in different security areas. The latter is the major assumption required to implement Fragmented Data Processing techniques.

If all the intruders belonging to other areas cooperate against a single one, the system can be considered from a security point of view as composed of only two different areas. In this case, it is obvious that a user cannot trust the other security area and must execute its security applications in his own area. (This is similar to the fault-tolerant systems that are composed of two machines; if one fails, the system is reduced to a non fault-tolerant one).

This general and rather common working environment being described, we would like to show, according to known solutions, that ensuring both secure and reliable processing in such distributed computing architectures is either very costly or very difficult in practice.

2.2. Security and networking

The free use of remote resources is one key feature of a distributed computing environment, not only with respect to files and applications which can be remotely accessed from any computer on the network, but also with respect to processing facilities that can be used to run subtasks belonging to a global application. Communication and synchronisation mechanisms are available to implement an application as a set of cooperating entities in order to improve parallelism and/or fault tolerance. This paper addresses only security attacks such as intrusions, discarding intentional design faults (the class of faults that are introduced during the design of the system and used later subversively, during system operation, e.g. Trojan Horses).

According to the working environment, which has been previously described, and with respect to security, a particular user (A) has to consider a Distributed Computing System (DCS) as divided into two disjoint subsets:

1. **Trusted part** (T_A) : this part of the system is composed of security areas in which user A can trust the other users as well as the processing and storage facilities, with respect to the security policy. Attacks against user A from inside the trusted part T_A of the system are considered to be impossible (the size of the trusted area can be reduced until this assumption is true, and Multi-Level Security may be implemented).

2. **Untrusted part** (U_A) : it is assumed the users in this part of the system may behave as opponents and realize any type of attack against A. Hardware and software in this subset of the system cannot be trusted by user A.

Figure 1 shows a network divided into two sub-areas T_A and U_A. The grey zone indicates user A's security area (T_A).

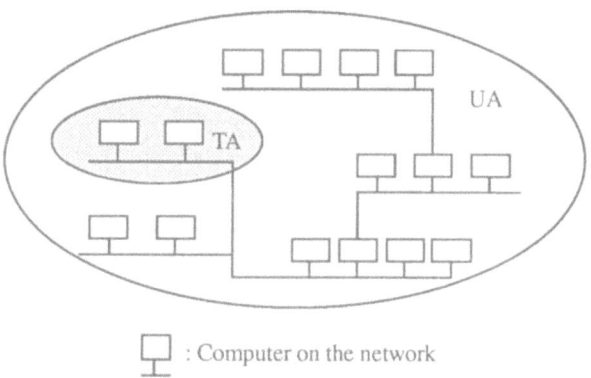

☐ : Computer on the network

Figure 1. Trusted and Untrusted network components for user A

According to the distributed computing environment described in section 2.1, $T_A \neq DCS$ i.e. $U_A \neq \emptyset$. If a user has to process sensitive data, T_A cannot be empty. As a matter of fact, inputs and ouputs must of course be intelligible and realized on the user's interface devices (terminal, workstation, printer...) where interactive activity takes place and/or on which results are presented. **Due to the intelligible form of inputs/outputs, the I/O devices must be located in the trusted part of the system.** Consequently, T_A cannot be empty.

According to this, we consider now the processors on which secure applications are executed.

Can secure processing take place in the Untrusted part of the system?

If it were the case, user A could initiate remote processing (for example on a set of powerful remote servers) in the U_A part of the system without loss of data security. An attempt to solve this problem had been proposed with the help of cryptography [15]. The principle was to process ciphered data without preliminary decipherment as shown in figure 2:

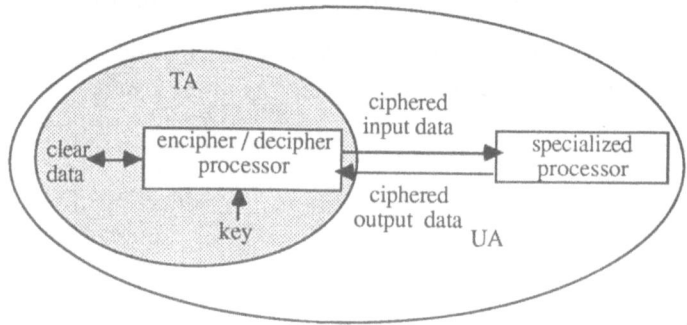

Figure 2. Processing ciphered data

The cryptographic technique able to make secure processing possible has to be consistent with the basic operations required for programming (arithmetic and logic operations). It was demonstrated that, whatever the method used, it is impossible to process ciphered data and to have a secure ciphering technique. The demonstration is as follows : suppose that the ciphered value of integer 1 is known, say 1'. Since arithmetic operations are available, it is easy to compute

2' (=1'+1'), 4' (=2'+2'), etc., and thus all $(2^n)'$, for all n. Using the available logic operations such as comparisons, the clear value of any ciphered value can be obtained by a dichotomic approach.

This shows that it is not possible to design a secure cryptographic code that, without preliminary deciphering, allows ciphered data processing with all the usual operations.

Another idea based on a hardware implemented security technique consists of "black boxes" in which ciphered input data is first deciphered before being processed, output data being ciphered as shown in figure 3.

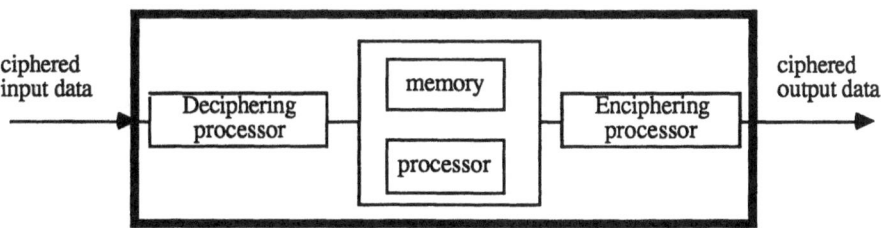

Figure 3. "Black box" processor

The "black box" should include a processor and sufficient memory space to store binary code and data to be efficient. To be really secure, this box cannot be opened without destroying its contents and thus must be of small size (no maintenance of this box is possible). To be able to communicate with other components of the system, the box should contain a cryptographic key such as a master key [10], this master key being installed once and never changed. The reduced size of the box implies multiple exchanges with the main memory whose content (and adresses) are ciphered.

This solution has major drawbacks. First, user A must be sure that the black box to which he sends his data is really closed and that he is not communicating with a white box which has the same functionalities (A can be sure of it if the box cannot be opened and if user A trusts both the design of the box as well as its installation). Secondly, key distribution and use are rather difficult to manage, due to the high number of such keys and their transmission. This solution is nevertheless possible, although it requires refined physical protections and complex data and key distribution.

Finally, from the point of view of data storage in untrusted environments, some satisfactory solutions are today of common usage, classically based on cryptography [3] or more recently at LAAS based on fragmentation-scattering [6]. We do not develop this aspect in this paper.

Figure 4 summarizes user A's view of the system with respect to security, for I/O, processing and storage resources in a distributed system. Only long term data storage can be located in the untrusted part (U$_A$) of the system. I/O devices, processing resources and storage devices for interactive accesses must stay in the Trusted part (T$_A$). The problem of processing sensitive information in the untrusted part of the system greatly restricts the usage of distributed resources.

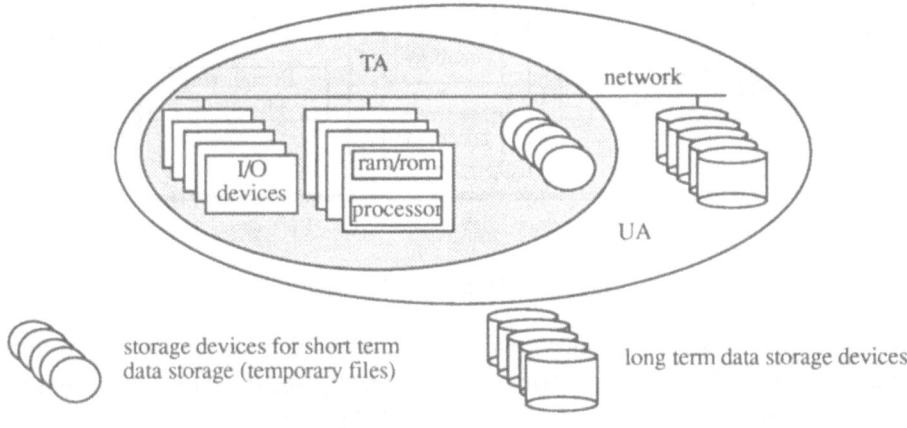

Figure 4. User A's view of the system

The conclusion of this section is that processing of sensitive information in untrusted environments is a very hard challenge.

2.3. Secure and reliable data processing

System reliability can be achieved by a number of techniques that are not detailed in this paper [9]. A large set of methods and tools have been developed or are under developement that enable a specified safety and reliability objective to be achieved. They are usually based on appropriate fault tolerance mechanisms implemented on redundant hardware components (processors and memory boards, busses, disks, machines, clusters...), to deal with hardware and/or software faults [1]. Companies like Stratus, August and Tandem propose solutions with specialized computers implementing fault-tolerant mechanisms; the techniques are hardware [18, 7] or software implemented [19]. This first approach based on specialized hardware is expensive and moreover previous investments cannot be integrated. More recent solutions consist in designing fault-tolerant mechanisms based on classical computers interconnected through broadcast networks and appropriate distributed algorithms. Communication mechanisms such as logical addressing, replicated communication endpoints and atomic multicasting can enhance the set of possible solutions, as for example demonstrated in the DELTA-4 project [11]. The granularity of various techniques for hardware/software fault tolerance in distributed systems can be the procedure (the block) [14], the object [11] or the process [2], [4], and so code and data are replicated on different hosts. For simplicity, we will consider that fault-tolerance is based on process replication.

According to the working environment defined in section 2.1, what are the possible solutions for reliable processing of sensitive information?

Since processing must be realized in the Trusted part of the system, the redundant components used to implement fault-tolerance mechanisms must seemingly also belong to T_A. Thus, fault tolerance cannot take advantage of all the available computers on the network, and so the fault-tolerant mechanisms, even based on flexible software-implemented techniques, are restricted to T_A, or implemented within expensive specialized computers located in that part.

Implementing secure and reliable processing in distributed computing systems seems either very costly, or very constraining since it is not possible to take full advantage of the flexibility provided by this distribution.

2.4. An original approach

The original approach which is proposed in this paper aims at providing a satisfactory solution to the problem of Secure and Reliable Processing in Distributed Computing Systems, i.e. that applications can take advantage of the available hardware redundancy to process sensitive information. The question is :

How to securely process sensitive information in a Untrusted environment?

Figure 5 presents the new distribution of processing and storage resources with respect to the T_A and U_A partition of the system. Actually, only I/O devices must be kept in T_A.

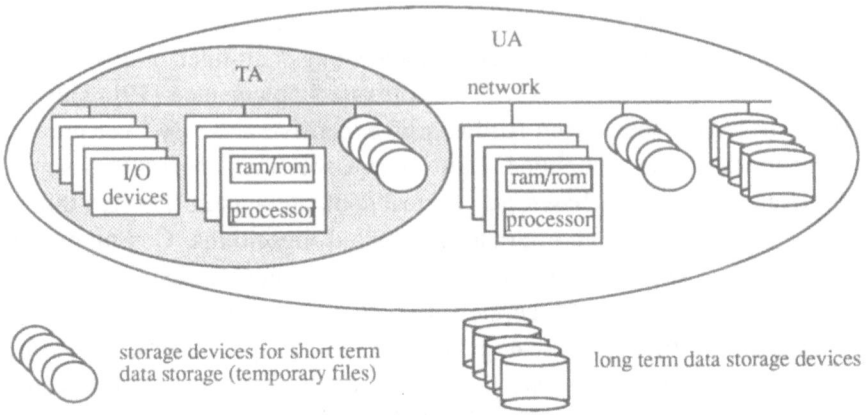

Figure 5. User A's new view of the system

More precisely, the idea is to process any type of application in the untrusted part of the system. The user does not care about the location where the processing takes place. In order to prevent conspiracy, the different processors used to run a given application (and its copies) must be located in different guarded security areas, as shown on figure 6.

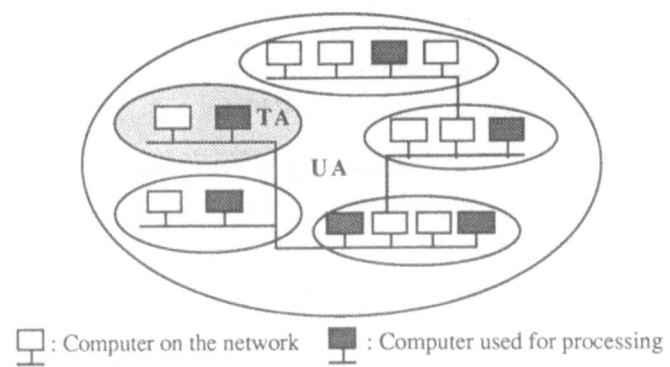

⊡ : Computer on the network   ■ : Computer used for processing

Figure 6. Involved processors in the application

Let P be the set of processors involved in the processing of the application. The idea of the FDP approach is to process only a subset of the code and/or the data on every processor P_i in P, so that each isolated subset cannot be considered as sensitive information. The underlying techniques are discussed in the next section.

3. Fragmented data processing

3.1. Principles of fragmentation

The fragmented data processing principle consists in scattering the sensitive information (code and data) among distinct non-conspiring areas of a distributed computing architecture. Consider as a first example, a simple application program AP composed of one code segment C and one data segment D (AP={C,D}). The fragmentation technique can be applied to C and D, such that AP is split into n fragments i.e. AP = { $(C_1,D_1),(C_2,D_2), ... C_n,D_n)$ }. Each application fragment $AP_i=(C_i,D_i)$, composed of a (slightly modified) subset of the code segment C_i and data segment D_i of the complete application, is processed on processors located in different areas, as summarized in figure 7.

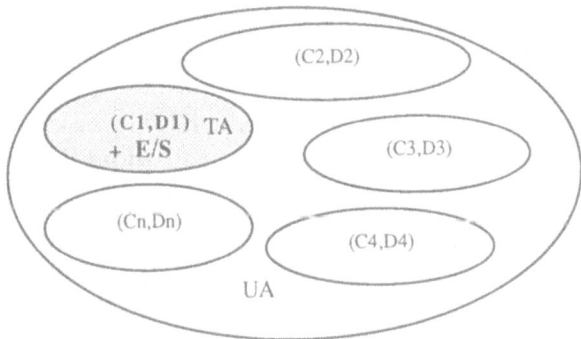

Figure 7. Scattering of fragmented processing

Security relies on the fact that in any area, the corresponding information (code and data) does not constitute a sensitive information in itself, although not ciphered. Data security can be ensured by reducing the amount of information in D_i , for all i in [1,n]. The key to the method consists in defining a fragmentation algorithm such that a set of D_i even located in the same area does not reveal any information. This rule should also be true for the code. Integrity and availability of the application AP can thus rely on redundant computation of application fragments.

Suppose that a given application AP is split into two application fragments $AP_1=(C_1,D_1)$ and $AP_2=(C_2,D_2)$, and that there are 4 processors available in the untrusted part of the system, located in 4 different security areas. According to the previous definitions, secure and reliable processing of application AP may be realized by running two copies of AP$_1$ and two copies of AP$_2$ on 2 out of the 4 processors, with appropriate fault-tolerance mechanisms. The solution consists in applying transformations at the application level (code and data) such that cooperation of application fragments is functionally equivalent to the initial application.

According to the system environment which has been defined in section 2.1, clear (non-fragmented) information only appears in T$_A$, and so input/output operations are realized in the trusted part of the system as shown in section 2.2. In U$_A$, the data is always in a fragmented form. Thus, fragmentation/reassembly mechanisms are invoked each time data crosses the border between T$_A$ and U$_A$. These functions are integrated in the system and could rely on specific hardware at the lower level of the system, for instance the communication unit. A secret key is given as a parameter for the fragmentation and reassembly operations. Secure data transmission between nodes in the U$_A$ part must also use classical security techniques, such as communication ciphering.

FDP is not only applied to data but can also be applied to the code segment of the application. In any case, transformations are realized on the code in order to process the local data and to synchronize with the rest of the application, i.e. the other application fragments. Thus, every code fragment is specific, but the global behaviour of all the cooperating application fragments is equivalent to the initial application. The application fragments can be generated by a pre-processor and a compiler, and linked to specific library functions. A problem is nevertheless to generate fragments that have no interesting value from a security point of view. This can only be achieved with help of the user (assisted possibly by purpose-built tools).

Different fragmentation modes must be investigated with respect to a given application. These modes could be used simultaneously and some parameters could be used to adapt the fragmentation process to a given application and its specific security requirements. As a first example, one can distinguish data which is fragmented and the code that can be used in its complete form, (with additional synchronization functions). The fragmentation can thus apply to data only, code only, or code and data segments. The last parameter that can be defined is the granularity level of fragmentation. Classical tradeoffs between security and performance must be taken into account in implementation. If a large number of fragments is used, security is improved at the expense of performance.

According to these principles, we propose to illustrate this form of data processing in sections 3.3 and 3.4.

3.2. Specific requirements

Fragmented Data Processing requires secure communications between the nodes of the network where processing takes place, as well as between processors and I/O devices. Loading of any node involved in the application can be realized in a first step, before the application is triggered, and be resident during execution of the application. On the other hand, data might be exchanged very often. Data messages are very short and so communication mechanisms should be very efficient; I/O traffic represents a small amount of information.

Execution code is mainly composed of arithmetic expressions, assignments, logical comparisons and conditional branches. Since information is in a fragmented form , one processor does not have sufficient local information for processing all instructions. Some information must be efficiently transferred between the processors involved in the computation. From a security point of view, transferred information must be limited, such that any given processor holds the minimum amount of information to process a given basic operation.

Even with a great number of nodes on the network, the number of security areas may be limited. It is thus possible to implement a complete interconnection of security areas. This solution, although more expensive than a classical broadcast network, is more efficient not only from the performance point of view but also from the security point of view, in the sense that a security area can only access information that is sent to it. Eavesdropping becomes more difficult in such a meshed network topology because of the number of links [8]. One possible topology is presented in figure 8.

Black links indicates the inter security area channels. The inter-area communication units are responsible for the communications with the other areas. For a given application, the inter-area communication unit belonging to the Trusted security area T_A, can also be responsible for the fragmentation and reassembly operations. Thus, fragmentation and reassembly operations are additional functions of the inter-area communication unit, that are only used by applications launched from the security area T_A.

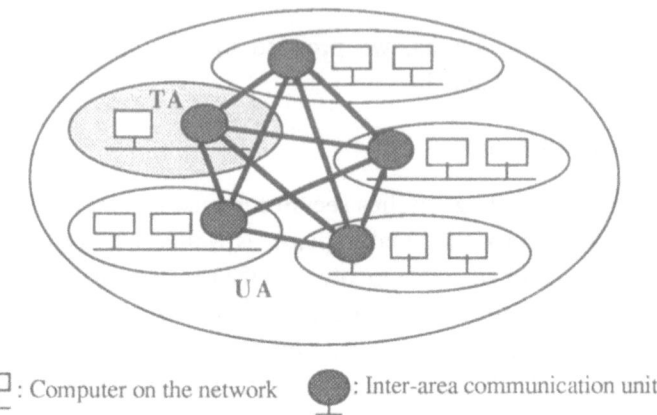

⊑ : Computer on the network   ● : Inter-area communication unit

Figure 8. Network topology

3.3. Fragmentation granularity

Fragmentation can be applied to data, code or both code and data. It does not seem that Fragmented Data Processing may be carried out only on the data, i.e. both data and code are modified and are both transformed in order to perform a fragmented execution. We can consider in a first step, that only data is the sensitive information. But even in this case code transformation is necessary for cooperation between the application fragments (synchronization and data exchange mechanisms).

Fragmentation operations may be carried out at various levels of granularity. We propose three examples of granularity that show the influence of this basic parameter.

First, fragmentation is possible at a very low level of data such as bit-slices. (For example, each 16-bit word may be fragmented into four-bit fragments). In these conditions, the operations are similar to those which are used in massively parallel computers and the work that has been done in this field could be of great help to design fundamental techniques of fragmented computing. An important constraint to take into account is data security and particularly data confidentiality, that must not be compromised due to the data exchanges.

An intermediate level of granularity could be defined by the elementary types of a programming language (characters, integers, reals, booleans...). The atomic granules (piece of information which is not fragmented) would be the variable of the programming language. If higher granularity does not endanger security, fragmentation can be based on structured data types such as arrays and records. Subsets of all the variables would be allocated to each processor. In this case, the granularity is compatible with the high level programming language structure. Moreover, only a subset of code is executed by each processor, thus enhancing code confidentiality and performance. Communication messages would be less numerous than in the previous case.

Finally, it is possible to design fragmentation at a high level, such as modules or procedures. The main advantages of a high level of granularity are the ease of implementation as well as a lower amount of communication required to synchronize the execution. Nevertheless, a large subset of data would be available on each processor, decreasing data confidentiality or integrity.

These examples of fragmentation granularity show the trade-off between security and performance. If fragmentation is carried out at a low level, security is enhanced while complexity of design and implementation increases. Another interesting observation is that the more data is fragmented at a lower level, the more the original code must be transformed in order to process the data at each machine. Moreover, if data is fragmented at a low level, all instructions are executed by each processor while at a high level of fragmentation, only a restricted subset of the code is executed by each processor (group of statements of a programming language, a single procedure, for example).

The principles presented in this section open a large field of investigations on basic techniques that could be used for fragmentation, taking into account security constraints. It would be interesting to further investigate the impact of fragmentation at different levels in order to offer different tradeoffs between security and performance: the quantification of the amount of information transmitted during arithmetic and logic operations for instance is of interest not only with respect to performance but also with respect to security. Fragmentation can also be user assisted in order to take into account the security semantics of application data structures. If a high level of fragmentation is considered (data types of a programming language), only the designer of the application can define the variable sets that do not reveal sensitive information. This can be done by means of development tools or by specific statements in a programming language.

3.4. Fragmentation examples

According to the various granularity levels defined in the previous section, we would like to illustrate fragmentation on two examples. Several rules must be observed during the fragmentation process, in order to apply code transformation such that the global behaviour of the cooperating application fragments is equivalent to the initial application. The code transformation can be applied to the source code (high level language or assembly language) by using library functions for communication and synchronization. Let's take two very simple examples to give an idea of simple fragmentation rules and code transformations which are nevertheless of interest with respect to security. In this first example, the granularity level is a byte.

Example 1: (Granularity: byte)

```
program SMALL;
        var A,B : integer;
        begin
        read (A,B);
        write (A+B);
        end.
```

This simple application program has two variables A and B. Each integer is represented by a 16-bit word (b1,b0) ; this application could be split into two application fragments (SMALL_0, SMALL_1). The two application fragments could be as follows:

```
SMALL_0 :
        D0 = { A.b0 , B.b0 } = {A0,B0}

        C0 = {  define X: temporary variable ;
                X=A0+B0;
                propagate_carry (to SMALL_1);
                send (X); (to source area TA)
                }
```

SMALL_1 :
>   D1 = { A.b1 , B.b1 } = {A1,B1}
>
>   C1 = { define X: temporary variable ;
>   X=A1+B1;
>   wait for carry from SMALL_0;
>   add_carry (X);
>   send (X); (to source area T_A)
>   }

In this example, the input/output operations are executed in the trusted part T_A. Fragmentation, compilation and linking, loading, I/O operations (read/write) and initialization of fragmented variables are performed in T_A. The A1+B1 and A0+B0 computation can be performed in the U_A part of the system, and SMALL_1 and SMALL_0 can be run with several copies. The initialization of the fragmented variables is such that A1 is the left-hand byte in A, B1 the left-hand byte in B; A0 and B0 are initialized with the right-hand byte of A and B, respectively. Synchronization is achieved by the *add_carry* operation in SMALL_1, waiting for *propagate_carry* in SMALL_0. Information transferred from SMALL_0 to SMALL_1 is 1 bit, giving 1 extra bit of information in addition to the initial values. Theoretically, some information about initial values is revealed by carry propagation, but practically it seems to be very difficult to trap and to use such information.

Example 2: (Granularity: integer)

> **program** NOT_QUITE_SO_SMALL (alias BIG);
>   **var** A,B,C,D : **integer**;
>   **begin**
>   read (A,B,C);
>   D:=A+B;
>   write (D*C);
>   **end.**

In this example, the granularity is the integer type and this application can also be split into two application fragments. User A could decide that the complete set of variables {A,B,C,D} represents sensitive information, but that each of the two isolated subsets {A,B} and {C,D} is not sensitive information according to the semantics of each individual variable. In this case, the two application fragments can be presented as follows:

BIG_0 :
      D0 = { A, B}
      C0 = { define X: temporary variable ;
           X=A+B;
           propagate(X); (to BIG_1)
           }

BIG_1 :
      D1 = { C,D}
      C1 = { define X: temporary variable ;
           assign (D); (with X from BIG_0);
           X=D*C
           send (X); (to source area T_A)
           }

Like in the first example, the input/output operations and initializations are executed in the trusted part T_A. BIG_1 and BIG_0 processing can be performed in the U_A part of the system. As in the first example, BIG_1 and BIG_0 can be run with several copies. Synchronization is achieved by the *assign* operation in BIG_1, waiting for *propagate* in BIG_0. Information transferred from BIG_0 to BIG_1 is A+B, giving information on the initial values A and B. However, the whole information (A,B,C,D) is never present on none of a single untrusted processor.

3.5. Is FDP technically and economically possible?

The answer to these questions is not easy. From a technical point of view, the main problem is to find fragmentation techniques that ensure a real dispersion of sensitive data. Efficient tools have to be designed to take into account what constitutes a acceptable dispersion from the user point of view. If a very low level of granularity is used, it may be possible to fragment the data independently of its value and meaning. Nevertheless, it is probably the most expensive solution especially in terms of CPU time and communications.

From an economic point of view, FDP must be compared to a solution where all the security areas have their own fault-tolerant machines. This simple solution is costly if many security areas are involved. Inversely, FDP is of increasing interest in this case.

4. Conclusions

This paper is an attempt to solve the problem of sensitive data processing in an untrusted environment, with an original approach: Fragmented Data Processing. This method would take advantage of available resources that exist in a distributed system, and particularly in order to perform reliable processing thanks to the natural redundancy offered by a local area network.

Implementation of the method would benefit from a specific network with point-to-point links interconnecting all the security areas, as well as hardware and software components on all the machines in order to execute the specific operations of fragmentation. Nevertheless, these components could be added to any type of machine allowing implementation to be carried out on existing networks. The main difficulty is to design fragmentation techniques that ensure effective security for data and/or code and that would be of acceptable performance. Some analogies can be found with coroutines and also with the data flow principle.

We have just presented here the basic idea of Fragmented Data Processing in this paper; many questions have to be investigated before concluding on the interest of the method. Specific techniques must be designed and validated. Nevertheless, we think that this idea offers a large field of investigation for future work and is presently one of the working direction of our research group.

Acknowledgements

The authors wish to thank very much Jean-Claude Laprie, head of the *Dependable Computing and Fault-Tolerance group* at LAAS, who originated the concept of "Intrusion tolerance" as a technique for ensuring security. Thanks must also go to several past and present members of the group who have contributed to the work in this area. Special thanks must go to Yves Deswarte for the many stimulating discussions we have had on the Fragmented Data Processing technique.

References

[1]   A. Avizienis, J.C. Laprie, "Dependable computing: from concepts to design diversity", *Proc. of the IEEE*, Vol. 74-5, pp. 629-638, May 86.

[2]   J. Bartlett, "A non-stop operating system", *Proc. of the 8th Symp. on Operating Systems Principles*, Pacific Grove, California, USA, pp. 20-29, Dec. 87.

[3]   D.E. Denning, *Cryptography and data security*, Addison Wesley, 1983.

[4]   J.C. Fabre, Y. Deswarte, J.C. Laprie, D. Powell, "Saturation: reduced idleness for improved fault-tolerance", *Proc. of the 18th Symp. on Fault-Tolerant Computing, FTCS-18*, Tokyo, Japan, June 1988, pp. 200-205.

[5]   J. Fraga, D. Powell, "A fault and intrusion-tolerant file system", *Proc. of the third Int. Cong. on Computer Security (IFIP/SEC'85)*, Dublin, Aug. 85, pp. 203-218.

[6]   J.M. Fray, Y. Deswarte, D. Powell, "Intrusion-tolerance using fine-grain fragmentation-scattering", *Proc. of the 1986 IEEE Symp. on Security and Privacy*, Oakland, April 1986, pp. 194-201.

[7]   A.L. Hopkins, T.B. Smith, J.H. Lala, "FTMP: A highly reliable fault-tolerant multi-processor for aircraft", *Proc. of the IEEE*, Vol. 66, n° 10, October 1978, pp. 1221-1239.

[8]   Y.Koga, E. Fukushima, K. Yoshihara, "Error recoverable and securable data communication for computer network", *Proc. of FTCS-12*, Santa Monica, June 1982, pp. 183-186.

[9]   J.C. Laprie, "Dependability: a unifying concept for reliable computing and fault-tolerance", in *Resilient Computing Systems*, Collins (UK) and Wiley (USA), T. Anderson editor, 1988.

[10] C.H. Meyer, S.M. Matyas, *Cryptography*, J.Wiley & Sons, 1982.

[11] D. Powell, G. Bonn, D. Seaton, P. Verissimo, F. Waeselynck, "The DELTA-4 approach to dependability in open distributed computing systems", *Proc. of the 18th Symp. on Fault-Tolerant Computing, FTCS-18*, Tokyo, Japan, June 1988, pp. 246-251.

[12] M.O. Rabin, "Efficient Dispersal of Information for Security, Load Balancing, and Fault-Tolerance", *Journal of the ACM*, Vol. 36, n° 2, April 1989, pp. 335-348.

[13] P.G. Ranéa, Y. Deswarte, J.M. Fray, D. Powell, "The Security Approach in DELTA-4", *Proc. of the European Telematic Conference (EUTECO 88) on Research into Networks and Distributed Applications*, Vienna, Austria, April 88, Ed. North-Holland, pp. 455-466.

[14] B. Randell, "Fault-tolerance and system structuring", *Proc. of the 4th Jerusalem Conference on Information technology*, Jerusalem, Israel, May 1984, pp. 158-169.

[15] R.L. Rivest, L. Adleman, M.L. Dertouzos, "On data banks and privacy homomorphisms", *Foundations of secure computations*, Academic Press 1978, pp. 169-179.

[16] J. Rushby, B. Randell, "A distributed secure system", *IEEE Computer Mag.,* Vol. 16-7, July 1983, pp. 55-67.

[17] A. Shamir, "How to share a secret", *Communications of ACM*, Vol. 22, n° 11, November 1979, pp. 612-613.

[18] T.B. Smith, "High performance fault-tolerant real-time computer architecture", *Proc. of the 16th Symposium on Fault-Tolerant Computing,* Vienna, Austria, July 1986, pp. 14-19.

[19] J. H. Wensley, L. Lamport, J. Goldberg, M.W. Green, K.N. Lewitt, P.M. Melliar-Smith, R.E. Shostak, C.B. Weinstock, "SIFT: design and analysis of a fault-tolerant computer for aircraft control", *Proc. of the IEEE,* Vol. 66, n° 10, October 1978, pp. 1240-1255.

Security and Fault Tolerance

Chair: R. Turn (California State University, Northridge, California, USA)

INTEGRATION PROBLEMS IN FAULT-TOLERANT, SECURE COMPUTER DESIGN

Mark K. JOSEPH
Los Angeles, CA - USA

Abstract

This chapter explores part of what has been termed "secure fault tolerance" [34]. Essentially, this involves ensuring that the fault-tolerance techniques used in a fault-tolerant, secure computer design do not accidentally nor deliberately violate its security policy. The impact that fault tolerance and computer security have on each other is discussed.

1. Introduction

This chapter explores several basic problems that can occur when integrating multiple dependability [25] concerns into one design. Here we address computer systems that have reliability, availability and security requirements, and where fault tolerance is used to achieve the reliability and availability goals. This type of study is viewed as one of the necessary steps toward the design of a fault-tolerant, secure computer.

Specifically, the contributions of this work are: (a) several examples of how fault-tolerance mechanisms can undermine the security of a system are presented, (b) major weaknesses of current integrated design approaches are discussed (section 4), and (c) in order to start to propose solutions to item "b" several issues in fault-tolerant, secure computer design are discussed (section 5). Item "a" is undertaken by addressing two fundamental security concerns: tamperproof design (section 2) and covert channels (sections 3 and 5).

"Tamperproof" (or tamper resistant) refers to a design attribute that guarantees that the most fundamental security mechanisms cannot be maliciously corrupted in a deployed system. Where as "covert channels are those paths not meant for communication but that can be used to transmit data indirectly" in violation of a system's security policy [30]. As a simplified example, TOP SECRET information could be illegally transmitted as a bit stream to an uncleared system user via deliberate fluctuations of the state of a globally visible data item. A program with legitimate access rights to the TOP SECRET information continually changes a file's name (i.e., the visible data item) to "covert" to transmit a binary 1, and to "leak" to transmit a binary 0. The uncleared user periodically reads the file's name in order to receive the signalled data [14]. Examples of covert channels in the context of a fault-tolerant, secure computer are given below.

Throughout this chapter we assume the following principle: it is impractical to trust--with regard to security concerns--all of the mechanisms used to provide fault tolerance. Thus, some of the fault tolerance mechanisms (e.g., recovery software) can themselves be malicious and/or be used to bypass security barriers.

It is recommended that the reader who is unfamiliar with the security concepts used in this chapter, with the importance of the security problems presented, and/or with the effort required for some of the mentioned solutions, to refer to [10, 14] for an introduction to these issues (also see the small glossary before the references).

2. Testability versus a tamperproof cesign

Problem Description

The reference monitor [10, 14] of a secure computer system has a fundamental requirement to be tamperproof. At the same time, all secure computer systems from classes C1 through to A1 [1] require some hardware and/or software features that periodically validate the proper operation of the hardware and firmware elements of the Trusted Computing Base (TCB) (i.e., diagnostic capabilities) [10].

In order to provide effective diagnostic capabilities (i.e., with reasonable coverage), hardware should be designed to be testable [13, 35]. This facilitates the location of hardware faults so that a failed hardware unit can be replaced or switched out. This is viewed as so important that all IBM mainframes, starting with the early 360 series, include extensive testability mechanisms as hardware standards.

Testability involves the observation and control of a computer system's state. Testability can be utilized for run-time diagnostics or in initial hardware checkout for manufacturing defects, and as such is also useful in non-fault-tolerant systems. A commonly used technique for testable processor design is Level Sensitive Scan Design (LSSD) [32]. Here, a set of registers in a processor can be connected into one long shift register. Observation of system state is accomplished by shifting register contents out, and control is accomplished by shifting in test patterns in order to force errors to manifest themselves. This mode of operation is invoked upon the detection of an error.

The issue here is obvious: how to allow the most trusted part of a secure computer system (i.e., the reference monitor) to be tested, by LSSD or similar techniques (e.g., [2]), without compromising its ability to maintain the security of a system? Obviously, if a reference monitor's state is subject to unauthorized modification via LSSD, then it is not tamperproof. The following is a discussion of this issue with LSSD specifically in mind.

---

[1] The Trusted Computer System Evaluation Criteria [10] provides a ranking of systems (i.e., D, C1, C2, B1, B2, B3, A1) based on the degree of trust (in regard to security concerns) that can be placed in computer systems. D class systems provide extremely limited to no security, and A1 class systems provide the most comprehensive security.

After the testing mode is entered, the current state in the affected area of a processor [2] is typically read out and stored for later analysis. For any secure computer system such state information will likely be sensitive. Therefore, its transfer and storage must be protected against unauthorized access (by analogy, if intentional design faults are at work, then treat this state information as audit data).

Control of a testable circuit via LSSD involves placing test patterns into the shift register. This implies the need for two distinct modes of hardware operation: test and normal. If this is not enforced adequately, then the reference monitor could be subverted by using LSSD to place it into an insecure state. For example, changing copies of access rights temporarily stored in a processor's register while data is being accessed (e.g., the Intel 80286 [18] has segment registers that cache access information when a segment is in use). Also, changing the program counter while in privileged mode can lead to untrusted code running with the privileges of the reference monitor.

Therefore, if the testability hardware, or any diagnostic software driving this hardware, or a special maintenance operator (if one is needed) is malicious, then the security of a system can be subverted.

Possible Solutions

One obvious solution to the above problem is to completely separate test modes with normal service delivery. After a system test is conducted a trusted system restart would be performed. The new reference monitor's state would be untouched by the previous test actions.

Another possibility is to provide assurances [10] (e.g., via formal verification) that all test hardware and software do not (accidentally nor intentionally) modify the reference monitor's state, and that all test logic is itself protected from subversion (i.e., make it part of the TCB). Note, that both of these solutions are fault avoidance techniques.

---

[2] Note, that if one failed hardware component in a system enters test mode this does not necessarily mean that the whole computer system stops providing service. Designs, such as the IBM 3081, can isolate the regions of hardware to be tested from the rest of the system.

3. Fault-tolerance mechanisms introducing
new covert channels

Problem Description

The added complexity of fault-tolerance mechanisms in a secure computer design is likely to result in new information flows and covert channels [34]. Here we present unique instances of both that can be directly attributed to fault-tolerance techniques.

Several new information flows are: (a) error reporting from error detection logic to recovery logic (this can be hardware or software), (b) health and status information including diagnostic results, and (c) reconfiguration and state restoration control signals from recovery logic to redundant hardware and state. One aspect determining if these information flows could be used for covert channels is whether or not system-defined and protected data objects are used for communication [22]. Low-level hardware signals are used in most fault-tolerant designs, and these would not currently be included under the security mechanisms control (e.g., serially transmitted error reporting messages can be too low level to be audited).

When analyzing computer systems for covert channels the following issues must be addressed: bandwidth of the channel, type of channel (storage or timing), its ease of use, determining the necessary shared resource used as the channel (i.e., without a shared resource no channel is possible), determining whether the sending and receiving subjects are from different security domains (i.e., different subjects at different security levels, otherwise no channel exists), determine how the sender uses the shared resource to transmit information, how the receiver senses the information sent, and how the sender and receiver synchronize the information transfer [22, 24, 30]. Additionally, it is interesting to determine natural causes of channel noise. This section presents two potential cases of covert channels due to fault-tolerance mechanisms.

Example 1 -- Covert Storage Channel

Consider a simple N-Modular Redundant (NMR) computation configuration with a disagreement detector. Two cases exist, one where all processors have identical copies of a process, or one where each processor has a diverse version.

The NMR configuration ensures process synchronization in order to enable voting on results. The voter compares the outputs of all processes at predetermined times, and the majority of all values is the output result. If any processor(s) disagrees with the majority decision, then it is recorded by the disagreement detector [23] and this information is sent to some recovery logic and/or error log.

Thus the shared resource for this storage channel is the N processors. The sending subject is the N copies of a process, and the receiver is the recovery logic or any subject that can read the error log. The storage variable, which is indirectly written by the sender, is the disagreement detector.

The following is the scenario for sending information on this channel. First, one or more processors force a disagreement, by the process running on it deliberately performing an incorrect calculation (i.e., the process contains an intentional software design fault). The disagreement detector will detect this and give the processor number(s) to the recovery logic. (For this channel to remain hidden the number of deliberately incorrect processors must be a minority so as not to interfere with normal processing.). There are many ways to encode information in this way: (a) each processor may have a value associated to it, or (b) for a 5MR configuration, for example, a 2 out of 5 and/or a 1 out of 5 code can be used to encode several values.

To signal information by the above method the N processes, which make up the sender, must produce errors at different times and in different patterns. For the case where all processors have identical process copies two subcases exist: (a) all input data to the NMR configuration is identical, and thus some small amount of visible state (e.g., initialization data) must be deliberately forced to be different in at least one processor in order to distinguish it from the others--otherwise the channel cannot be used, and (b) diverse data [1] is used as input, which can also be used to differentiate identical process copies in order to transmit several failure patterns over the channel. Lastly, when diverse process versions exist [4], the visible state and input can be identical, and the channel can be used with just one version deliberately failing at different times.

Additionally, the channel receiver must be given the failure pattern to signalling code relation in order to correctly read the transmitted data. This can be done by a calibration sequence, where for the first few covert transmissions the sender runs through the entire code in some predetermined order.

The bandwidth of this channel may be high, since it does not interrupt normal system functioning and the recovery logic is the receiver. Thus the recovery logic can ignore the generated errors when it knows they are being used for transmission purposes. This last point is important, because permanent faults are distinguished from transient faults by their persistence over a time interval. If the recovery logic was not the receiver or at least a "cooperating scheduler" [30], then the covert channel would force many permanent faults to appear, likely closing off the channel and raising suspicion.

Channel noise will occur naturally due to all types of accidental errors that force some of the processes to occasionally produce an incorrect output when they are not scheduled to fail. This reduction in bandwidth is dependent on: (a) the rate of occurrence of accidental faults, and (b) the probability that accidental faults occur in processors that are not scheduled to fail, versus hitting processors that are already scheduled to fail.

If accidental errors should cause graceful degradation [3] of the number of processors in the redundant configuration, then an extra channel, used to synchronize the sender and receiver, can be used by the recovery logic to inform the sender of this condition. There is no way for the sender to notice a degradation by itself, and thus either the channel would be closed or channel noise would increase. So upon receipt of the degradation information, the sender could recover by using a simpler signalling code with lower bandwidth.

The sending subject can be at any security level, and over time will change as new processes run on the processors. The recovery logic and any subject that can read an error log can thus, at some time, be at a different security level than the sending subject. Additionally, this is not an overt channel for the following reasons: (a) the disagreement detector is not meant as a communication path between a process and the recovery logic [30], and (b) the disagreement detector is not a system protected data object, and is not normally viewed as a data container [22].

Example 2 -- Covert Storage & Timing Channel

A watchdog processor is a simple coprocessor that provides system-level, concurrent error detection by monitoring the behavior of the CPU [26]. The

---

[3] A specified and designed transfer to lower *levels of service.*

watchdog can be added to the design of any computer system, at a reasonable cost, without major changes. Thus, it may be desirable to retrofit existing non-fault-tolerant, secure computers with a watchdog in order to add some error detection capability. However, doing this requires careful design, since a watchdog processor that utilizes derived signatures or assigned tokens as program control flow markers can[4] add both a storage and timing channel to a secure computer design.

For these channels the sending subject is the process running on the CPU being monitored by the watchdog. The receiving subject is the watchdog itself--which can contain its own writable memory. The main idea is that the sender can transfer information by modifying its control flow at specific times. The watchdog will easily detect this by recognizing specific signatures.

The following is the scenario for sending information on the storage channel. Special loops (i.e., code sequences) in a process are associated with a signalling code to be transmitted (e.g., a binary 0). The correspondence between the signature generated by the code loops and the signalling code is done as in the previous covert channel (i.e., by a calibration sequence). To send a code number a process simply runs a special loop. The receiving watchdog processor dynamically generates the signature and compares it with its known signalling code to signature relation. Additionally, data can be encoded by the number of times a particular signature is seen in a row, or by the order of a sequence of signatures.

The following is the scenario for sending information on the timing channel variant. The only difference from the storage channel is that the receiving watchdog senses the information transfer by the length of time it takes to receive the next signature (e.g., the actual length of the loop) [30].

In order to accomplish the signalling of information via this channel, the extra executions of signalling code loops must not affect the final result of a process. These signalling code loops can be routines that perform some check function, such as reading through a link list to find a special entry; or they can be specially implanted code that perform useless functions and have no effect on the program's results.

---

[4] This type of watchdog processor is also known as a Program Flow Monitor (PFM).

Bandwidth can be restricted if a process has a time limit on it, since the use of the covert channel incurs added execution time. Additionally, as in the channel presented in example 1 above, channel noise can occur due to accidental errors.

Possible Solutions

The covert channels described above can be removed from a fault-tolerant, secure computer design, but to do so requires making trade-offs. Five approaches seem immediately possible: (a) place covert channel receivers (e.g., the watchdog processor in example 2) at the system-high sensitivity level[5] ; this can be done since both channels have few receivers, (b) place channel receivers into the TCB, (c) replace the chosen fault-tolerance mechanisms with mechanisms that do not contain covert channels--however, they must still meet other requirements (e.g., coverages, latencies), (d) attempt to remove the channel or limit its bandwidth by careful redesign of the same fault-tolerance mechanism(s) [10, p.81], or (e) audit potential use of storage channels [11, 14].

All the approaches listed above are essentially covert channel avoidance techniques; approach "e" tries to discourage channel use by introducing the possibility of detection. In addition, approach "a" does not work for the recovery logic receiver discussed in example 1. Placing it at system high will restrict its actions (e.g., prevent writing, or moving information [data and programs] residing at lower sensitivity levels), and thus can prevent it from performing its task. A possible alternative is to use separate instances of a recovery process at each sensitivity level [34]. Lastly, the disadvantage with approach "b" is that placing a receiver into a TCB makes it harder to ensure the TCB's correctness (e.g., via formal verification for A1 class systems, justification for a minimal TCB for B3 and A1 class systems).

4. Problems with current integrated design approaches

Presented in [27] is a design approach for critical systems where criticality encompasses the following system properties: human safety, fault tolerance, high availability, security, privacy, integrity [5, 6], and timely responsiveness for the entire life cycle of a computing system. Neumann proposes that a hierarchical layering based on current multi-level security (MLS) and multi-level integrity

---

[5] The highest sensitivity level defined for a specific computer system.

(MLI) technology with allocations of safety, security, and fault-tolerance mechanisms at each layer (depending on the function of that layer), will provide a much better system than is achievable with current design methods.

The following are two shortcomings with this MLS/MLI hierarchical approach. For this discussion we limit our concerns to fault tolerance and security aspects of a design. First, a lower layer in a MLS/MLI hierarchy cannot rely on a higher layer to perform error detection and/or recovery for it. This is for security reasons. In typical MLS hierarchies, the lower the layer the more *trusted*[6] it usually is--the lowest layer (i.e., the operating system kernel) being the most trusted. Thus it does not make sense for a trusted, lower layer to depend on an untrusted, higher layer for anything. If error detection and/or recovery does progress up in a hierarchy, then the result can be some form of denial of service [15].

As an example, Hamming codes for memory would reside in the lowest layer of Neumann's hierarchy. Reasonable tests, on the values of data stored in memory, appearing in an application would normally reside in one of the untrusted, higher layers. An error undetected by a single error correcting, double error detecting Hamming code (e.g., certain burst errors) would depend on the untrusted reasonable tests to detect the incorrect values from memory. However, if these reasonable tests were *malicious* they could ignore the error entirely, or even cause false alarms. Likewise, malicious recovery logic (at less trusted, higher layers) could perform the wrong recovery action, perform only partial recovery, or do no recovery at all. Thus Neumann's scheme does not provide protection against integrity-based denial of service, which should be a major concern for critical military systems.

Reliance on untrusted, higher hierarchy layers for error detection can also be found in [28]. Here the concept of a "recovery layer" (a more trustworthy, lower layer) dedicated to fault tolerance is described. In order to support a recovery block mechanism [29] the recovery layer imports application layer acceptance tests. That is, less trusted, possibly malicious acceptance tests are used for error detection in the more trusted recovery layer. Thus, this example is equivalent to the Hamming code example given above.

---

[6] To perform its defined function as specified (e.g., in a non-malicious manner). Trusted not to violate some stated (security) policy.

Second, at each layer in Neumann's hierarchy multiple levels of sensitive information can exist. Thus any recovery mechanism at such a layer may have to deal with this information and critical state. This would require recovery logic to be trusted not to mix information at different sensitivity and integrity levels [34] (i.e., not to violate the star [*-]property). Neumann's scheme does not address this problem.

Due to the amount of logic to be trusted, and its potential complexity, these last two points indicate that such a hierarchy could easily require more trust than is practical to rely on. (Trust is supported by assurances [e.g., fault avoidance techniques such as formal verification] that a design is correctly implemented.)

Currently, this author knows of no existing solution approaches that could effectively deal with the problems described so far in this section. However, a research direction that seems promising involves the creation of a unified design and evaluation paradigm for computer systems required to possess multiple dependability attributes. Section 5 discusses several design issues that would have to be captured by such a unified paradigm.

A basic principle in dependable computer design is that a reliable computer can be built from unreliable parts. Presented in [9] is a design approach for building a reliable secure computer from unreliable insecure parts. However, if a secure system were composed of insecure layers, then all an attacker needs to do is to penetrate the lowest, insecure layer in order to overcome the entire system [27, p. 908]. This was certainly one motivation behind the creation of the reference monitor and TCB concepts.

To justify this criticism consider the following three points. First, the scheme in [9] relies on distributed systems with design diversity, replication, and adjudication to achieve security. However, each insecure component of such a distributed system could be overcome one at a time. The effects of each penetration on the functioning of the components can be designed to be delayed (e.g., a time bomb, a computer virus). After enough of the system has been subverted, all the malicious logic left after each penetration could overcome the adjudication and replication provisions.

Second, the Internet Worm attack is an example of an intentional fault successfully attacking diverse computers. The worm carried two different binary versions of itself [31], one to run on a VAX and one on a SUN 3, and it was designed to carry more. In addition, it had several methods of attack some of which were specific to a particular operating system. Thus, it was designed to deal with limited diversity in the instruction set and operating system of its victims.

Third, notice that higher layers (which could include security mechanisms) typically rely on lower layers for certain basic functions and state information. In addition, lower layers (e.g., an operating system kernel) typically have more access privileges that allow them to access--and thus subvert--almost any piece of data or program in a system. Attacking and corrupting the lower layers of a system can prevent the higher system layers from functioning correctly, thus resulting in an insecure system.

As an example of this third point, consider supplementing the security of a weakly defended system (e.g., a class C2 system [10]) with an intrusion-detection mechanism [7]. Such a design could effectively deal with the unsophisticated attacker. However, the advanced attacker would aim at subverting the event gathering logic, of the detection mechanism residing on the monitored machine, immediately after the initial penetration. The subverted event gathering logic would thus include new provisions to *edit out* all events, which would normally be sent to the bulk of the intrusion-detection mechanism, that point to the attacker's presence in the system. This attack is possible since the add-on detection mechanism does not directly protect the event gathering logic.

5. Toward fault-tolerant, secure computer design

The purpose of this section is to motivate consideration of issues necessary for the formation of a design paradigm for dependable systems that includes security concerns. One potential benefit of such a unified paradigm is the discovery of superior solutions to the integration problems presented in this chapter.

First, in [17, 34], it is stated that the redundant storage of sensitive information increases its chance for exposure (i.e., unauthorized access). However, this does not appear to be the case when employing fragmentation scattering [8, 12]. Here a file is broken into pieces, encrypted, rearranged and replicated, and these fragments are scattered to different sites on a local area network. No master directory exists that indicates locations of a file's fragments. It is not possible to determine which fragments belong to a file simply by observing fragments on storage devices, and redundant copies of fragments does not change this situation. Thus, redundantly storing sensitive information does not always result in a less secure system.

Second, in [9], it is observed that N-Version Programming[7] (NVP) could detect and prevent the use of some covert channels[8]. For this to work for storage channels, a consensus decision is required for every version's action that can result in a change to some *globally visible* (or at least visible to both the sender and receiver of the channel), internal machine state. Otherwise, the version utilizing a storage channel will produce normal data for existing voting points while--in the background, unseen--its other (non-voted) actions send sensitive information over a channel. For example, the size of a memory buffer pool can be a globally visible, internal machine state. Thus, the system calls, which any version can make, to obtain or release a buffer must be voted on [21].

Additionally, NVP does not seem effective against many covert timing channels, because a version's effect on another process's real response time would often be hard to vote on. For example, how could a meaningful consensus decision be performed on the amount of CPU time spent once a version is scheduled (since this can transmit a binary 0 or 1)? Timing channels that use system input/output [30] could be prevented, since access to such facilities can easily be voted on.

Third, standard fault-tolerance techniques cannot be used against security threats without some modifications (if at all). For example, redundancy in data structures [33] may enable detection and correction of invalid data structure modifications. However, an attacker (human, software or hardware) will probe until it has determined the correct number of pointers that have to be changed for a valid data structure modification. Once this is determined, malicious deletions and additions to a data structure (e.g., which is used for firing a missile) can be done without a trace! As another example, the Program Flow Monitor used in [20] required several basic modifications to ensure its correct operation against well-planned, malicious attacks.

Fourth, a unified design paradigm should include a characterization of the origin of faults and their basic properties. Figure 1 presents a classification that distinguishes accidental from intentional human-made faults. (Note, this figure is an extension from one appearing in [3].) Two differences are evident. First, *intrusions* are intentional interaction faults by users without legitimate access to

---

[7] As an example of one design fault-tolerance technique

[8] Assuming that the disagreement detector covert channel, presented in section 3, is closed.

a computer system (i.e., attacks originating from outside a system), or by authorized users attempting to extend their privileges *without* authorization (i.e., attacks originating from inside a system).

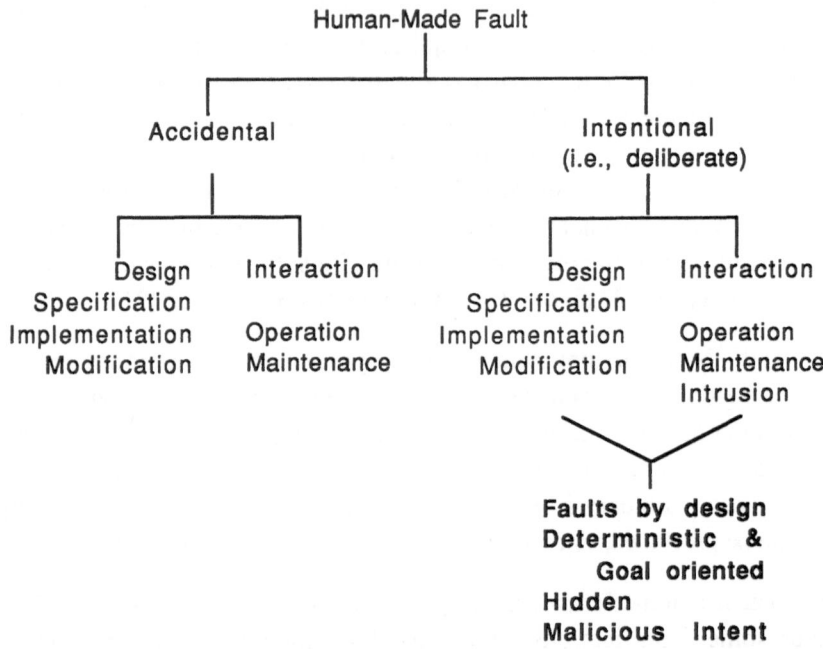

Figure 1. Classification of accidental and intentional human-made faults.

In Figure 1, operation, maintenance, and intrusion are all listed as intentional interaction faults. At first it may be hard for the reader to see the distinction between them, since both operation and maintenance faults can directly lead to intrusions. However, there are cases of intentional, operation and maintenance, interaction faults that have nothing to do with intrusions. For example, a computer system's "superuser" can have privileges that allow him to delete any other user's files. If he does so by accident, then this is a case of an accidental, operational, interaction fault. However, if he deletes someone's files to be malicious, then this is a case of an intentional, operational, interaction fault. (In addition, this example pertains to systems without an all powerful superuser) In *neither* of these cases does an intrusion occur, since the offending user had the privileges to delete files. The malicious file deletion is an *abuse of rights*, since the offending user is given the privileges and misuses them. Other examples of misuse of rights can be found in [6].

Second, four basic characteristics of intentional faults are listed in bold face in Figure 1. *Faults by design* is used to emphasize that intentional faults are well planned at a system's weaknesses. They are *smart faults* in that they are designed to overcome mechanisms used for protection, detection, and recovery. Intentional faults are also *deterministic* and *goal oriented,* since their results are usually previously known and specified (e.g., unauthorized disclosure of sensitive information, denial of service, or just to demonstrate that an attack can be done). They are *hidden* in that a great deal of effort is typically invested to conceal the faults themselves, and their behavior (i.e., they don't leave any fingerprints around). And finally, there is usually a *malicious intent* behind their use (included is the goal to annoy others!).

Note, that in Figure 1 the terms *design* and *interaction* are general categories. More precisely, specification, implementation and modification faults are all design faults, and operation, maintenance and intrusion are all interaction faults.

Classifying Several Security Threats

With the concepts that have been presented in this section we can start to classify several standard security threats. To begin with, the *presence* of a covert channel in a design can either be an accidental or intentional design fault. Some instances of accidental covert channels can be removed from a design, or their bandwidth significantly reduced. However, other instances of accidental covert channels are inherent to computer systems and cannot be removed, while reducing their bandwidth can sometimes seriously affect system performance.

The *act of using* a covert channel by a subject is an intentional implementation or modification fault, while the act of using a covert channel by a user is an intentional interaction fault. In any secure system the continuity of confidentiality is a major requirement in its specification of proper service. Thus, use of a covert channel results in improper service, because it violates confidentiality.

Denial of service instances can be classified as errors, since they (a) can be caused by a wide range of faults (e.g., intrusions, design faults), and (b) can lead to a failure. However, the *act of causing* denial of service can be an intentional design or interaction fault. In [16], it was shown that a denial of some service can be used as a covert channel. Thus, in some cases, the act of causing a denial is the same as the act of using a covert channel.

Computer viruses and Trojan horses can be classified as intentional implementation or modification faults, because they add functionality to programs that were not defined in their specifications. And finally, a trap door, can also either be an intentional implementation or modification fault. But in addition, it can be the result of either an intentional operational fault (e.g., a privileged system operator adds a special password to the password file so that he will be able to enter the system after being fired), or an intrusion.

6. Conclusion

It should not be deduced from this work that all fault tolerance mechanisms would necessarily lead to insecure systems. Or that the above problems could not be eliminated with careful design. However, it should be clear that lack of consideration of the possible effects of fault tolerance on security would be a mistake.

The work presented in this chapter has been part of a larger effort at UCLA concerned with design issues in fault-tolerant, secure computers, and the extension of fault-tolerance techniques to address unsolved problems in computer security [19, 20, 21].

Glossary

Covert Storage Channel -- A covert channel that involves the direct or indirect writing of a storage location by one subject and the direct or indirect reading of the storage location by another subject [10]. Typically, this storage location describes the status of some shared resource (e.g., a file's name, or whether it is locked or not).

Covert Timing Channel -- A covert channel in which one subject signals information to another by modulating its own use of system resources (e.g., CPU time) in such a way that this manipulation affects the real response time observed by the second subject [10].

Subject -- An active entity, generally in the form of a person, process, or device that causes information to flow among objects or changes the system state [10].

Object -- A passive entity that contains or receives information (e.g., pages, segments, files, directories, programs, video displays, keyboards, printers) [10].

References

[1] P.E.Ammann, and J.C.Knight, "Data Diversity: An Approach to Software Fault Tolerance", *17th Int'l Symp. on Fault-Tolerant Computing*, July 1987, pp.122-126.

[2] H.Ando, "Testing VLSI with Random Access Scan", *Proc. COMPCON 1980*, pp.50-52.

[3] A.Avizienis, "Fault-Tolerant Computing Systems", UCLA Class Notes, Computer Science Department, Jan. 1985.

[4] A.Avizienis, "The N-Version Approach to Fault-Tolerant Software", *IEEE Trans. on Soft. Eng.*, Vol. SE-11, No. 12, Dec. 1985, pp.1491-1501.

[5] K.J.Biba, "Integrity Considerations for Secure Computer Systems", Mitre Technical Report TR-3153, Mitre Corp., Bedford, MA., April 1977.

[6] D.D.Clark, and D.R.Wilson, "A Comparison of Commercial and Military Computer Security Policies", *IEEE Symp. on Security and Privacy*, April 1987, pp.184-194.

[7] D.E.Denning, "An Intrusion-Detection Model", *IEEE Symp. on Security and Privacy*, April 1986, pp.118-131.

[8] Y.Deswarte et al., "A Saturation Network to Tolerate Faults and Intrusions", *IEEE 5th Symp. on Reliability in Distributed Software and Database Systems*, Jan. 1986, pp.74-81.

[9] J.E.Dobson, and B.Randell, "Building Reliable Secure Computing Systems out of Unreliable Insecure Components", *IEEE Symp. on Security and Privacy*, April 1986, pp.187-193.

[10] Department of Defense Trusted Computer System Evaluation Criteria, DoD 5200.28-STD, Dec. 1985.

[11] A Guide to Understanding Audit in Trusted Systems, NCSC-TG-001, Version-2, June 1988.

[12] J.M.Fray, Y.Deswarte, and D.Powell, "Intrusion-Tolerance Using Fine-Grain Fragmentation-Scattering", *IEEE Symp. on Security and Privacy*, April 1986, pp.194-201.

[13] S.Funatsu et al., "Designing Digital Circuits with Easily Testable Consideration", *Proc. Int'l Test Conf.*, 1978, pp.98-102.

[14] M.Gasser, *Building A Secure Computer System*, Van Nostrand Reinhold, New York, 1988.

[15] V.D.Gligor, "A Note on the Denial-of-Service Problem", *IEEE Symp. on Security and Privacy*, April 1983, pp.139-149.

[16] V.D.Gligor, "Denial-of-Service Implications for Computer Networks", *Proc. DoD Computer Security Center Invitational Workshop on Network Security*, March 1985, pp.9-33 -- 9-48.

[17] D.K.Hsiao, D.S.Kerr, and S.E.Madnick, *Computer Security*, Academic Press, New York, 1979.

[18] Intel, *iAPX 286 Programmer's Reference Manual*, Santa Clara, California, 1983.

[19] M.K.Joseph, "Towards the Elimination of the Effects of Malicious Logic: Fault Tolerance Approaches", *10th National Computer Security Conf.,* Sept. 1987, pp.238-244.

[20] M.K.Joseph, and A.Avizienis, "A Fault Tolerance Approach to Computer Viruses", *IEEE Symp. on Security and Privacy,* April 1988, pp.52-58.

[21] M.K.Joseph, "Architectural Issues in Fault-Tolerant, Secure Computing Systems", *Ph.D. dissertation,* University of California, Los Angeles, CA., Technical Report CSD-880047, June 1988.

[22] R.A.Kemmerer, "Shared Resource Matrix Methodology: An Approach to Identifying Storage and Timing Channels", *ACM Trans. on Computer Systems,* Vol. 1, No. 3, August 1983, pp.256-277.

[23] R.E.Kuehn, "Computer Redundancy: Design, Performance, and Future", *IEEE Trans. on Reliability,* Vol. R-18, No. 1, Feb. 1969, pp.3-11.

[24] B.W.Lampson, "A Note on the Confinement Problem", *Comm. of the ACM,* Vol. 16, No. 10, Oct. 1973, pp.613-615.

[25] J.C.Laprie, "Dependability: A Unifying Concept for Reliable Computing and Fault Tolerance", Chapter 1, *Dependability of Resilient Computers,* T.Anderson editor, BSP Professional Books, Boston, Mass., 1989, pp.1-28.

[26] A.Mahmood, and E.J.McCluskey, "Concurrent Error Detection Using Watchdog Processors--A Survey", *IEEE Trans. on Computers,* Vol. C-37, No. 2, Feb. 1988, pp.160-174.

[27] P.G.Neumann, "On Hierarchical Design of Computer Systems for Critical Applications", *IEEE Trans. on Soft. Eng.,* Vol. SE-12, No. 9, Sept. 1986, pp.905-920.

[28] B.M.Ozaki, E.B.Fernandez, and E.Gudes, "Software Fault Tolerance in Architectures with Hierarchical Protection Levels", *IEEE MICRO,* Vol. 8, No. 4, August 1988, pp.30-43.

[29] B.Randell, "System Structure for Software Fault Tolerance", *IEEE Trans. on Soft. Eng.,* Vol. SE-1, No. 2, March 1975, pp.220-232.

[30] M.Schaefer et al., "Program Confinement in KVM/370", *Proc. ACM National Conf.,* Oct. 1977, pp.404-410.

[31] E.H.Spafford, "The Internet Worm Program: An Analysis", Purdue Univ., Dept. of Computer Science Technical Report CSD-TR-823, Nov. 1988.

[32] L.A.Stolte, and N.C.Berglund, "Design for Testability of the IBM System/38", *Proc. Int'l Test Conf.,* 1979, pp.29-36.

[33] D.J.Taylor, D.E.Morgan, and J.P.Black, "Redundancy in Data Structures: Improving Software Fault Tolerance", *IEEE Trans. on Soft. Eng.,* Vol. SE-6, No. 6, Nov. 1980, pp.585-594.

[34] R.Turn, and J.Habibi, "On the Interactions of Security and Fault Tolerance", *9th National Computer Security Conf.,* Sept. 1986, pp.138-142.

[35] M.J.Y.Williams, and J.B.Angell, "Enhancing Testability of Large-Scale Integrated Circuits via Test Points and Additional Logic", *IEEE Trans. on Computers,* Vol. C-22, No. 1, Jan. 1973, pp.46-60.

CONCURRENT ERROR DETECTION

USING SIGNATURE MONITORING

AND ENCRYPTION

Kent WILKEN - John Paul SHEN
Center for Dependable Systems
Department of Electrical & Computer Engineering
Carnegie Mellon University - Pittsburgh, PA 15213 - USA

Abstract

This paper presents an efficient approach to concurrent detection of program execution errors that combines signature monitoring with program encryption. Sources of detectable errors include permanent and transient hardware faults, software and hardware design faults, and computer viruses. Errors are detected by a simple monitor that uses signatures embedded in a compatibly encrypted program. The monitor concurrently decrypts the program using the processor control-bit sequences that are included in the signatures. Computer virus attacks are difficult because details of the processor's internal operation are needed to attach compatibly encrypted code. Encryption and a small signature cache added to the monitor allow the lowest memory overhead of any proposed signature-monitoring technique. Encryption and the program memory's error correction/detection code are combined to reduce signature-error detection latency by more than 60 times, while maintaining memory error correction/detection.

1. Introduction

Complete computer dependability requires detection of errors from all sources. Since the earliest computers, much attention has been focused on detecting errors caused by hardware faults. As system complexity increased, detection of errors caused by software and hardware design faults became important. Although faults are often assumed to be inadvertent, deliberate faults (e.g. computer viruses) cause errors that must be detected. The potential for deliberate faults becomes greater as computer use and computer communication increases. Detection of errors caused by deliberate faults, a problem traditionally considered by computer security researchers, is emerging as a fault-tolerant computing research topic [10].

This paper proposes an efficient behavior-based approach to detecting errors caused by certain hardware, design, and deliberate faults. In the behavior-based approach, a program's behavior is abstracted and the abstraction is monitored for run-time violations. No fault model is assumed, any fault (hardware, design, or deliberate) that causes incorrect program behavior is potentially detectable. To be efficient, the selected abstraction must provide high error-detection coverage at a low cost. Schmid, et al., [15] studied several program abstractions and found that program control flow offers the most error detection potential. Researchers [9, 13, 14, 16, 18, 22] have proposed techniques that monitor program control flow using signatured programs and a simple hardware monitor, a general approach we call signature monitoring. This paper proposes a new approach to *signature monitoring* that increases its efficiency and effectiveness.

1.1. Signature monitoring

To provide efficient error detection, signature monitoring exploits a common program redundancy: few instructions alter control flow. This redundancy allows program segments containing many instructions to be coded and later checked, as a unit. Figure 1 shows an elementary signature-monitoring technique. The signature compiler divides the assembly code into *basic blocks* [1] and computes a *reference signature* for each block using the function V. The compiler embeds the reference signature at the end of the block, and sets the *indicator bit* in an added memory column at the corresponding location.

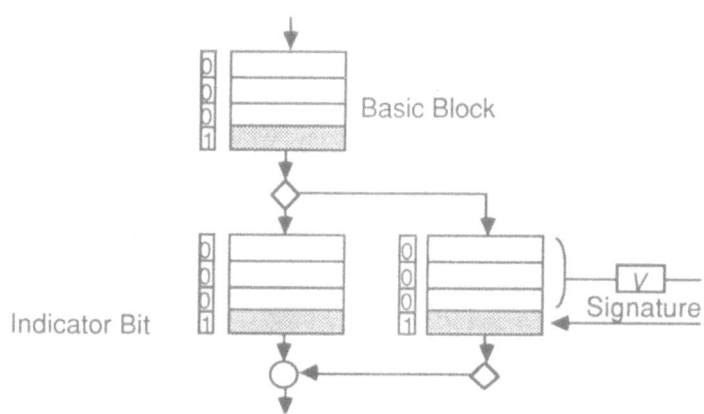

Figure 1. Elementary signature-monitoring technique.

During execution, the monitor generates the block's run-time signature as it observes the executed instructions. At the set indicator bit the monitor compares the run-time signature with the reference signature and declares an error if they differ. Detectable errors include *control-bit errors* and *control-flow errors*. A control-flow error occurs when the instruction execution sequence is incorrect. Control-bit errors result when instructions are executed in the correct order but one or more of the signatured control bits is incorrect. The signature can include control bits from assembly code, microcode, and hardware control lines.

Signature monitoring techniques have been proposed that reduce memory overhead for storing signatures by allowing a *path* containing more than one basic block to be encoded into each signature [13, 16, 22]. The technique proposed by Wilken and Shen [22], *Continuous Signature Monitoring* (CSM), was shown to reduce memory overhead to a theoretical lower bound by partitioning the program into the minimum number of paths. This result is based on the assumptions that the monitor contains no memory other than a register for accumulating the run-time signature, and that control-flow error detection coverage is $1-2^{-w}$ for a w-bit signature [22]. In Section 2 of this paper, these assumptions are relaxed and a technique is proposed that uses encryption and a small signature cache added to the monitor to further lower memory overhead without impacting coverage.

The CSM technique proposed in [22] significantly reduces the latency for detecting signature errors by using the program memory's parity column to store an encoded bit that allows detection of parity and signature errors at each program location. In Section 3, an encryption-based technique is proposed for use with single-error-correcting/double-error-detecting (SEC/DED) program memory. This technique exploits the SEC/DED code to produce a dramatic reduction in signature-error detection latency, while preserving the code's error correction/detection capability.

1.2. Computer virus detection

Recently, the computing community has experienced numerous computer virus attacks [5]. Cohen [3] showed that computer viruses can be created with modest skill and effort, can spread rapidly, and pose a significant security threat. As computers proliferate, the number and severity of computer virus attacks is likely to increase. Effective and efficient virus-detection techniques are needed.

Joseph and Avizienis [9] propose extending signature monitoring to include concurrent virus detection. Signature monitoring can detect a virus, unless the virus is properly signatured. Proper signaturing of a virus may be easy for earlier techniques because they use a single signature function that the attacker might know or easily deduce [9]. Joseph and Avizienis propose using multiple signature functions, one of which is randomly selected by the signature compiler for each program. Using a technique proposed by Namjoo [14], signatures are linked to form a graph that is isomorphic to the program flow graph. While the processor executes the program, the monitor traverses the graph and checks the signatures [14]. Joseph and Avizienis proposed encrypting the signature graph and a vector that represents the function. The decryption key is securely stored, and later delivered to the monitor when the program is loaded. The monitor decrypts the graph and the function's vector, and stores the *plaintext* [2] in its local memory, which is not readable externally. Attacks are averted because a virus cannot easily attach segments to the program that conform to the existing signature graph, or easily alter the program and the encrypted signature graph, without detection.

Although innovative, the encrypted signature-graph approach has limitations. The decryption overhead precludes this approach if process context switches are frequent [9]. For systems that use virtual memory, the monitor's memory is large because it must contain the entire signature graph, even though only a fraction of the program may reside in the processor's real memory. Moreover, the signature graph is large because it contains the signatures plus the links that form the graph. For microprocessors that use an on-chip cache, the monitor must be located on-chip to observe the program's behavior [22]. For this approach, an on-chip monitor requires a separate address and data bus (and possibly its own cache) for accessing the signature graph, to avoid reducing processor performance, and to ensure the privacy of the graph's plaintext.

In Section 4, an alternative approach to concurrent virus detection is proposed that uses signatures embedded in the program. This approach provides significant resistance to virus attacks, and avoids the limitations of the encrypted signature-graph approach. Section 5 summarizes the paper's contributions.

2. Basic encryption and signature caching

This section introduces the basic approach to combining signature monitoring and program encryption. The improvements provided by this approach include reductions in both memory overhead and error detection latency.

2.1. Basic encryption

Figure 2a shows a path that is signatured using the conventional approach. A path's signature is the result of a series of intermediate calculations performed at each word in the path:

$$S_k = V (S_{k-1}, C_{k-1}) \tag{1}$$

where k ranges from 1 to j, V is the signature function, S_0 is a specified initial value, and C_{k-1} represents the values of the monitored control bits during execution of word W_{k-1} . Location k is associated with *intermediate signature* [21] S_k , which for $k>0$ is the signature of the sub-path [0, k-1]. The last intermediate signature, S_j , is the path's signature.

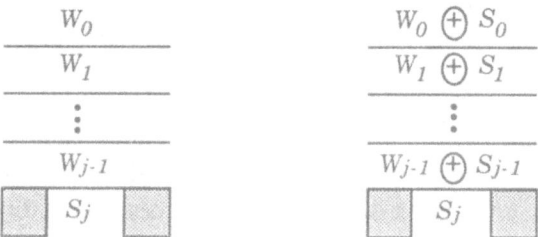

Figure 2. Program signaturing.

The conventional approach's only alteration to the assembly code is the embedded reference signatures. For the proposed approach illustrated in Figure 2b, reference signatures are embedded as before, but the signature compiler also encrypts each word using that location's intermediate signature as the key. Figure 2b shows an efficient encryption function, the exclusive-or (XOR) operator. The monitor generates the run-time intermediate signature, decrypts the word, and delivers the result to the processor for execution.

Schuette and Shen [16] proposed a related technique called *branch address hashing* (BAH) that eliminates reference signatures that follow branch instructions. Each branch address is replaced by the branch address hashed (XORed) with its intermediate signature. Following a signature error, the unhashed branch address becomes a pseudo-random value, and a jump is taken to an arbitrary location, where the error may be detected. The basic encryption approach proposed here can be viewed as a generalization of BAH because all instruction words are hashed, not just branch addresses. Thus, this approach is termed *instruction hashing*.

Instruction hashing provides several improvements to signature monitoring's efficiency. A computer system that uses signature monitoring will generally include numerous other hardware and software mechanisms to detect such errors as illegal opcodes, address or capability violations, etc. Experiments show that such mechanisms can detect a large fraction of processor errors [16]. Following an error that produces an incorrect signature, the intermediate signatures and hence the unhashed instructions are pseudo-random. Execution of pseudo-random instructions will trigger numerous error detection mechanisms, resulting in reduced error-detection latency. Moreover, these mechanisms provide a redundant (and diverse) means for detecting signature errors should the monitor fail in a mode that prevents it from detecting or reporting errors.

Instruction hashing also provides improvements where BAH would otherwise be applied. Signatures that follow branch instructions can be removed using instruction hashing, as with BAH. However, BAH must expand a branch instruction that contains a short branch address into a branch instruction followed by a full-word address [16]. Instruction hashing reduces memory overhead because address expansion is not needed. Also, instruction hashing's unhashing circuit is less complex and adds less delay to instruction fetches because all words are unhashed, compared with BAH's selective unhashing.

An advantage can result from the pseudo-random instruction distribution that instruction hashing produces following a signature error. The CSM technique proposed in [22] uses BAH to eliminate the signature that would otherwise follow each subroutine CALL. After an error, if a CALL is executed before the error is detected, and if the arbitrary destination address (produced by unhashing the branch address) is the beginning of any subroutine, detection of the error is delayed until that subroutine completes [22]. Combining instruction hashing with CSM decreases the probability of these long-latency events, because the fraction of CALLs in a pseudo-random distribution of instructions is generally significantly less than the fraction of CALLs in a program.

In [22], CSM was shown to be more efficient than previous signature-monitoring techniques because it has less memory overhead, lower error-detection latency, higher error-detection coverage, a less complex monitor, and low processor-performance loss. Adding instruction hashing to this technique further increases its efficiency because CSM benefits from all of the aforementioned instruction-hashing improvements. Moreover, CSM combined with instruction hashing is the basis for the signature-caching technique proposed later in this section, the latency reduction technique proposed in Section 3, and the virus-resistant technique proposed in Section 4.

2.2. Justifying signatures

This subsection reviews *justifying signatures* [13] as background for the signature-caching technique proposed in the next subsection. A justifying signature is a word embedded in a path that sets (justifies) the path's signature to a particular value. Namjoo's [13] *Path Signature Analysis* (PSA) technique uses justifying signatures to reduce memory overhead. In Figure 3, a simple program is represented by a program graph, a directed graph that represents each basic block by a node and each possible transition between basic blocks by an arc. PSA constructs sets of paths that cover all legal sequences of nodes in the graph. All paths in a set start at the same node, and share a common reference signature, which is embedded at the beginning of the starting node. PSA adds justifying signatures to selected nodes so that all paths in a set produce the same signature. The path sets for Figure 3 are {ABD, ABC} and {BD, BC}. Reference signatures are embedded at the beginning of nodes A and B. A justifying signature is embedded in node C or D so that these nodes (and hence the paths in the two path sets) produce the same signature. Thus, PSA uses three signatures for this program, compared with four used by the basic technique.

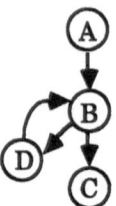

Figure 3. Example program graph.

The CSM technique [22] further reduces memory overhead by using justifying signatures to create a program that produces the same signature along any route from entry to exit. CSM could be viewed as a generalization of PSA that needs only one path set and one reference signature to cover an entire program. CSM partitions a program graph into *maximal paths* [22]. The first maximal path is formed starting at the program's entry node. Each remaining maximal path begins at the "branch taken" arc of a conditional-branch node that is already included in a maximal path. A maximal path is formed by adding a node (from a contiguous location if possible) and the connecting arc until the maximal path merges with another maximal path or itself, or a program exit is reached. A justifying signature is embedded in the maximal path so that the path's signature equals the merge location's intermediate signature. The program's reference signature is the signature of one maximal path that ends at a program exit. A justifying signature is embedded in other maximal paths that end at a program exit so that the path signatures equal the reference signature. CSM requires only two signatures for the program graph in Figure 3 the program reference signature from maximal path ABC and a justifying signature embedded in maximal path D.

2.3. Signature caching

In [22], CSM was shown to achieve a theoretical lower bound for the number of signatures that must be added to a program, assuming that the monitor contains no memory other than a register for accumulating the run-time signature, and that control-flow error detection coverage is $1-2^{-w}$ for a w-bit signature. Memory overhead can be further reduced by relaxing these assumptions and adding a small signature cache to the monitor. Within each program loop, a CSM justifying signature provides the monitor with the correct intermediate signature when the processor returns to the loop's first location. However, the monitor previously calculated and discarded this intermediate signature. The proposed monitor stores each calculated intermediate signature and its corresponding address in a small cache. For simplicity, a direct-mapped cache [19] can be used. When a branch instruction is executed, and the instruction is not followed by a justifying signature, the monitor compares the branch's destination address with the addresses in its cache, and copies the corresponding intermediate signature into its signature register when a match occurs. Given the signature-cache size, the signature compiler can determine which of the CSM justifying signatures can be removed from program loops. For the program in Figure 3, a cache equal to or larger than the size of node B plus the size of node D allows node D's justifying signature to be eliminated. Thus, signature caching uses only one signature for this program, the reference signature from maximal path ABC.

To avoid reduced error detection coverage, signature caching must be used with instruction hashing. Without instruction hashing, an error that occurs in a loop where the justifying signature was removed, and that is not detected at the loop's end, is undetectable. This occurs because the error-free intermediate signature of the loop's first location is copied from the cache after the branch instruction at the loop's end is executed. Using instruction hashing, a valid intermediate signature is copied from the cache following an error only if an executed pseudo-random instruction is a branch to a location contained in the cache. Because the cache is small, only a few instructions from the vast instruction space (e.g. 2^{32}) cause such an event. Thus, the decrease in error-detection coverage is negligible.

A technique's memory overhead can be estimated using high-level language control-flow constructs, by determining the number of signatures required for each construct type and determining each type's average size [22]. CSM requires one justifying signature for each construct that contains a conditional branch: IF, IF-ELSE, SWITCH, FOR, WHILE, and DO [22]. Signature caching can remove the signature from those constructs that contain a loop: FOR, WHILE, and DO. For the program statistics used in [22], FOR, WHILE, and DO contain 22% of the total signatures. Thus, signature caching can reduce CSM memory overhead from a range of 3 to 7% [22], to as low as 2.3 to 5.5%.

The optimum cache size depends on a few factors. A larger cache allows the signature compiler to eliminate more justifying signatures, but increases the monitor's cost. Context-switch time increases with cache size because the signature cache is part of the process state that must be saved when a context switch occurs.

3. Short error-detection latency

Computers used in critical applications generally contain single-error-correcting/double-error-detecting (SEC/DED) memory. This section presents an instruction hashing technique that dramatically reduces signature-error detection latency by exploiting the SEC/DED code.

3.1. A new SEC/DED code

Figure 4 illustrates an SEC/DED *code word* [11] $c_k(X)$ at location k that consists of the w-bit instruction word W_k and $2+\log_2 w = m$ check bits. The code word $c_k(X)$ is the matrix product of W_k and the code's *generator matrix* [11]. k k The SEC/DED *Hamming code* [11] is well suited for bit-serial communication because the decoder can use a simple linear-feedback shift register (LFSR). For computer memory, Hsiao's [8] SEC/DED code is widely used because it optimizes parallel decoding.

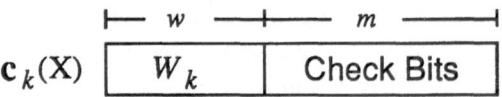

Figure 4. SEC/DED code word.

A new code is proposed that uses instruction hashing and an SEC/DED code to reduce error-detection latency. An even *weight* [11] $w+m$ bit hashing vector $s_k(X)$ is formed using the intermediate signature S_k . For example, the vector can be formed by dividing the w-bit intermediate signature into m groups, calculating an even parity bit for each group, and then appending the m parity bits to the intermediate signature. Any even-weight vector is a multiple of $(1+X)$ [11]. Thus:

$$s_k(X) = (1+X)\, q_k(X) \qquad\qquad (2)$$

The new code word, $C_k(X)$, is the SEC/DED code word XORed with the hashing vector:

$$C_k(X) = c_k(X) + s_k(X) + s_k(X) \qquad\qquad (3)$$

Figure 5 illustrates the decoding organization of the monitor, memory, SEC/DED decoder, and instruction execution unit. During program execution, the monitor generates the run-time intermediate signature S_k' , and forms the even-weight unhashing vector $s_k'(X)$. Using $s_k'(X)$, the monitor unhashes the vector read k k ' from memory, $C_k'(X)$.

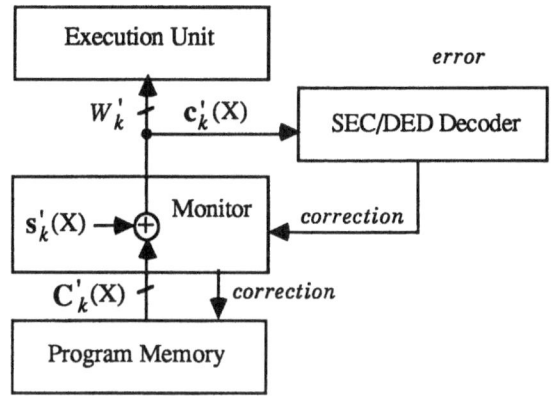

Figure 5. Decoding organization.

The vector $c_{k'}(X)$ is received by the SEC/DED decoder:

$$c_{k'}(X) = C_{k'}(X) + s_{k'}(X) \tag{4}$$

3.2. New code's performance

The correction/detection performance of this code is analyzed assuming a single fault. If a memory error $e_k(X)$ occurs at location k, $C_k'(X) = c_k(X) + s_k(X) + e_k(X)$. From the single fault assumption, the intermediate signature is error-free because it is derived from previous instructions, i.e. $s_k'(X) = s_k(X)$. Substituting these expressions into equation (4) produces the vector received by the SEC/DED decoder:

$$c_{k'}(X) = [c_k(X) + s_k(X) + e_k(X) + s_k(X)]$$

$$c_{k'}(X) = c_k(X) + e_k(X)$$

This same vector would be received by the decoder if the memory error occurred without instruction hashing. Thus, the SEC/DED capability still exists for this new code.

Using the new code, the SEC/DED decoder also detects signature errors. If the run-time intermediate signature contains an error at location k, that error is included in the even weight vector formed by the monitor. The unhashing vector $s_k'(X)$ contains an error of the form:

$$s_k'(X) = (1+X)\, q_k'(X)$$

$$s_k'(X) = (1+X)\, [q_k(X) + e_k(X)]$$

Expanding the terms, and using equation (2):

$$s_k'(X) = s_k(X) + (1+X)\, e_k(X)$$

From the single fault assumption, the memory at location k is error-free, thus $C_k'(X) = C_k'(X)$. The vector received by the SEC/DED decoder, $c_k'(X)$, can be determined by substituting these expressions for $s_k'(X)$ and $C_k'(X)$ into equation (4):

$$c_k'(X) = C_k(X) + s_k(X) + (1+X)\, e_k(X)$$

Substituting the expression from equation (3) for $C_k(X)$

$$c_k'(X) = c_k(X) + s_k(X) + s_k(X) + (1+X)\, e_k(X)$$

$$c_k'(X) = c_k(X) + s_k(X) + (1+X)\, e_k(X) \tag{5}$$

A Hamming code word is a multiple of $(1+X)\mathbf{p}(X)$, where $\mathbf{p}(X)$ is a *primitive polynomial* of degree m-1 [11]. A Hamming SEC/DED decoder divides the received vector by $(1+X)$ and by $\mathbf{p}(X)$ to produce the remainders (syndromes) r_1 and r_2, respectively [11]. For the new code, both terms of the received vector in equation (5) are multiples of $(1+X)$, thus $r_1 = 0$. The SEC/DED code word $c_k(X)$ is a multiple of $\mathbf{p}(X)$. Because $(1+X)e_k(X)$ is a random value with respect to $\mathbf{p}(X)$, the syndrome r_2 is uniformly distributed over the 2^{m-1} possible values. With probability 2^{m-1}, $r_2 = 0$ and the vector is assumed to be error-free. With probability $1-2^{m-1}$, $r_2 \neq 0$ and the decoder reports an uncorrectable error [11].

For Hsiao's code, each column of the *parity check matrix* [11] has odd weight [8]. For the even-weight error vector from equation (5), Hsiao's decoder produces an even weight syndrome, because the syndrome is the sum of an even number of odd weight columns. All but one of the 2^{m-1} even weight syndromes correspond to an uncorrectable error. The remaining even weight syndrome, the all zeros syndrome, indicates no error. Thus, when used with the new code, Hsiao's decoder detects signature errors with the same probability as the Hamming code.

For a 32-bit processor, an SEC/DED decoder detects the signature error with probability 63/64 before the first program word is executed following an error, and if the error remains undetected, with probability 63/64 before subsequent words are executed. The average error detection latency for this geometric series is $(1-63/64)/(63/64) = 0.016$ program memory cycles. This average latency is more than 60 times shorter than the best existing latency-reduction technique [22]. Moreover, this short latency is achieved without increasing memory overhead.

The new code facilitates recovery from transient errors. When the decoder reports an uncorrectable error, the processor assumes that the error is a transient signature error, and invokes a rollback procedure (e.g. [20]) that restores the processor and monitor to a previous state. The error is deemed uncorrectable only if the rollback fails. The new code's short latency significantly reduces the size and complexity of the rollback buffers. For example, saving a single state for a 32-bit processor allows >98% of transient signature errors to be tolerated. The short, predictable recovery time is well matched to the needs of real-time systems. In addition to transient processor errors, transient monitor errors can be detected and corrected, without duplicating the monitor as required by previous techniques [12].

This approach requires the memory bus width to increase to $w+m$ bits from w bits so that the encoded check bits can be written and read along with the encoded instruction. In addition to the approach's aforementioned benefits, single-bit bus errors are correctable, even-weight bus errors are detectable immediately, and other bus errors are detectable with high probability after a short latency.

4. Computer virus resistance

This section proposes extensions to instruction hashing that provide significant resistance to computer virus attacks.

4.1. Modifications to CSM

Instruction hashing combined with CSM provides some resistance to computer virus attacks because a program's assembly code is encrypted using CSM's pseudo-random intermediate signatures. However, this combination contains weaknesses that must be eliminated for the virus resistance to be significant.

First, the typical CSM signature function, a cyclic-redundancy check (CRC) polynomial, allows the program's plaintext to be readily deduced from the hashed program using well known methods [2]. Instead, this function must be replaced by a cryptographic function. The external structure of the cryptographic function is illustrated in Figure 6. The inputs to the function V are the instruction word W_k, the other monitored control bits c_k that occur during the execution of W_k, the function's previous output S_k, and a program key K_p

that the signature compiler selects at random for each program. The p resulting intermediate signature S_{k+1} is used for hashing and unhashing W_{k+1}.

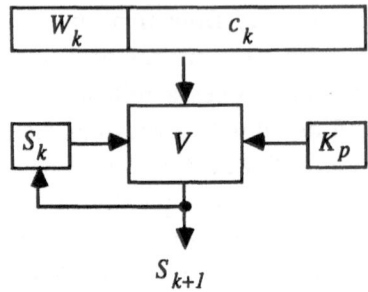

Figure 6. Cryptographic signaturing function.

To avoid reducing processor performance, a hardware signature generator should execute the function in real-time, i.e. the result S_{k+1} should be available W_{k+1} when the memory fetch of W_{k+1} completes. Also, the key should be difficult to deduce by examining the function's outputs for a known set of inputs, a _known plaintext attack_ [2]. In practice these requirements are likely to conflict, with performance taking precedence for most applications. The performance requirement may restrict the possible functions to those for which the program is not _theoretically secure_ [2] against a general cryptanalytic attack. However, practical security against a virus attack can still exist. A virus has limited resources: its size and hence its ability to attack are constricted; its computation facility is limited to the host computer; excessive execution time can make the virus conspicuous.

Second, the CSM technique (summarized in Section 2.2) derives all intermediate signatures from the program's instructions and from the initial intermediate signature assigned to the program entry-node's first location [22]. As proposed in [22], the program's initial intermediate signature is the same for all programs, e.g. 0. For virus resistance, the signature compiler selects each program's initial intermediate signature at random.

Third, the CSM technique uses BAH to eliminate justifying signatures following CALLs [22]. The intermediate signature of each subroutine's first location is a fixed value, e.g. 0. For virus resistance, these constant intermediate signatures must be eliminated. This can be done by reinstating a justifying signature after some CALLs, using the technique illustrated in Figure 7. A subroutine is signatured as a part of a maximal path that contains a CALL to the subroutine. A justifying signature is not needed after this CALL. A justifying signature is embedded after other CALLs to the subroutine so that the path's signature equals the intermediate signature of the subroutine's first location. A stack is added to the monitor to save the intermediate signature at a CALL's justifying signature, which is used to derive the intermediate signature at the return (next) location.

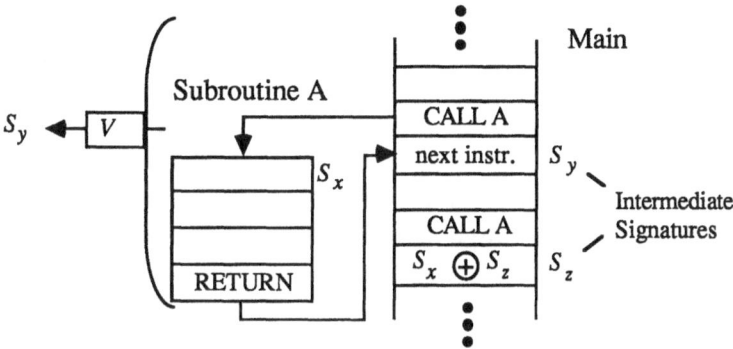

Figure 7. Signaturing subroutines.

Fourth, the program's initial intermediate signature, reference signature, and key must be securely stored, and then be available to the monitor when the program executes. Also, when a context switch occurs, the contents of the signature register, signature cache, and stack must be securely stored, and then restored when the process resumes. In both cases, these data can be encrypted and decrypted using a cryptographic function contained in the monitor. This function need not execute in real time and might be unrelated to the real-time cryptographic function. This function's key can be loaded during system generation and stored in non-volatile memory that is not readable by the processor. The trusted and encrypted signature compiler writes to the monitor the plaintext of the program's key, initial intermediate signature, and reference signature. The ciphertext of these data is read by the compiler and stored with the encrypted program. When the program executes, the processor delivers to the monitor the ciphertext, which is decrypted and used internally. Similarly, when a context switch occurs, the ciphertext of the contents of the signature register, cache, and stack are read and stored by the processor, and later decrypted by the monitor. The processor cannot read the plaintext of the various signatures or the key stored inside the monitor.

With these changes, instruction hashing poses a significant barrier to virus attacks. Additional virus resistance can be had by exploiting the processor's complex internal behavior.

4.2. Monitor-assisted signature compilation

Previous signature-monitoring papers suggest using a software signature compiler to generate a program's signatures. Although the signature compiler's complexity and execution time are important, they have not been considered in the literature. Error detection coverage can be increased significantly by including the processor's internal control sequences (from microcode and hardware control lines) in the signature along with the assembly code [16]. However, generating such signatures using software requires that the signature compiler include a model of the processor's control section. Using this model, signature generation is equivalent to simulating the response of the processor's control section to each program instruction. Besides adding significant complexity to the signature compiler, compiler throughput is significantly reduced.

Minor modifications to the monitor and processor can create a new instruction-execution mode that allows the monitor's hardware signature- generator to assist with signature compilation. ENTER MODE and EXIT MODE instructions are added to the instruction set. While the new mode is enabled the following occurs: (1) the monitor does not unhash instructions, (2) instruction output (store to register or memory) is blocked, and (3) the program counter (PC) always increments, i.e. PC = PC + 1. The compiler constructs a code segment that includes the instruction sequence that is the target of signature generation, bracketed by delimiters as illustrated in Figure 8. The leading delimiter is a hashed ENTER MODE followed by a justifying signature that equals the intermediate signature of the sequence's first instruction. The trailing delimiter is EXIT MODE.

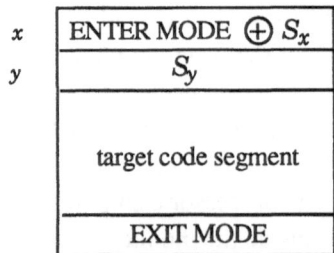

Figure 8. Signature-generation code segment.

After the segment is constructed, ENTER MODE is unhashed and executed, and then the monitor copies the justifying signature into its signature register. Next the target sequence is executed. During execution, each intermediate is calculated and saved in the monitor's cache. After EXIT MODE is executed, the plaintext intermediate signatures are read by the compiler and hashed with the corresponding instructions in the target sequence to form the encrypted instructions.

Using monitor-assisted signature compilation, signatures that include a large number of the internal control sequences are feasible. For maximum error detection coverage, all possible control sequences are included in the signatures. These signatures can capture a significant portion of the processor's complex internal behavior. This complexity provides an instruction-hashed program with significant additional virus resistance.

The proposed instruction-execution mode must be secured so that it can only be enabled by the trusted and encrypted signature compiler during program installation. Thus, a virus cannot receive monitor assistance when attempting to decrypt a program or encrypt itself. The virus must carry its own software decryption/encryption mechanism. Because the complexity of the processor's internal operation must be reflected in this mechanism, it is likely to be large (which makes the virus conspicuous), to be difficult to construct, and to execute slowly.

4.3. Comparison with previous approaches

Instruction hashing has many advantages compared with previous approaches to computer virus detection. Various techniques have been proposed that use cryptographic checksums to ensure program integrity when a program is loaded, e.g. [4]. Unlike these techniques, instruction hashing also ensures integrity during program execution. For example, instruction hashing could detect (and preclude) the "fingerd" attack used by the Internet Worm [17]. Cryptographic checksums must be securely stored, otherwise a virus can attach to a program and substitute a new checksum that reflects its changes. Instruction hashing requires no secure storage outside the monitor.

Contrasted with the encrypted-graph approach [9] (summarized in Section 1.2), instruction hashing can be applied where context switching is frequent, it works in a memory hierarchy without complication, and it requires less memory overhead. Using instruction hashing, a monitor failure does not make the system insecure because the failure can be detected by other processor detection mechanisms. The encrypted-graph approach must duplicate the monitor to create a fail-safe system. Unlike the encrypted-graph approach, instruction hashing provides an additional element of security: privacy. Instruction hashing is a deterrent against unauthorized examination or use of a program.

Techniques have been proposed that use program encryption and managed key distribution to prevent software piracy, e.g. [7]. By employing an error-detection code, these techniques can detect modifications to the assembly code, including those caused by computer viruses. Unlike these techniques, instruction hashing can detect control-flow errors, and control-bit errors from all levels in the control hierarchy, including assembly-code modification. By adding managed key distribution, instruction hashing could also be used to deter software pirates.

Unlike the aforementioned approaches, instruction hashing's cryptographic function must execute in real-time. This may limit instruction hashing's effectiveness against a general cryptanalytic attack. The other approaches can be based on non real-time functions that have established effectiveness, e.g. DES [6]. Instruction hashing's security may be derived in part from the concealed details of the processor's internal implementation, which could be exposed. Also, code that is encrypted using internal control sequences is implementation specific and will not execute on different implementations of the same instruction set architecture. This can be a limitation for heterogeneous multi-processor/multi-computer systems.

5. Summary

The paper presents a new approach to concurrent error detection that combines signature monitoring with encryption. The new approach, called instruction hashing, is shown to be robust because it allows several signature-monitoring advances. Following a signature error, instruction hashing produces pseudo-random instructions, which can trigger various processor error-detection mechanisms. This reduces error detection latency and provides a redundant and diverse means for detecting errors should the monitor fail in a mode that prevents it from detecting or reporting errors. A signature cache is proposed that reduces memory overhead to below that of the best existing technique by eliminating justifying signatures from loops. Instruction hashing prevents signature caching from reducing error-detection coverage. Computers used in critical applications generally use SEC/DED memory. A new code is proposed that uses instruction hashing to exploit the SEC/DED code. The new code retains memory error correction/detection capability, and provides an average signature-error detection latency of 0.016 program memory cycles, 60 times shorter than the best existing technique. This short latency facilitates low-cost recovery from transient hardware faults. Basic instruction hashing provides some resistance to computer virus attacks because the program is hashed with CSM's pseudo-random intermediate signatures. Minor modifications to basic instruction hashing significantly increase virus resistance. Monitor-assisted signature compilation is proposed, which allows numerous internal control sequences to be included in a program's signatures. Hashing instructions with these signatures further increases virus resistance because the virus requires inferred details of the processor's internal operation. Instruction hashing is a low-cost approach that is well suited for applications that must tolerate transient hardware faults, detect permanent hardware faults, and resist attacks by computer viruses.

Acknowledgement

This work was supported by the Office of Naval Research (ONR) under contract N00014-86-K-0507.

References

[1]   A. Aho, R. Sethi and J. Ullman, *Compilers: Principles, Techniques, and Tools*, (Addison-Wesley, 1985).

[2]   H. Beker and F. Piper, *Cipher Systems: The Protection of Communications*, (John Wiley, 1982).

[3]   F. Cohen, "Computer Viruses: Theory and Experiments", pp. 240-263, *7th National Computer Security Conf.*, (Sept. 1984).

[4]   F. Cohen, "A Cryptographic Checksum for Integrity Protection", *Computers & Security* 6, 6 (Dec. 1987), 505-510.

[5]   P. Denning, "Computer Viruses", *American Scientist 76*, (May-June 1988), 236-238.

[6]   National Bureau of Standards, *Data Encryption Standard*, FIPS Publication 46, U.S. Department of Commerce, (1977).

[7]   A. Herzberg and S. Pinter, "Public Protection of Software", *ACM Transactions on Computer Systems 5*, 4 (November 1987), 371-393.

[8]   M. Hsiao, "A Class of Optimal Minimum Odd-Weight-Column SEC-DED Codes", *IBM Journal of Research & Development 14*, 4 (July 1970), 395-401.

[9]   M. Joseph and A. Avizienis, "A Fault Tolerance Approach to Computer Viruses", pp. 52-58, *Proc. Symp. on Security and Privacy*, IEEE, (1988).

[10]   M. Joseph, "Architectural Issues in Fault-Tolerant, Secure Computing Systems", Ph.D. Dissertation, T.R. #CSD-880047, UCLA Computer Science Dept., (1988).

[11]   S. Lin, *An Introduction to Error-Correcting Codes*, (Prentice Hall, 1970).

[12]   A. Mahmood and E. McCluskey, "Concurrent Error Detection Using Watchdog Processors - A Survey", *IEEE Transactions on Computers 37*, 2 (February 1988), 160-174.

[13]   M. Namjoo, "Techniques for Testing of VLSI Processor Operation", pp. 461-468, *Proc. 12th ITC*, IEEE, (1982).

[14]   M. Namjoo, "Cerberus-16: An Architecture For a General Purpose Watchdog Processor", pp. 216-219, *Proc. 13th FTCS*, IEEE, (1983).

[15]   M. Schmid, R. Trapp, A. Davidoff and G. Masson, "Upset Exposure by Means of Abstraction Verification", pp. 237-244, *Proc. 12th FTCS*, IEEE, (1982).

[16]   M. Schuette and J. Shen, "Processor Control Flow Monitoring Using Signatured Instruction Streams", *IEEE Transactions on Computers C-36*, 3 (March 1987), 264-276.

[17]   E. Spafford, "The Internet Worm: Crisis and Aftermath", *Communications of the ACM 32*, 6 (June 1989), 678-687.

[18]   T. Sridhar and S. Thatte, "Concurrent Checking of Program Flow in VLSI Processors", pp. 191-199, *Proc 12th ITC*, IEEE, (1982).

[19]   H. Stone, *High-Performance Computer Architecture*, (Addison-Wesley, 1987).

[20]   Y. Tamir, M. Tremblay and D. Rennels, "The Implementation and Application of Micro Rollback in Fault-Tolerant VLSI Systems", pp. 234-239, *Proc. 18th FTCS*, IEEE, (1988).

[21]   K. Wilken and J. Shen, "Embedded Signature Monitoring: Analysis and Technique", pp. 324-333, *Proc. 17th ITC*, IEEE, (1987).

[22]   K. Wilken and J. Shen, "Continuous Signature Monitoring: Efficient Concurrent-Detection of Processor Control Errors", pp. 914-925, *Proc. 18th ITC*, IEEE, (1988).

Fault-Tolerant Distributed Systems

Chair: M. Dal Cin (J.W. Goethe University, Frankfurt, Germany)

A LINGUISTIC APPROACH TO FAILURE HANDLING IN DISTRIBUTED SYSTEMS

Richard D. SCHLICHTING
Department of Computer Science
The University of Arizona Tucson, Arizona 85721 - USA

Flaviu CRISTIAN
IBM Almaden Research Center
650 Harry Rd. San Jose, CA 95120 - USA

Titus D.M. PURDIN
Department of Management Information Systems
The University of Arizona Tucson, Arizona 85721 - USA

Abstract

Distributed computer systems are increasingly being used for controlling critical applications. An important aspect to constructing dependable systems for such use is ensuring that the system software is robust to failures in the underlying computing platform. One property that makes failures difficult to handle in this environment is that they can occur concurrently with other system events. This paper describes a language-based approach for constructing system software that can cope with such asynchrony in a systematic manner. The basic idea is to treat failures as just another class of events that are handled similarly to normal events. Linguistic

The first and third authors have been sponsored by the National Science Foundation under Grant CCR-87-01516 and the Air Force Office of Scientific Research, Air Force Systems Command, USAF, under Grants AFOSR-84-0072 and AFOSR-88-0147.

constructs that can be added to distributed programming languages with minimal impact are then proposed to handle such failure events. To make our ideas precise, we use the SR distributed programming language as a basis for incorporating these constructs. The approach is illustrated by a detailed presentation in the extended SR language of a replicated directory management program.

1. Introduction

Critical applications are increasingly being controlled by distributed computer systems, that is, systems based on a collection of processors connected by a local-area network. There are numerous reasons why such systems are especially appropriate for this type of application. One is that many critical applications are process-control situations in which the components being controlled are physically dispersed. Another reason is that the multiple processors afforded by such an architecture provide a natural basis for the redundancy required to construct fault-tolerant systems. Moreover, it seems clear that the use of distributed systems in this area is a trend that is likely to accelerate.

For a distributed system to provide the dependable service required for critical applications, the system software itself must be robust to failures in the underlying computing platform. Unfortunately, programming such software has proven to be a difficult task: in addition to the complications induced by normal distributed execution, the programmer must also deal with component failures that can occur asynchronously during program execution. Although abstractions such as atomic transactions [9] are often used at the application level to handle this extra complexity, such high-level techniques are inappropriate for building system software.

In this paper, we describe how flexible mechanisms oriented towards handling processor failures in system software can be integrated into concurrent or distributed programming languages such as Ada [1], CSP [8], Mesa [11], or SR [3]. The design of the mechanisms is based on the observation that the occurrence of failures can be considered as events (i.e., state transitions) caused spontaneously by the *adverse environment* of the system [5]. With this view, the failure of a processor is treated logically within the software as a concurrent event that is generated and signaled in real-time by an underlying *processor membership* service that detects the failure [7]. The interprocess synchronization and communication mechanism provided by the language (e.g., semaphores, condition variables, messages) is then used to wait for the occurrence of a failure signal and synchronize its activity with normal processing.

Although our approach is general, the SR (Synchronizing Resources) distributed programming language is used in this paper as the basis for failure handling extensions because of its intended use as a systems programming language and because of the wide variety of synchronization constructs provided by the language [3]. An overview of the language is provided in Appendix 1. Also, we restrict our attention here to the failure and recovery of *fail-stop processors*, processors whose only failure is a detectable crash [12]. The approach can easily be generalized to handle other types of failures, however.

2. Extensions to handle failures

Our approach to handling failures is to view their occurrence as events generated by concurrently executing adverse environment processes. We assume that occurrences of processor failures and restarts are detected by a processor membership service implemented in the language run-time kernel. This service then makes use of the interprocess communication and synchronization mechanisms provided by a distributed programming language to notify other processes that the event has occurred. In SR, *operation invocation* is the standard way in which processes communicate and synchronize, so operations and their invocations serve as the basis for our failure-handling mechanisms. In particular, we propose language extensions to allow a programmer to specify an *event handling operation* (or *event handler*) for an event; such an operation is invoked automatically by the SR run-time kernel whenever the event occurs.

Before proceeding, we point out several specific objectives that guided the development of these mechanisms and influenced their final form. First, the mechanisms should be designed to mesh with SR so as to have minimal impact on the existing language. In particular, our goal is not to develop a *new* language for fault-tolerant computing, but rather to show how existing languages can be adapted for such purposes. Second, it should be possible to use the same mechanism to handle the failure or restart of any processor. For example, the means used within an SR *resource* (i.e., module) for establishing a handler for the restart of its own processor (i.e., restart code to recover following a processor crash) should be the same as establishing handlers for other processors. Third, it should be possible to isolate code for dealing with failures from other code so that it is easier to understand the program's normal flow of control. Lastly, we would like the mechanisms we design to be expressive enough to be applicable in many different kinds of situations, yet very simple to use in most common situations. This philosophy is most apparent in the way in which event handlers are bound to events. Specifically, we have chosen to do such association dynamically for expressiveness, while providing essentially a static equivalent that is applicable in most common situations. This is explained more fully in Section 2.2 below.

2.1. Events

An *event* is defined as a state change of interest in the system. Event occurrences can be caused by humans, the adverse environment processes, or simply the passage of time. The main focus of this paper are those events related to the transition of processors between functioning and failed states. Specifically, two kinds of events are considered: a processor *crash* and a processor *restart*. These events are described using the *event descriptions* "crash(machine_denotation)" and "restart(machine_denotation)", where the argument denotes a particular processor. This denotation can take one of three forms: a machine name from a built-in **machine** enumeration type, a *resource capability* identifying a resource instance executing on that machine, or an *operation capability* identifying an operation within a resource instance executing on that machine.

For instance, let rescap be a resource capability for a resource instance executing on machine *mach* and *opcap* be a capability for an operation within that instance. We describe the event that occurs when *mach* goes from a functioning to a failed state by any of the event descriptions crash(*mach*), crash(*rescap*), or crash(*opcap*). Similarly, the transition that occurs when the processor goes from a failed to a functioning state is described by restart(*mach*), restart(*rescap*), or restart(*opcap*). Note that each group of event descriptions are aliases that describe exactly the same event: a machine failure or restart. Thus, it is not accurate to interpret, for example, crash(*opcap*) as the failure of the *operation* denoted by *opcap* since that operation may not in fact be executing when the failure occurs. An event description always refers to a change in the state of a processor, with *opcap* being used simply to denote the processor on which the operation *opcap* is located.

The ability to describe processor crash and restart events at various abstraction levels provides useful flexibility. For example, a server responsible for routing messages in a distributed system might wish to know which machines are functioning and which are down so that it can avoid routing messages through failed machines. In such a situation, the logical component of interest with respect to failures is the processor itself, and so using the explicit machine name in the description is most appropriate. On the other hand, consider the case where a process in some resource is interacting with a process in another resource instance. The name of the machine on which the instance is running may not be readily available and is in any case largely irrelevant--the logical unit of interest here is the resource instance itself. Thus, in this kind of situation, it is more appropriate and convenient to use resource capabilities in the event description rather than explicit machine names. Similarly, a proc passed a reply capability upon invocation may not know what resource instance is implementing the corresponding operation or on what machine that instance resides. Use of an operation capability in the event description would be most appropriate in this situation.

Although a processor failure or restart occurs by definition at a specific instant in time, it is the *detection* of the event by the membership service on a processor that actually triggers invocation of event handlers. In keeping with the semantics of fail-stop processors, we assume that each such event is detectable by all functioning processors. We also assume that the order in which the transitions associated with a single processor are detected by the membership service on each processor is consistent with their occurrence in real time (i.e., a failure is always detected before a restart) and that all failure and restart events in the system are detected in the same order on all functioning processors. Implementation of a processor membership service that meets this requirement is described in [7].

Note that the use of capabilities as arguments allows events involving any processor to be described uniformly. For example, it is simple for the programmer of a resource to describe events involving the processor on which a resource instance is executing by using a capability for that instance. In SR, the predefined function 'myresource()' returns this capability. This is most useful for the restart event (i.e., restart(myresource())), which occurs when the processor containing the instance is rebooted following a failure.

2.2. Handler bindings

The association of a given event with an event handling operation to be invoked when the event occurs is accomplished by the introduction of the new data type **binding**. Variables of this type ("bindings") are declared as follows:

var binding_var : **binding**

Declaration of such variables can appear in the same location as other variable declarations in SR. That is, they can be declared to be global to a resource, local to a proc, or even local to a block within a proc.

The value assigned to a binding variable has two components—an event description and a *handler invocation*—and is constructed using the new constructor **when**:

> **when** event_descrip **send** event_hand_denot(arg_list)

This expression specifies that the operation denoted by event_hand_denot is to be invoked asynchronously with the argument list arg_list when the corresponding event is detected. The handler denotation and argument list are evaluated when the constructor is evaluated, not when the invocation is actually performed. For example, consider the expression

> **when** restart(myresource()) **send** *retry*(*count*)

This expression associates the restart of the machine on which the executing resource resides with an invocation of the operation *retry*. The argument to *retry* in the handler invocation is the value of *count* when the constructor is evaluated.

Evaluation of an expression containing the **when** constructor only associates a handler with an event and does not initiate monitoring for the occurrence of the event by the system. The actual monitoring is started when the value of such an expression is assigned to a binding variable. The monitoring then continues until the variable is assigned another value or the scope of the variable is exited. The second property means that, for example, the monitoring associated with a binding variable declared local to a proc ends when the proc terminates. In sum, then, at any point during the execution of a program, the events to be monitored and the invocations to be generated should the occurrence of such an event be detected are determined by the current values of all active binding variables.

To illustrate the use of bindings, consider a client resource that obtains service from a server resource. Suppose that a capability for the server is passed to the client upon creation, and that the client performs some remedial action should the server fail. Then, one possible code outline for the client resource is shown in Figure 1. Upon creation of the client, the process *do_work* assigns to the variable *sf* ("server failure") a value binding the handler *fail_handler* to the failure of the machine executing the server serv. This assignment initiates monitoring by the system for the failure of the designated processor. Should that machine subsequently crash, *fail_handler* is invoked with an argument consisting of the capability for the failed server. The process created by this invocation could then perform some remedial action; any such action would, of course, require coordination with the main process *do_work* using standard SR synchronization mechanisms.

```
resource client(serv : cap server)
    ...
    var sf : binding
    ...
    process do_work
        sf := when crash(serv) send fail_handler(serv)
        ... use operations implemented by serv ...
    end

    proc fail_handler(downserver)
        ... handle failure of downserver ...
    end

end
```

Figure 1. Client resource

The binding variable *sf* in the above example could also have been assigned its initial value in the declaration using standard SR syntax:

var binding_var : **binding** := initial value

If a handler invocation is not assigned as an initial value, a binding variable takes on the special value **null**, indicating that no monitoring is currently to be associated with this binding variable. This **null** value can also be assigned to a binding during execution to cancel monitoring for a particular event.

2.3. The 'state' predefined function

One problem with the client/server code shown above is that the failure of the server will go unnoticed by the client should it be detected by the local membership service before execution of the assignment to *sf*. This is in fact a general problem with associating handlers with events: the assignment to a binding only ensures that *future* detections of the event generate invocations of the corresponding handler. We complement this by requiring that state information about the monitored component be available as of the assignment occurrence. Specifically, execution of an assignment to a binding also assigns the current perceived state of the processor of interest to a third, "hidden" field of the variable. The value assigned is the appropriate member of the enumeration type *status*:

$$\textbf{type } status = \textbf{enum}(\text{up,down,undefined})$$

An undefined status applies only when the value **null** is assigned to the binding, i.e., when no monitoring is to be associated with that variable. The determination of component status and the establishment of component monitoring is done atomically; that is, it is guaranteed that any event detected before the monitoring starts is reflected in the processor status assigned to the binding variable.

Once an assignment to a binding has taken place, the value of the hidden status field can be retrieved using a new predefined function 'state'. This function takes a binding as its argument [1] . For example, given the assignment

$$binding\_var := \textbf{when } crash(res\_cap) \textbf{ send } fail()$$

the function '*state(binding_var)*' returns 'down' if the machine denoted by *res_cap* was in a failed state when the assignment took place. If the machine was functioning, then 'state(*binding_var*)' returns 'up'.

Given the 'state' predefined function, the client/server example above might be coded as illustrated in Figure 2. In this implementation, the handler *fail_handler* is explicitly invoked at the beginning of *do_work* should the server be down when the assignment to sf took place. If the server fails subsequently, *fail_handler* will be invoked implicitly by the system.

The need for performing all of the actions that comprise an assignment to a binding as an atomic unit is one of the motivations for making association of event handlers with events dynamic instead of static. By using assignment to a binding variable as the means for establishing this association, the moment at

---

[1] The use of a hidden field whose value is retrieved by a predefined function is consistent with the way SR provides status information following invocation.

which the current state is determined and the monitoring commenced is clearly visible to the programmer. This moment would be less obvious had the association been done statically, say by enabling a handler upon block entrance. Nevertheless, we point out that this ability to establish and change handlers with a simple assignment statement is a powerful feature that is best utilized in a disciplined fashion. For example, in many programs it is sufficient to have a global failure handler for a given event; this can be done easily using our mechanisms by assigning the appropriate value to a binding variable during resource initialization and not changing it thereafter.

```
resource client(serv : cap server)
    var sf : binding
    ...
    process do_work
        sf := when crash(serv) send fail_handler(serv)
        if state(sf) = down → send fail_handler(serv) fi
        ... use operations implemented by serv ...
    end

        proc fail_handler(downserver)
        ... handle failure of downserver ...
        end

end
```

Figure 2. Client using the 'state' predefined function

2.4. Using handler bindings

Bindings provide the programmer with a very flexible mechanism for dealing with failures. For example, it is easy using arrays of binding variables to establish a single parameterized handler for a collection of related events. In the above example, suppose that there were multiple servers whose capabilities were passed in an array to the resource instance. Then, the client could establish a single event handler for the failure of any of these servers as shown in Figure 3.The code for *do_work* associates the handler invocation *fail_handler(serv[j])* with the failure of the server whose capability is *serv[j]*. Thus, when a given server fails, the handler proc is invoked with the corresponding capability being passed as argument. Note that in this implementation, multiple failures are handled concurrently since a new process is created every time the proc is invoked. If this leads to synchronization problems, the handling of failures could be serialized by implementing *fail_handler* using an input statement that is repeatedly executed by a single process.

```
resource client(serv[0:N-1] : cap server)
    var sf[0:N-1] : binding              # server failures

    process do_work
        fa j := 0 to N-1          # for all j from 0 to N-1
            sf[j] := when crash(serv[j]) send fail_handler(serv[j])
        af
        ...
    end

    ...

    proc fail_handler(downserver)
        ... handle failure; downserver is cap of failed server ...
    end

end
```

Figure 3. Parameterized failure handler

The above example illustrates how binding variables can be used to establish a single handler for multiple events. It is also quite easy to do the opposite: write a program that contains multiple handlers for a single event. To do this, all that is required is multiple binding variables that have been assigned the same event description but different handler invocations. These variables would normally be found in different resource instances to give each instance its own handler for a given event; for example, a client and a server allocator that both interact with servers would probably have separate event handlers to be invoked should a server fail. Note, however, there is no reason why multiple handlers for a single event could not be established in the same resource instance if that is convenient. This should be done with care since multiple event handlers will be initiated in parallel when the event is detected.

2.5. Stable storage

In the course of a crash and subsequent restart, all information stored in volatile storage is lost. To avoid the loss of information essential to recovery, provision must be made in SR to have such information placed in *stable storage*, that is, storage whose contents are unaffected by failure [9].

Our approach to this problem is to allow variables identified by the user as necessary for recovery to be explicitly designated as stable variables and then provide run-time access to stable storage in the implementation of SR. A variable or parameter is declared as stable using the keyword **stable** before the type designation as follows

$$\textbf{var } \text{variable\_name} : \textbf{stable } \text{variable\_type}$$

where variable_type is any built-in type, including **binding**. The stable designation is allowed only for resource parameters and variables since these are the only entities whose scope is reestablished upon restart. In particular, it would do no good to declare a proc's parameters or local variables to be in stable storage—the only process that could access such values ceases to exist when a failure occurs.

Once declared, a stable variable can be manipulated just as any other SR variable. Writes to a stable variable are performed using assignment statements, and reads are performed implicitly when the name of the variable is used in an expression. A single access to a simple stable variable is implemented as an atomic action by the SR run-time kernel.

Our use of stable variables can be viewed as a middle road between most current programming languages and applications-oriented languages such as Argus [10]. The former force the user to implement stable storage, while the latter typically have higher level operations whose execution implicitly writes information in stable storage.

2.6. Processor restart

When a processor restarts following a failure, the system goes through certain steps in order to reestablish the appropriate execution environment. First, those resource instances that contain stable variables or event handlers associated with the restart of the processor are reestablished. Since the contents of volatile storage is lost when a failure occurs, certain SR run-time support values are implicitly maintained in stable storage so that there will be enough information available to reestablish the appropriate resource instances. For example, the code for these resources and certain run-time tables associated with their instances are kept in stable storage.

It is important to emphasize that resource instances are reestablished and *not* recreated. This means, for example, that the **initial** code for the resource instance is not re-executed. It also means that the capability for the resource instance and the capabilities for any operations it implements remain the same: other resources that possessed capabilities for the resource prior to the crash can continue to use those capabilities after the restart.

Once the resource instances exist, the second and final step is taken: generating those handler invocations necessitated by the occurrence of a restart event. If P is the restarting processor, the relevant bindings are those whose event descriptions are equivalent to 'restart(P)'. If the variable is located on a processor other than P, the run-time support on that processor will generate the invocation when its processor membership service detects that P has restarted. On processor P, the invocations to be performed are determined by examining all bindings in stable storage. Each invocation is then generated asynchronously by the run-time support using the equivalent of a **send** statement.

After generating all of the handler invocations, the run-time support of the restarted processor resumes normal processing, including monitoring of processors as dictated by the current values of the bindings. However, whether or not user activity is resumed depends entirely on the programmer, i.e., the system takes no action other than sending handler invocations. Thus, it is the responsibility of the programmer to ensure that these invocations actually reinitiate execution. This is achieved by servicing the restart event with a proc rather than an input statement.

3. Implementing a replicated directory

In this section, we present an example to illustrate how these failure-handling mechanisms can be used to structure a portion of a fault-tolerant distributed system. Our program solves a problem abstracted from a distributed operating system design: maintenance of a replicated directory that maps names of system services to the capabilities of the servers that implement those services. Under the hypothesis that directory updates are much less frequent than lookups, it is beneficial to maintain this mapping in replicated directory tables managed by directory managers executing on all operational processors. In particular, such replication provides both fast read access and maximum resilience to failures. Consistency between the copies of the table can be guaranteed through the use an atomic broadcast [6]. That is, if all updates to the replicated directory are broadcast atomically, then all copies of the table will be identical at identical local clock times, assuming that all clocks in the network are synchronized.

The typical operations exported by the service registry to clients are the addition of a new service and its server to the table, deletion of a service and its server, and lookup of the server currently implementing a given service. Hence, the replicated service registry undergoes state transitions in response to the following events: installation of a service SV and a capability for its implementing server S, removal of a service, failure of a processor executing a server S, and the restart of a processor. When service SV is added, the correspondence between SV and its implementing server S is added to the registry tables on all functioning processors. We assume that there is only a single server implementing a given service at any moment; thus, if two add operations are initiated simultaneously for identical services, one will succeed while the other will fail. Should a service SV provided by server S be revoked, the correspondence between SV and S is removed from the tables on all functioning processors. When a processor P crashes, a mapping from a service SV to server S is removed from all tables if S was executing on P. When a processor restarts, it joins with an empty set of resident servers.

To simplify our presentation, we assume that there is at least one functioning processor at all times, i.e., that there is no complete system failure. In addition, we assume that no service additions, removals, or processor failures occur while a newly restarted processor joins a set of functioning processors. For a treatment of the on-the-fly processor join problem, see [7]. Finally, we assume that no directory table overflow occurs.

Our implementation of this collection of interacting directory managers consists of a single resource from which one instance is created for each of N machines. The interface of this resource is shown in Figure 4. The first three op declarations specify the operations exported by the directory manager resource to the client: *add*, *remove*, and *lookup*. The operation *add* is invoked to enter a service name and the capability of its implementing server into the replicated directory. The operation *remove* is invoked to remove a service from the replicated directory; execution has no effect is the service name if not in the table. The operation *lookup* is invoked to obtain the capability of the server implementing a given service. Each of these three operations takes as argument a string value corresponding to the service name. In addition, *add* takes the capability of the implementing server (*ser*) and the capability for a reply operation (*reply*) in the client that the directory manager invokes to indicate whether the service has been successfully added. For *lookup*, the capability of the server that provides the named service is returned as the functional value. If there is no such service, the null capability is returned.

---

spec *direc_manager*

    # operations invoked by clients

        op *add*(*name* : **string**(*),

            ser : **cap** *server*,

            reply : **cap**(*success* : **bool**))

        op *remove*(*name*: **string**(*))

        op *lookup*(*name* : **string**(*)) **returns** *ser*: **cap** *server*

    # operations invoked by directory managers

        op *insert*(*name* : **string**(*);

            ser : **cap** *server*,

            reply : **cap**(success : **bool**);

            orig : **cap** *direc_manager*)

        op *delete*(*name* : **string**(*))

    # interface of main proc; *dm* is array of capabilities for

    # all directory managers in the system

        op *manager_body*(*dm*[0:N-1] : **cap** *direc_manager*)

resource *direc_manager*(*tsize*)

Figure 4. Interface of directory manager resource

The next two operations—*insert and delete*—modify the local table in response to a request that has been broadcast by some directory manager. The arguments to *insert* are the name of the service to be added and three capabilities: one for the implementing server, a second for the reply operation in the client, and the third for the directory manager that broadcast the request. This last capability is entered into the table along with the service name and implementing server for use in failure handling (described below). The argument to *delete* is the name of the service to be deleted.

Finally, the spec for *direc_manager* contains an operation declaration for *manager_body*, the main process of the resource, and the specification of the resource parameters. The maximum number of entries to be stored in the table (*tsize*) is the only resource parameter, while an array containing capabilities for each of the N directory managers is the argument to *manager_body*. This array of capabilities is necessary to allow each manager to invoke the operations *insert* and *delete* in other managers.

The body of the resource, shown in Figure 5, consists of global declarations and two procs: *manager_body* and the directory manager recovery protocol *myrecovery*. The main proc *manager_body* is structured as an infinite loop that repeatedly waits for an invocation of any one of seven operations. These operations include the operations exported to clients and other directory managers, as well as the event handlers *failure* and *recovery*. The program could have been structured so that some or all of these operations were procs, but having a single process avoids potentially concurrent access to the table and the resulting critical section problems.

The code for the client and nameserver operations is relatively straightforward. When *add* or *remove* is invoked, the implementing process simply initiates an atomic broadcast to the *insert* or *delete* operations, respectively, of *all* managers, including itself [2]. The use of atomic broadcast to invoke such local operations not only simplifies the code, it also ensures that insertions and deletions into the table are performed in the same order by all managers. These invocations of *insert* and *delete* are then translated into actual modifications of the local table. In *insert*, an indication of the success of the addition is sent to the client by the resource initiating the broadcast using the forwarded reply capability.

---

[2] For this example, we assume a built-in SR primitive '**abcast** *caps*' that does an atomic broadcast to each operation whose capability is in the array *caps*. A low-cost algorithm to achieve an atomic broadcast is described [6].

```
body direc_manager
    # directory type declaration
        type directory = rec(name : string(*); ser : cap server;
                             orig : cap direc_manager)
    op failure(man : cap direc_manager)        # directory manager failures
    op recovery(man : cap direc_manager)          # directory manager restarts
    op myrecovery()
    op accept_table(t[0:*] : directory; dman[0:N-1] : cap direc_manager)
    var table[0:tsize-1] : directory
    var myr stable binding := when restart(myresource()) send myrecovery()
    proc manager_body(dm)
        var fail[0:N-1] : binding          # manager failures
        var recov[0:N-1] : binding         # manager restarts
        # initialize handler bindings
            fa j:=0 to N-1 →
                fail[j] := when crash(dm[j]) send failure(dm[j])
                recov[j] := when restart(dm[j]) send recovery(dm[j])
            af
    do true →
      in add(name,ser,reply) →
            abcast dm[*].insert(name,ser,reply,myresource())
        [] remove(name) → abcast dm[*].delete(name)
        [] lookup(name) returns ser →
            if name present in table → ser := associated server cap.
    [] name not present → ser := null            # return null capability
        fi
        [] insert(name,ser,reply,orig) →
            ... insert name, ser, orig into table if name not already present ...
            if orig = myresource() → send reply(...) fi
        [] delete(name) → ... delete name if present ...
        [] failure(man) → ... delete names associated with manager man...
        [] recovery(man) → send man.accept_table(table,dm)
        ni
      od
    end
    proc myrecovery()
      var received : bool := false
      do true →
        in accept_table(t, dman) →      if ~received →
                                          received := true; table := t
                                          send manager_body(dman)
                                        fi
        ni
      od
    end
end
```

Figure 5. Body of directory manager resource

The remaining operations implemented by *manager_body* are the two event handlers *failure* and *recovery*, which are invoked when the crash or restart of a processor executing another manager is detected. The handler bindings corresponding to these particular events are the arrays *fail* and *recov* declared in proc *manager_body*. The values assigned to these arrays when the main process is created specify that the argument passed to *failure* is the capability of the directory manager on the failed processor, while the argument passed to *recovery* is the capability for the manager on the newly recovered machine. Note that these event arrays are declared local to *manager_body* and are not in stable storage. These characteristics together ensure that a recovering directory manager can obtain an up-to-date table before monitoring resumes.

The last proc in *direc_manager* is the recovery operation *myrecovery*. As specified by the value of the global binding *myr*, an invocation of this operation is generated whenever the processor executing the manager recovers following a failure. Once this invocation occurs, the newly-created process waits for another directory manager to invoke the operation *accept_table*; this is performed by every other functioning manager in operation *recovery* as a response to the restart. The process then initializes the local table using the directory contents transmitted in the invocation and restarts the main process. To avoid cluttering up the run-time tables with undeliverable messages, the restart process repeatedly accepts subsequent invocations of *accept_table* and discards them.

4. Conclusions

In distributed programs, it is common for a process to wait for a signal from another process that a particular event has occurred. The events themselves may vary—for example, the release of a resource or permission to enter a critical section—yet the techniques involved and the language mechanisms employed remain the same. Our approach to doing failure handling in distributed systems software capitalizes on this fact by treating a failure as another type of event. By reducing the handling of a failure to the handling of just another type of concurrent event, the systems programmer that needs to deal with the highly asynchronous nature of failures can rely on well-known and well-understood techniques instead of being forced to develop new ones. This characteristic is especially appropriate for programming critical applications, where the need for dependability dictates a conservative strategy.

Our approach also has the benefit of being quite general, in two different ways. One is that it is easy to extend the type of events that can be considered; to handle new events—for example, failure events of models other than fail-stop—all that needs be done at the language level is to define new events and event descriptions. We are currently exploiting this generality by extending our approach to handle exception conditions in distributed programming languages. The second way in which our approach is general is that it is relatively language independent. We used SR in this paper, yet the general approach would be the same no matter what concurrent programming language was used. For example, in a monitor-based language, the occurrence of an event could cause an entry procedure in a specified monitor to be invoked; for a language that uses message-passing, the event could trigger the transmission of a message.

Future work will concentrate on two efforts. One is completing the implementation of an extended version of SR containing the mechanisms described in Section 2. By doing so, we hope not only to refine our approach, but also to provide a useful language for designing and implementing fault-tolerant distributed systems. The initial test for this extended SR will consist of re-implementing portions of the Saguaro distributed operating system [2].

The other area of future research is the development of a new language for constructing fault-tolerant distributed systems. While many of the constructs can be borrowed directly from languages such as SR, it is clear that the freedom to design features without being concerned about minimizing the impact on an existing language will result in a more suitable language for constructing fault-tolerant applications. For example, one approach that we considered for this paper was to have two predefined language operations, **crash**(*pname*) and **restart**(*pname*), where pname is the name of the failed or restarted processor; these operations would then be invoked automatically with the appropriate argument value when the corresponding event occurred. The problem with this approach is that it is inconvenient to specify which processors should be monitored without some notion of process groups (or resource groups in the case of SR) such as found in the V system [4]. Adding such a facility to SR was deemed too substantial a modification for the purposes of this paper, but it appears to be a promising foundation upon which to base a new language.

Acknowledgments

Discussions with R. Olsson and especially G. Andrews were very beneficial in developing the extensions presented in Section 2. G. Andrews, S. Manweiler, R. Olsson, and several anonymous reviewers read earlier drafts and provided many helpful comments.

References

[1] *Reference Manual for the Ada Programming Language.* ANSI/MIL--STD--1815A, 22 January 1983.

[2] G.R. Andrews, R.D. Schlichting, R. Hayes and T. Purdin, "The design of the Saguaro distributed operating system", *IEEE Transactions on Software Engineering* SE-13,1 (Jan. 1987), 104-118.

[3] G.R. Andrews, R. Olsson et al., "An overview of the SR language and implementation", *ACM Trans. on Prog. Lang. and Syst. 10*, 7 (Jan. 1988), 51-86.

[4] D.R. Cheriton and W. Zwaenepoel, "Distributed process groups in the V kernel", *ACM Trans. on Comp. Sys.* 3,2 (May 1985), 77-107.

[5] F. Cristian, "A rigorous approach to fault-tolerant programming", *IEEE Transactions on Software Engineering* SE-11, 1 (Jan. 1985), 23-31.

[6] F. Cristian, H. Aghili, R. Strong and D. Dolev, "Atomic broadcast: From simple message diffusion to byzantine agreement", *Proc. 15th Annual International Symposium on Fault-Tolerant Computing*, Ann Arbor, Michigan (June 1985), 404-409.

[7] F. Cristian, "Agreeing on who is present and who is absent in a synchronous distributed system", *Proc. 18th Annual International Symposium on Fault-Tolerant Computing*, Tokyo (June 1988), 206-211.

[8] C.A.R. Hoare, "Communicating Sequential Processes," *Comm. ACM* 21, 8 (Aug. 1978), 666-677.

[9] B.W. Lampson, "Atomic transactions", In *Distributed Systems--Architecture and Implementation.* Lecture Notes in Computer Science Vol. 105, Springer-Verlag, New York, 1981, Chapter 11.

[10] B. Liskov and R. Scheifler, "Guardians and Actions: Linguistic support for robust, distributed programs", *Proc. 9th Symp. on Prin. of Programming Languages*, Austin, TX (Jan. 1983), 7-19.

[11] J. Mitchell, W. Maybury and R. Sweet, "Mesa language manual", Version 5.0. Report CSL-79-3, Xerox PARC, April 1979.

[12] R.D. Schlichting and F.B. Schneider, "Fail-stop processors: An approach to designing fault-tolerant computing systems", *ACM Trans. on Comp. Sys.* 1,3 (Aug. 1983), 222-238.

Appendix 1:

The SR Distributed Programming Language

An SR program consists of one or more *resources*. These resources can be thought of as patterns from which resource instances are created dynamically. Each resource is composed of two parts: an interface portion called the *spec*, which specifies the interface of the resource and the *body*, which contains the code to implement the abstract object. The specification portion contains descriptions of objects that are to be exported from this resource—i.e., made available for use within other resources—as well as the names of resources whose objects are to be imported. Of primary importance are the declaration of *operations*—actions implemented by sequences of statements that can be invoked. These declarations specify the interface of those operations that are available for invocation from other resources. For example,

op example1(**var** *x*: **int**; **val** *y*: **bool**)

declares an operation, *example1*, that takes as arguments an integer *x* that is passed with copy-in/copy-out (**var**) semantics and a Boolean y that is copy-in only (**val**). Result parameters (**res**) are also supported, as are operations with return values.

The declaration section in the resource body together with the **spec** define the objects that are global to the resource, i.e., accessible to any process within the resource. All of the usual types and constructors are provided. In addition, there are *capability variables*. Such a variable functions either as a pointer to all operations in a resource instance (a *resource capability*), or as a pointer to a specific operation within an instance (an *operation capability*). A variable declared as a resource capability is given a value when a resource instance is created, while an operation capability is given a value by assigning it the name of an operation or from another capability variable. Once it has a value, such variables can be used to invoke referenced operation(s), as described below.

An operation is an entry into a resource. An SR operation has a name, and can have parameters and return a result. There are two different ways to implement an operation: as a proc or as an alternative in an input statement. A proc is a section of code whose format resembles that of a conventional procedure:

> **proc** opname(parameters) **returns** result
>   op_body
> **end**

The operation body *op_body* consists of declarations and statements. Like a procedure, the declarations define objects that are local to the operation *opname*. Unlike a procedure, though, a new process is created, at least conceptually, each time *opname* is invoked. It is possible to get standard procedure-like semantics, however, depending on how the proc is invoked (see below). The process terminates when (if) either its statement list terminates or a **return** is executed.

An operation can also be implemented as an alternative of an input statement. An input statement implementing a collection of operations $opname_1$, $opname_2$, ..., $opname_n$ has the following form:

> **in** $opname_1$(parameters) $\rightarrow$ op_body_1
> [] $opname_2$(parameters) $\rightarrow$ op_body_2
> ...
> [] $opname_n$(parameters) $\rightarrow$ op_body_n
> **ni**

A process executing an input statement is delayed until there is at least one alternative $opname_i$ for which there is a pending invocation. When this occurs, one such alternative is selected non-deterministically, the oldest pending invocation for the chosen alternative is selected, and the corresponding statement list is executed. The input statement terminates when the chosen alternative terminates.

An operation is invoked explicitly using a **call** or **send** statement, or is implicitly called by its appearance in an expression. The explicit invocation statements are written as

> **call** op_denotation(arguments)
> **send** op_denotation(arguments)

where the operation is denoted by a capability variable or by the operation name if the statement is in the operation's scope.

Execution of a **call** terminates once the operation has been executed and a result, if any, returned. Its execution is thus synchronous with respect to the operation execution. Execution of a **send** statement is, on the other hand, asynchronous: a **send** terminates when the target process has been created (if a proc), or when the arguments have been queued for the process implementing the operation (if an input statement). Thus, the effects of executing the various combinations of **send/call** and **proc/in** are described by the following table.

Invocation	Implementation	Effect
call	**proc**	procedure call
send	**proc**	process creation
call	**in**	rendezvous
send	**in**	asynch. message passing

To illustrate how the individual pieces of the language fit together, consider the implementation of a bounded buffer shown in Figure 6. Two operations are exported from this resource: *deposit* and *fetch*; *deposit* places a value in the next available slot if one exists, while *fetch* returns the oldest value from the buffer. A depositing process is delayed should the buffer be full. Similarly, a fetching process is delayed whenever the buffer is empty. Note also that the resource has a parameter *size*; its value determines the number of slots in the buffer. The use of resource parameters in this way allows instances to be created from the same pattern, yet still vary to a certain degree. Finally, note the single input statement to implement both the deposit and fetch operations, and the use of a send statement in the initialization code to initiate the main (parameterless) proc *buff_loop*. Creating a process in this manner is so common that the keyword **process** can be used instead of **proc** as an abbreviation for the **send** in the resource initialization code and corresponding **op** declaration.

```
spec buffer
      op fetch() returns value : int
      op deposit(val newvalue : int)
resource buffer(size : int)

body buffer
      var first, last : int := 0, 0
      var slot[0:size - 1] : int

      initial
            send buff_loop()
      end
      proc buff_loop()
            do true →
                in deposit(newvalue) and first ≠ (last + 1) % size →
                      slot[last] := newvalue
                      last := (last + 1) % size
                [] fetch() returns value and first ≠last →
                      value := slot[first]
                      first := (first + 1) % size
                ni
            od
      end

end
```

Figure 6. Bounded buffer

FAULT-TOLERANT MEMBERSHIP SERVICE IN A SYNCHRONOUS DISTRIBUTED REAL-TIME SYSTEM

H.KOPETZ, G. GRÜNSTEIDL, J. REISINGER
Institut für Technische Informatik
Technical University Vienna, A-1040 Vienna - Austria

Abstract

In many hard real-time applications, timely knowledge about the operational state of the nodes in a distributed computer system is of significant importance. This paper presents a fault-tolerant distributed membership protocol for the determination of the set of active nodes in a synchronous distributed real-time system. After a discussion of the system architecture and the fault hypothesis, which is supported by experimental data, the membership protocol is described in detail. In the final section the application of this protocol for the design of selfchecking nodes, for the solution of the atomic multicast problem and for the remote monitoring of nodes is discussed.

Keywords: Membership Protocol, Atomic Multicast, Real-Time, Fault-Tolerance, Distributed Systems.

This work has been sponsored in part by the Austrian Ministry of Science and Research under contract Nr.GZ.605.504/3-26/87 and by the Austrian Foundation for Basic Research (FWF) under contract P6251P.

1. Introduction

In the domain of hard real-time applications, distributed computer architectures are well accepted. In such a distributed architecture the set of active nodes, i.e. the membership set, will change over time. An active node may fail or be turned off while an inactive node may join the set of active members at a given point in real-time.

Many distributed algorithms, e.g. algorithms for the execution of a coordinated action of a number of nodes, require a timely and consistent knowledge about the membership set.

It is the objective of this paper to present a distributed membership protocol to determine the set of active member nodes in a synchronous distributed real-time system.

Since time critical actions of a real-time system may have to be delayed until the membership is known, the time interval between the point in time t_{past} of the membership and the point in time $t_{observe}$ when the consistent knowledge about this membership at t_{past} is available to all active partners in the ensemble, must be bounded and should be kept as small as possible. The minimum amount of information required to establish the membership set at t_{past}, even if a very simple fault model is assumed, requires information from every node of the set, i.e. a complete round of information exchanges.

The problem of reaching agreement among group members on group membership in a synchronous distributed system in the present of different faults was first studied in [6]. There it is assumed that the communication network provides an atomic broadcast service [2], [5]. The problem is also related to the area of system diagnosis in distributed systems [9], [10], [15].

This paper is organized as follows. In the first part the characteristic properties of the real-time environment are introduced, the architecture is described and a detailed discussion of the fault hypothesis, supported by experimental data, is presented. In the second part, the core section of this paper, a protocol for the solution of the membership problem in a distributed real-time system is first described informally and then analyzed in some detail. The final section is concerned with the application of this protocol for the solution of a number of important problems in distributed real-time systems, such as the atomic multicast problem or the remote monitoring of distributed actions.

2. System architecture and fault hypothesis

2.1. Characteristics of the real-time environment

In a typical real-time system a controlled object (the controlled environment) and a control system (the computer) are connected via sensor- and actuator-based interfaces. Within predefined intervals of real-time the control system has to accept data from the sensors, has to process these data and deliver the results to the control object via the actuators. The output data influence the control object such that the effects of these outputs can be observed via the sensor inputs thus closing the loop as shown in Fig.1.

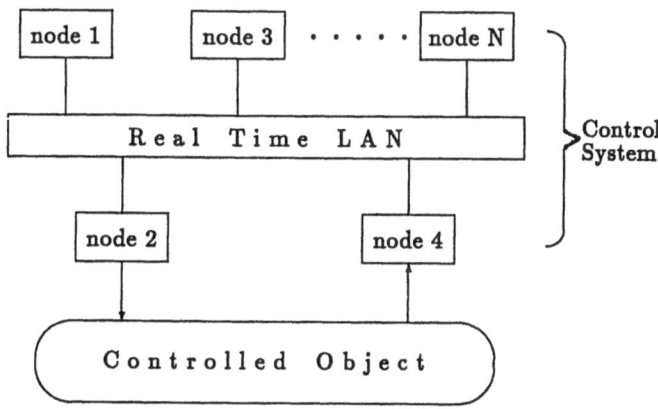

Figure 1. Architecture of a distributed real-time system

The following properties are characteristic of the real-time environment:

- The control system must respond to a stimulus from the controlled object within an interval dictated by the environment, called "response time". This response time must be guaranteed, even under the anticipated fault conditions (fault hypothesis) and under the specified peak load conditions (load hypothesis). Normally it is impossible to exercise explicit flow control in order to reduce the input rate from the environment.

- Many real-time systems have to operate for extensive periods of time (up to years) without shutdown for maintenance or functional enhancements. Thus the requirements on reliability and ease of maintenance are very high. Only fault-tolerant architectures can meet this challenge.

- If a serious failure occurs, either in the control system or the control object, the system must shut down in a controlled and consistent predetermined manner (failsafe operation).

In order to achieve the required level of performance and fault-tolerance, distributed architectures have become common in real-time applications. A distributed real-time system consists of a real-time local area network (RT LAN) connecting a number of node computers which have to perform a coordinated action. If anyone of the partners is missing the action should not be performed. This facility is sometimes referred to as "atomic multicast" [3], [4], [5]. Either all selected partners perform the coordinated action or no irrevocable activity is initiated by any of the partners.

In a real-time environment, a control system can fail in two domains, the value domain and the time domain. In the rest of this article we will use the following definitions:

Def.:   Correct result
        A result is correct if its values correspond to the intentions of the user.

Def.:   Timely result
        A result is timely if it is produced within the intended interval of real-time.

Def.:   Valid result
        A result is valid if it is correct and timely.

In the "non real-time world" we are only concerned with the value domain, i.e. with correctness. The inclusion of the real-time domain adds a new dimension to the "validity" problem in real-time systems, since the speed of processing in a given interval, i.e. the system performance becomes an essential property.

Architectural assumptions

The architectural assumptions are based on the MARS-project Maintainable Real-Time System) [8]. Therefore, in the rest of this paper, we will assume that a distributed real-time system consists of a set of autonomous nodes, i.e. selfcontained computers including the application software, which are loosely coupled by a Real-Time Local Area Network (Fig. 1). Every node consists of a single board computer, an incoming link and an outgoing link. Thus the network is reduced to a passive channel only, e.g. a bus. Communication among the nodes is achieved by the exchange of broadcast messages, i.e. a message sent by any one node can be received by all other nodes.

By a synchronous distributed system we mean that every node contains a local real-time clock. At any point in time, all clocks of good nodes are synchronized to within a known constant Δ by a fault-tolerant synchronization algorithm [7], thus establishing a global time base of known accuracy. This global time is used to control the access to the communication channel by a synchronous time division multiple access strategy (TDMA), i.e. the time for a message to be delivered is bounded. We call a time interval on the channel a slot and assume that the slot length is constant.

Given N nodes in the system, a complete TDMA round, i.e. a TDMA cycle, will use up N slots. In order to facilitate the analysis we will refer to a particular slot as $t^{i,k}$, where i denotes the cycle number (from the set of positive integers) and k denotes the slot number (from the set 1 to N). It is evident that if k reaches N a new TDMA cycle is initiated (and vice versa). It follows that at time $t^{i,k}$ node k has the right to send. We call this point in time $t^{i,k}$ the membership point for node k in cycle i.

We further assume that an active node k will send a broadcast message in its assigned slot $t^{i,k}$. In case no other message has to be transmitted, the node will broadcast an empty message as a life sign. It is a property of the TDMA strategy that empty slots cannot be used by other nodes anyway.

2.2. Fault-tolerance in distributed real-time systems

Fault-Tolerance is concerned with the continuation of acceptable service of a system despite the occurrence of faults within the stated fault hypothesis. It can only be achieved if there are enough "redundant" resources in the system to detect and circumvent the consequences of these faults [1]. In the context of real-time systems it is necessary to examine if the fault handling techniques which have been successful in the "non real-time world" can be transferred to the real-time environment.

If we analyse the time characteristics of faults, we can distinguish the following cases:

(i)   Transient Faults: A fault starts at a particular point in time and remains in the system for some period after which it disappears again. Examples of transient faults are hardware deficiencies caused by external interference (e.g. electric, mechanical, radioactive etc.). After the external disturbance disappears, the fault disappears as well.

(ii)   Permanent Faults: A fault that starts at a particular point in time and remains in the system until it is removed by some external activity (i.e. the completion of a repair action). Examples of permanent faults are a broken wire in the hardware or a design fault in a software program.

(iii) Intermittent Faults: A transient fault that occurs repeatedly within some given interval. An example of an intermittent fault is a marginally working IC component, which operates some of the time and does not operate some other time.

As has been discussed in the previous section, time is a limited resource in a Real-Time Environment. In the following, we will thus concentrate on a technique increasing the fault-tolerance of distributed real-time systems which does not invalidate the time characteristics of the system. This technique is based on the redundant operation of selfchecking nodes.

Def.: Selfchecking Node (SN)
       A node with a given functionality, which at any point in time produces either valid results or no results, i.e. all internal failure modes are mapped into a single external failure mode, i.e. the termination of service.

If and when a selfchecking node operates, it operates according to its specification, i.e. it produces valid results. It is up to the implementation of the selfchecking node to provide the necessary error detection mechanisms [11], [18] to guarantee the selfchecking property under the fault model specified. We are aware that this is a strong assumption with has to be justified by providing the appropriate mechanisms within the nodes.

In order to be able to detect failures of the incoming and outgoing link of a node, the node has to have knowledge about the traffic on the channel and the behavior of the other nodes. This knowledge is contained in the membership protocol which thus forms an integral part of a selfchecking node. Otherwise a node, which does not hear any traffic on the channel, can never decide if there is no traffic on the channel or if his incoming link has failed.

The introduction of the SN makes it possible to abstract from the complexity of the detailed inner structure and failure modes of a node. A SN can be defined at an abstract level by the messages it accepts, the messages it produces, its internal state space, its function, and its performance characteristics.

2.3. Fault hypothesis

We assume that the nodes can fail in the selfchecking mode (permanent crash failures) and messages can get lost (transient omission failures). Based on our experimental results outlined in the next section it does not seem to be necessary to include other types of failures (e.g. value failures or timing failures [5], [19] in our fault model.

Since a loss of the message on the channel has the same effect as the loss of the message in the outgoing link, the channel failure rate can be subsumed in the outgoing link failure rate. In order to reduce the number of omission failures, every message is sent twice within its sending node's TDMA slot. As long as any one of the two identical messages arrives at the receiver, the transmission is considered as successful.

We consider that if a sent message is lost by two receivers but received by the rest, then two receiver errors have occurred. The occurrence of two errors has been excluded from the fault hypothesis. This assumption has been confirmed by our experiments, which have shown that these two receiver errors are statistically independent and therefore extremely unlikely [16].

The following parameters are considered in our fault hypothesis:

Failure Mode	Failure Probability
transient node failure	λ_{trans}
permanent node failure	λ_{perm}
sender loss	λ_{send}
receiver loss	λ_{rec}

Experimental results

Our experimental system consists of 12 nodes coupled by a thin wire Ethernet controlled by a collisionfree synchronous TDMA strategy. We used a message length of 1500 bytes and a TDMA slot length of 4 msec (to send two redundant messages), i.e. a full TDMA cycle will consume 48 msec. This system has been in operation for more than a year. Based on the observed failure rates we assume a permanent node failure rate of 1/10000 *hours*$^{-1}$ and a transient node failure rate of 1/1000 *hours*$^{-1}$.

In order to obtain the transient message failure probabilities λ_{send} and λ_{rec} we performed extensive experiments on our experimental system. During our experiments we also determined whether multiple failures (failures effecting more than one node) and failures in double sending (where one message is sent immediately after the other) are statistically independent.

In our experiments we have sent two billion messages across the network. The experimental results gave us no reason to reject our hypothesis of statistical independence of the failure modes mentioned above [16].

As will be shown later, the membership protocol has to be executed after every slot. In order to keep the protocol simple (and fast) we assume that only one failure--either a transient or permanent node failure or a sender loss or a receiver loss--may occur within two TDMA cycles.

Based on our experimental data, we thus have the following failure probabilities within the two needed TDMA cycles in a system with 12 nodes, where every message is sent twice:

transient node failure (1 of 12 in 100 msec)	3.10^{-7}
permanent node failure (1 of 12 in 100 msec)	3.10^{-8}
sender loss (1 of 24 outgoing messages)	10^{-14}
receiver loss (1 of 24 incoming messages)	10^{-14}
probability of 1 failure in 100 msec	$< 10^{-6}$
probability of 2 independent failures	$< 10^{-12}$

Our fault hypothesis will be violated if two or more failures occur within two TDMA cycles. Assuming statistical independence the mean time to violate the fault hypothesis is more than 1000 years.

3. A membership protocol

We call the set of active nodes at a given point in time t the membership set. The membership set will change over time because an active node may depart from the membership set (e.g. node failure) or an inactive node may join the membership set.

The *aim of this protocol* is to guarantee that every active node at any point in time $t_{observe}$ has a consistent knowledge about the membership set at a past point in time t_{past}. The time interval between $t_{observe}$ and t_{past} has to be bounded and should be kept as small as possible.

3.1. Informal protocol description

Whenever a node k broadcasts a message at its membership point $t^{i,k}$ it includes the information about all messages it has received in the last $N-1$ slots in the message header. This information, called the view vector $VV^{i,k}$ of node k in cycle i is received by all other nodes and is used to build a view matrix $VM_l^{i,k}$ in every node l. Because messages can get lost, these view matrices can differ.

Immediately before node k sends its message at its membership point $t^{i,k}$ all nodes (including node k which is supposed to send next) analyse the local view matrix and determine the membership value $mval(i-1,k)$ (a boolean variable) of node k in TDMA cycle $i-1$. The protocol has to guarantee that all nodes arrive at the same membership value $mval(i-1,k)$ as long as the fault hypothesis is not violated. If $mval(i-1,k)$ is false, then node k has not been an active member at its last membership point $t^{i-1,k}$. Node k then has to turn itself off and does not send a message at the membership point $t^{i,k}$. A node which has failed must remain in the failed state for a protocol defined constant d_{fail}, which has to be longer than 2 TDMA cycles.

If a node k has been active at two successive membership points $t^{i-1,k}$ and $t^{i,k}$, it is concluded that this node has been active at any point in time during the interval $< t^{i-1,k} , t^{i,k} >$. If a node k has been active at the membership point $t^{i-1,k}$, but has been inactive at $t^{i,k}$ it is concluded that it has become inactive after the point in time $t^{i-1,k}$ and before the point in time $t^{i,k}$.

The membership set at time t is given by the set of all active nodes at time t.

The membership history of a node k is the sequence of membership values $mval(.k)$ in the interval of interest.

3.2. Analysis of the protocol

As mentioned in section 2, every active node broadcasts a message in each TDMA cycle (life sign). Under the given fault hypothesis, two situations may occur if a failure causes the loss of a message:

(i)   this message is not received by exactly one other active node

(ii) this message is not received by any of the other nodes

If the message is not received by two or more other active nodes, but received by the remainder of the active nodes, then two or more message losses have occurred, a case which is not covered by our fault hypothesis.

In case (i), the failure is associated with this receiver node (node failure or receiver interface failure). In case (ii) the failure is associated with the sender (node failure or sender interface failure). Hence it follows that for each message loss exactly one node is responsible. We further assume that, according to the membership protocol, a node shuts itself off immediately after it detects that it is responsible for the failure (provided it has not experienced a crash failure at some earlier time and is already silent). In order to diagnose the node responsible for the message loss, we need the information carried in the view vectors:

Def.:   View Vector
        The view vector of node k is the vector $VV^{i,k}$ containing information about the messages it has received in the last $N-1$ slots.

A boolean element of the view vector $VV^{i,k}$ is denoted by vv_j, whereby $vv_j = true$ if node k has received the last message from node j and false otherwise. In our implementation of this protocol, in a system with up to 32 nodes, the view vector will occupy 4 bytes. Considering our fixed message length of 1500 bytes this view vector length is less than 0.3 % of the message length.

In a fault free system, in every TDMA cycle each active node receives the view vector from every other active node. By using this view vectors every active node l constructs its view matrix $VM_l^{i,k}$ at every point in time $t^{i,k}$.

Def.:   View Matrix $VM_l^{i,k}$
        A matrix where each column q is constructed from the recent view vector received (or not received if there has been a message loss) from the node q. Each row refers to a sender p. An element of this matrix, denoted by vm_{pq}, represents l's knowledge about q's receipt of the last message sent

by p. The boolean values (true and false) of the view vectors are mapped on to the values 1 and -1 in the representation of the view matrix.

$$VM_l^{i,k} = \begin{pmatrix} vm_{1,1} & \cdot & \cdot & vm_{1,q} & \cdot & vm_{1,N} \\ \cdot & \cdot & \cdot & \cdot & \cdot & \cdot \\ \cdot & & \cdot & vm_{p,q} & \cdot & \cdot \\ \cdot & \cdot & \cdot & \cdot & \cdot & \cdot \\ \cdot & \cdot & \cdot & \cdot & \cdot & \cdot \\ vm_{N,1} & \cdot & \cdot & vm_{N,q} & \cdot & vm_{N,N} \end{pmatrix}$$

where

$$vm_{p,q} = \begin{cases} 1 \text{ if q has received the last message from p} \\ 0 \text{ if no information is available from q at 1 (message loss or} \\ \quad \text{not yet sent)} \\ -1 \text{ if q has not received the last message from q} \end{cases}$$

We will now analyse the effects of the different failure modes on the membership decision immediately before the membership point $t^{i,k}$, i.e. the point in time when the membership for node k is determined for its membership point $t^{i-1,k}$. We can assume that the membership set for the time $t^{i-2,k}$ has been established and is available. According to the fault hypothesis, only one failure may occur in the interval $<t^{i-2,k}, t^{i,k}>$.

Crash failures

In case of a permanent crash failure, a node is silent. After a transient crash failure, the implementation must guarantee that the node remains silent for at least two TDMA cycles.

If a crash failure has occurred prior to sending the message at $t^{i-1,k}$, then node k will not have sent any message at $t^{i-1,k}$. This will be recognized by all other nodes and $mval(i-1,k)$ is set to F consistently.

If a crash failure has occurred after sending the message at $^{i-1,k}$, then node k has been active at $^{i-1,k}$ and $mval(i-1,k)$ is set to T consistently.

Incoming link failures

If node k has two or more incoming link failures in the interval $<t^{i-2,k}, t^{i-1,k}>$, then it will fail silently before sending the message at $t^{i-1,k}$ and all other nodes will realize this failure by the missing message from node k at $t^{i-1,k}$.

If a transient incoming link failure occured at node k before time $t^{i-1,k-1}$ then the situation is easy to analyse, since the view vector of node $k-1$ indicates that the message was sent by node $k-2$. Node k can thus turn itself off before sending the message at $t^{i-1,k}$ and all other nodes will recognize the failure of node k by the missing message at $t^{i-1,k}$.

The most critical case is a transient incoming link failure immediately before sending. Let us assume that such a failure occured at $t^{i-1,k-1}$ at node k. At $t^{i-1,k}$, when node k has to send a message, it cannot determine if this failure is an incoming link failure of node k or an outgoing link failure of node $k-1$. Node k therefore has to send its message at $t^{i-1,k}$ with the attribute "suspect", and at the same time enters into the state "suspect". If, within the next TDMA cycle, i.e. the cycle stretching to immediately before $t^{i,k}$ no other failure is detected, then node k must conclude that it was responsible for the failure at $t^{i-1,k-1}$ (transient incoming link failure) and turn itself off immediately. All other nodes will come to the same conclusion by analysing the view vector of node k at $t^{i-1,k}$, indicating that node k did not hear the message sent by node $k-1$ at $t^{i-1,k-1}$ (which was heard by all other nodes) discard the suspect message sent by node k at $t^{i-1,k}$, and set $mval(i-1,k) = F$.

Outgoing link failure

If the outgoing message at $t^{i-1,k}$ is lost in the sender interface (or on the channel), then no node will hear the message and will set its view vector accordingly. At the membership decision point before $t^{i,k}$ it is then straightforward to recognize the outgoing link failure of node k. Node k will turn itself off and all other nodes will set $mval(i-1,k) = F$.

3.3. Join of new members

If a new node l wants to join the membership this node has to go through three phases:

- Initialization Phase (I-Phase)

- Trial Phase (T-Phase)

- Full Membership Phase (F-Phase)

In the following we describe only the actions inside these phases which are important for the membership protocol.

In the I-Phase the node l has to synchronize its local clock and to establish itself as a trial member. It receives and analyses all view vectors sent by the active nodes and broadcasts its own view vector (as a life sign) as soon as it has heard all other nodes. If the members of the last membership set receive this new view vector of the new node l they will admit node l as a new trial member. As soon as node l sees that it has been recognized by all other active nodes it enters the T-Phase.

During the T-Phase, which lasts for a time interval determined by some protocol parameter d_{stable}, the node behaves as a full member. It runs the membership protocol and participates in all actions as specified. The only difference to the full membership state is that no node may conclude that node l has been a stable member for a past period of at least d_{stable}.

Since such a trial phase has been introduced in the membership join protocol, every full member can conclude about every other full member, that, if a node has been an active full member at time t it has also been an active full member in the past interval starting with $t-d_{stable}$. This protocol property is very helpful for reasoning within real-time applications.

3.4. An example

We assume that a system consists of 4 active nodes. The membership set for the time $t^{i-2,k}$ has been established consistently at these nodes and contains all 4 nodes. As an example we have chosen the critical case of a transient incoming link failure immediately before sending.

In the following we describe the view matrices of the 4 nodes after different points in time $t^{i,k}$, whereby $VM_l^{i,k}$ denotes the view matrix of node l after $t^{i,k}$, i.e. the view vector at $t^{i,k}$ is already considered in this matrix. Consequently the matrix contains information about the last 4 slots including slot $t^{i,k}$.

We assume that node 2 suffers a transient incoming link failure at $tt^{i-1,1}$, i.e. the last message sent by node l is lost by node 2. This situation is reflected in the following matrices.

$$VM_1^{i,-1,1} = \begin{pmatrix} * & 0 & 0 & 0 \\ 1 & * & 1 & 1 \\ 1 & 0 & * & 1 \\ 1 & 0 & 0 & * \end{pmatrix}$$

$$VM_2^{i,-1,1} = \begin{pmatrix} * & -1 & 0 & 0 \\ 0 & * & 1 & 1 \\ 0 & 1 & * & 1 \\ 0 & 1 & 0 & * \end{pmatrix}$$

$$VM_3^{i,-1,1} = \begin{pmatrix} * & 0 & 1 & 0 \\ 1 & * & 1 & 1 \\ 1 & 0 & * & 1 \\ 1 & 0 & 1 & * \end{pmatrix}$$

$$VM_4^{i,-1,1} = \begin{pmatrix} * & 0 & 0 & 1 \\ 1 & * & 1 & 1 \\ 1 & 0 & * & 1 \\ 1 & 0 & 0 & * \end{pmatrix}$$

After $t^{i-1,1}$ the view matrices contain information about the last 4 slots ($t^{i-1,1}$, $t^{i,2,4}$, $t^{i-2,3}$, $t^{i-2,2}$), i.e. row 1 of a view matrix, which refers to the sender 1, corresponds indirectly to $t^{i-1,1}$, row 2 to $t^{i-2,2}$, row 3 to $t^{i-2,3}$ and row 4 to $t^{i-2,4}$. After $t^{i-1,1}$ the membership for node 2 has to be established for its membership point $t^{i-2,2}$. The message loss (loss of the view vector $VV^{i-1,1}$ of node 1) does not affect this decision, because node 2 can conclude from the other values of row 2 of its view matrix $VM_2^{i-1,1}$ that no outgoing link failure of node 2 has occured.

But node 2 cannot determine if it is responsible for the loss at $t^{i-1,1}$. Therefore it has to send at $t^{i-1,2}$ its message and enters into the state "suspect". In the following we describe the view matrices after $t^{i-1,4}$ when the membership for node 1 is determined for its membership point $t^{i-1,1}$.

$$VM_1^{i,-1,4} = \begin{pmatrix} * & -1 & 1 & 1 \\ 1 & * & 1 & 1 \\ 1 & 0 & * & 1 \\ 1 & 0 & 0 & * \end{pmatrix}$$

$$VM_2^{i,-1,4} = \begin{pmatrix} * & -1 & 1 & 1 \\ 0 & * & 1 & 1 \\ 0 & 1 & * & 1 \\ 0 & 1 & 0 & * \end{pmatrix}$$

$$VM_3^{i,-1,4} = \begin{pmatrix} * & -1 & 1 & 1 \\ 0 & * & 1 & 1 \\ 0 & 0 & * & 1 \\ 0 & 0 & 1 & * \end{pmatrix}$$

$$VM_4^{i,-1,4} = \begin{pmatrix} * & -1 & 1 & 1 \\ 0 & * & 1 & 1 \\ 0 & 0 & * & 1 \\ 0 & 0 & 0 & * \end{pmatrix}$$

After $t^{i-1,4}$ all nodes (inclusive node 2 itself) establish the membership of node *1* at $t^{i-1,1}$. Additionally by analysing row 1 of their view matrices every node can recognize that only node 2 has not received the message from node 1 at $t^{i-1,1}$. Hence its follows that node 2 turns itself off immediately and all other nodes set *mval(i-1,2)=F* consistently.

In the following the view matrices after $t^{i,1}$ are shown.

$$VM_1^{i,-1,1} = \begin{pmatrix} * & 0 & 0 & 0 \\ 1 & * & 1 & 1 \\ 1 & 0 & * & 1 \\ 1 & 0 & 0 & * \end{pmatrix} \qquad VM_2^{i,-1,1} = \begin{pmatrix} * & 1 & 0 & 0 \\ 0 & * & 1 & 1 \\ 1 & 1 & * & 1 \\ 1 & 1 & 0 & * \end{pmatrix}$$

$$VM_3^{i,-1,1} = \begin{pmatrix} * & 0 & 1 & 0 \\ 1 & * & 1 & 1 \\ 1 & 0 & * & 1 \\ 1 & 0 & 1 & * \end{pmatrix} \qquad VM_4^{i,-1,1} = \begin{pmatrix} * & 0 & 0 & 1 \\ 1 & * & 1 & 1 \\ 1 & 0 & * & 1 \\ 1 & 0 & 0 & * \end{pmatrix}$$

If only these matrices are used to calculate *mval(i-1,2)* a loss of the membership of node 2, caused by a transient incoming link failure, cannot be recognized. It follows that a node can send a message, even if it is no member at this membership point, i.e. if a node sends a message, which is received by all other nodes, it cannot be concluded that the node is a member at this membership point.

4. Application of the membership protocol

In the following section we will outline some possible applications of this membership protocol for the design of distributed real-time systems.

Selfchecking nodes

As has been pointed out in section 2.2, every selfchecking node has to contain as an integral part, a membership protocol which includes some knowledge about the intended behaviour of the node environment. Without such a knowledge it is impossible to detect link failures. The commonly used 'watchdog timer' is a primitive, but limited, mechanism which is introduced for this purpose.

In our opinion, selfchecking nodes are effective building blocks for the design of fault-tolerant distributed real-time systems. The simple external failure mode of selfchecking nodes facilitates the error detection, error diagnostics and error handling at the architectural level.

Atomic multicast

In many real-time applications a set of cooperating nodes has to perform a coordinated action. It has been argued convincingly that the atomic multicast communication is a powerful abstraction for this purpose [2], [3], [5]. Atomic multicast requires that a message is accepted either by all or by none of a group of selected nodes.

If the membership at $t^{i-1,k}$ has been established, it is evident who received the message sent at $t^{i-1,k}$ and who did not receive the message. If the group of selected nodes is a subset of the membership set at $t^{i-1,k}$, then the multicast at $t^{i-1,k}$ has been successful, otherwise it has not been successful.

It is straightforward to append a protocol for atomic multicast to the membership protocol. Since the membership view is consistent, every node has to eliminate the message sent at $t^{i-1,k}$ if anyone of the selected node was not member at $t^{i-1,k}$.

Clock synchronization

The degree of fault-tolerance of a distributed clock synchronization protocol, as has been published in [7], can be improved if an effective membership protocol is available. The active number of nodes N, which has to be used in the clock synchronization protocol is generated by the membership protocol. Thus every resynchronization interval can tolerate the number of failures specified in the clock synchronization protocol anew.

Remote action monitoring

The membership protocol can also be used to perform remote monitoring of actions of a node in a distributed system. Provided the system is based on selfchecking nodes it can be concluded that every node operates as intended as long as it is in the active membership. In case a node does not perform as intended, the selfchecking mechanisms will turn off the node and the remote monitor realizes immediately that the node in question has failed.

It has to be pointed out that such a remote action monitoring can be realized without increasing the communication traffic by additional messages. The information supplied by the membership protocol is fully sufficient for this purpose.

5. Conclusion

The membership protocol presented in this paper provides a consistent view of the membership in a synchronous distributed system without increasing the number of messages. Only the length of each message has to be increased by a bitvector of size N, (i.e. the number of nodes). Considering the performance characteristic of high speed local area networks, this increase in the message size is of no concern.

The CPU load generated by this membership protocol, even under the restricted (but we feel realistic) fault hypothesis considered in this paper, is significant. At every slot the protocol has to be executed at every node. In a future implementation of our distributed system MARS [8] we plan to include a separate coprocessor for the execution of the membership protocol and the generation of the membership history in each node.

Our membership protocol is not restricted to an architecture uses the synchronous TDMA access strategy. A token protocol, as proposed e.g. in [12], [13], [14], forms also a good basis for the implementation of this protocol.

We feel that many high level acknowledgement messages can be eliminated if up to date membership information is provided as a basic operating system service. These acknowledgement messages, particularly if they are related to protocols of the PAR (Positive Acknowledgement or Retransmission) type, can be the cause for performance breakdowns under peak load conditions in a hard real-time environment. The elimination of these messages will significantly improve the timing properties of a distributed real-time architecture. Application specific end to end protocols [17] and an efficient membership service are in our opinion the basic primitives for the design of fault-tolerant hard real-time distributed systems.

Acknowledgements

We would like to thank the members of the Mars Research Group at the Technical University of Vienna and the referees for their useful comments on an early version of this paper.

References

[1] T. Anderson, P.A. Lee, *Fault Tolerance, Principles and Practice*, Prentice Hall International, Englewood Cliffs, 1981, pp. 63-91.

[2] O. Babaoglu, R. Drummond, "Streets of Byzantium: Network Architectures for Fast Reliable Broadcasts", *IEEE Trans. Software Engineering*, Vol. 11, No. 6, June 1985, pp. 546-554.

[3] K. B. Birman, T. A. Joseph, "Reliable Communication in the Presence of Failures", *ACM Trans. Computer Systems*, Vol. 5, No. 1, Feb. 1987, pp. 47-76.

[4] Jo-Mei Chang, N. F. Maxemchuk, "Reliable Broadcast Protocols", *ACM Trans. Computer Systems*, Vol. 2, No. 3, Aug. 1984, pp. 251-273.

[5] F. Cristian, H. Aghili, R. Strong, D. Dolev, "Atomic Broadcast: From Simple Message Diffusion to Byzantine Agreement", *Proc. 15th Int. Symposium on Fault-Tolerant Computing (FTCS-15)*, Ann Arbor, Michigan, 1985, pp. 200-206.

[6] F. Cristian, "Agreeing on Who is Present and Who is Absent in a Synchronous Distributed System", *Proc. 18th Int. Symposium on Fault-Tolerant Computing (FTCS-18)*, Tokyo, 27-30 June 1988, pp. 206-211.

[7] H. Kopetz, W. Ochsenreiter, "Clock Synchronization in Distributed Systems", *IEEE Trans. Computers*, Vol. 36, No. 8, Aug. 1987, pp. 933-940.

[8] H. Kopetz, A. Damm, Ch. Koza, M. Mulazzani, W. Schwabl, Ch. Senft, R. Zainlinger, "Distributed Fault-Tolerant Real-Time Systems: The MARS Approach", *IEEE Micro*, Feb. 1989, pp. 25-40.

[9] J. Kuhl, S. Reddy, "Distributed fault-tolerance for large multiprocessor systems", *Proc. 7th Symposium on Computer Architecture*, May 1980, pp. 23-30.

[10] J. Kuhl, S. Reddy, "Fault-diagnosis in fully distributed systems", *Proc. 11th Int. Symposium on Fault-Tolerant Computing (FTCS-11)*, June 1981, pp. 100-105.

[11] A. Mahmood, E.J. Mc Cluskey, "Concurrent Error Detection Using Watchdog Processors - A Survey", *IEEE Trans. on Computers*, Vol. 37, No. 2, Feb. 1988, pp. 160-174.

[12] MAP/TOP 3.0 Network Management Requirements Specification, MAP Chapter C11, 1987.

[13] D. Powell, G. Bonn, D. Seaton, P. Verissimo, F. Waeselynck, "The Delta-4 Approach to Dependability in Open Distributed Computing Systems", *Proc. 18th Int. Symposium on Fault-Tolerant Computing (FTCS-18)*, Tokyo, 27-30 June 1988, pp. 246-251.

[14] D. Powell, "Delta-4 - Overall System Specification", Report of the ESPRIT Project Delta-4, Jan. 1989.

[15] F.P. Preparata, G. Metze, R.T. Chien, "On the connection assignment problem of diagnosis systems", *IEEE Trans. Electron. Computers*, Vol. EC-16, Dec. 1967, pp. 848-854.

[16] J. Reisinger, "Failure Modes and Failure Characteristics of a TDMA driven Ethernet", Research Report 8/89, Institut für Technische Informatik, Technical University Vienna, 1989.

[17] J.H. Salzer, D.P. Reed, D.D. Clark, "End-To-End Arguments in System Design", *ACM Trans. Computer Systems*, Vol. 2, No. 4, Nov. 1984, pp. 277-288.

[18] R.D. Schlichting, F.B. Schneider, "Fail Stop Processors: An Approach to Designing Fault-Tolerant Computing Systems", *ACM Trans. Computer Systems*, Vol. 1, No. 3, Aug. 1983, pp. 222-238.

[19] R. Strong, "Problems in Maintaining Agreement", *Proc. 5th Symposium on Reliability in Distributed Software and Database Systems*, Los Angeles, January 1986, pp. 20-27.

AUTHOR INDEX

Peter Alan Lee, Thomas Anderson

Fault Tolerance

Principles and Practice

(Dependable Computing and Fault-Tolerant Systems, Vol. 3)

Second, revised edition
1990. 36 figs. XV, 320 pages.
Cloth DM 145,–, öS 1015,–
ISBN 3-211-82077-9

Prices are subject to change without notice

This classic text on the fundamentals of fault-tolerant computing system design is being reprinted, with revisions and updated references, in the Springer series "Dependable Computing and Fault-Tolerant Systems"; it was the first book which dealt exclusively with the important topic of fault tolerance. The treatment of the subject is comprehensive, but is structured by means of a general framework. This framework greatly facilitates the reader's understanding of the basic principles, and enables them to be applied to both the hardware and software aspects of computing systems. Furthermore, the emphasis placed on the principles of fault tolerance ensures that the book will continue to be relevant for many years and future technologies, rather than being dependent on specific, present-day techniques.

This book is relevant to anyone interested in reliable computers (and who isn't!). It has been used as the basis for graduate and undergraduate courses in fault tolerance, and is suitable for any student of computer architecture. The book is also appropriate for the designers and builders of computer systems, as well as for existing fault tolerance practitioners.

Springer-Verlag Wien New York

A. Avižienis, H. Kopetz, J. C. Laprie (eds.)

The Evolution of Fault-Tolerant Computing

In the Honor of William C. Carter

(Dependable Computing and Fault-Tolerant Systems, Vol. 1)

1987. 52 figures, 35 portraits and
1 frontispiece. X, 465 pages.
Cloth DM 118,–, öS 830,–
ISBN 3-211-81941-X

This book contains contributions from a group of eminent computer scientists and engineers from several countries. It covers the evolution, the state of the art and the future perspectives of the field of fault-tolerant computing. Historic developments in academia and industry are described by those people who themselves have actively been involved in bringing them about.

U. Voges (ed.)

Software Diversity in Computerized Control Systems

(Dependable Computing and Fault-Tolerant Systems, Vol. 2)

1988. 41 figs. VII, 216 pages.
Cloth DM 75,–, öS 530,–
ISBN 3-211-82014-0

This book contains the state of the art of software diversity. Some relevant experimental work and the use of software diversity in industrial application with high dependability requirements is described. An extensive annotated bibliography on the topic completes the book.

Prices are subject to change without notice

Springer-Verlag Wien New York